Nitric oxide has a tantalizing rol… many of its wide-ranging effects are well known, there remains much more to explore and to learn about the interactions of this fascinating molecule in physiological and pathophysiological processes. The volume reviews the myriad of effects of nitric oxide as a chemical messenger in the central nervous system, peripheral nervous system, immune system and cardiovascular system. Furthermore, it provides a very practical introduction to the procedures and experimental protocols necessary to work with and study nitric oxide and its synthesizing enzyme, nitric oxide synthase, in the laboratory.

In this respect the volume is unique, providing as it does a complete single-volume review of the role of nitric oxide in health and disease, and a very practical introduction to the methods and protocols involved in this intriguing and active area of biomedical research.

Nitric oxide in health and disease

BIOMEDICAL RESEARCH TOPICS

Series Editor

Professor J. A. Lucy

The Babraham Institute, Babraham, Cambridge

This new series provides an essential introduction to the theory and practice of undertaking research in some of the most exciting and rapidly advancing areas of biomedical research.

Each volume provides a framework for understanding the scientific principles before going on to outline experimental approaches and then to describe in practical detail the relevant laboratory techniques and protocols. This integrated approach will provide a very sound introduction for those students and graduates about to embark on a new area of biomedical research.

Key features - each volume covers:

- Basic Principles and Recent Advances
- Experimental Approaches
- Protocols and Techniques

Nitric oxide in health and disease

J. Lincoln, C. H. V. Hoyle & G. Burnstock

Department of Anatomy and Developmental Biology and Centre for Neuroscience, University College London, UK.

PUBLISHED BY THE PRESS SYNDICATE OF THE UNIVERSITY OF CAMBRIDGE
The Pitt Building, Trumpington Street, Cambridge CB2 1RP, United Kingdom

CAMBRIDGE UNIVERSITY PRESS
The Edinburgh Building, Cambridge CB2 2RU, United Kingdom
40 West 20th Street, New York, NY 10011-4211, USA
10 Stamford Road, Oakleigh, Melbourne 3166, Australia

First published 1997

Printed in the United Kingdom at the University Press, Cambridge

Typeset in 9.5/14pt Concorde

A catalogue record for this book is available from the British Library

Library of Congress Cataloguing in Publication data

Lincoln, J. (Jill), 1953–
Nitric oxide, health, and disease / by J. Lincoln, C. H. V. Hoyle &
G. Burnstock
p. cm. – (Biomedical research topics series: 1) Includes bibliographical
references and index.
ISBN 0 521 55038 6 (hardback). – ISBN 0 521 55977 4 (pbk.)
1. Nitric oxide – Physiological effect. 2. Nitric oxide – Pathophysiology.
I. Hoyle, Charles H. II. Burnstock, Geoffrey.
III. Title. IV. Series. V., 1955–
[DNLM: 1. Nitric Oxide. QV 126 L737n 1997]
QP535.N1L56 1997
612'.01524–dc21 96-29492 CIP
DNLM/DLC
for Library of Congress

ISBN 0 521 55038 6 hardback
ISBN 0 521 55977 4 paperback

Contents

Preface

Our aim in writing this book has been to provide a full account of the current understanding of the biology of nitric oxide (NO) and its role both in normal physiological processes and in disease. It is intended to be useful as a reference for clinicians, postgraduate students and postdoctoral researchers starting out in research on NO and, for established researchers, to provide information outside their own area of expertise.

Since 1987, when NO was first identified as a biological messenger in mammalian systems, over 11 000 papers have been published on NO. While this represents remarkable scientific effort and progress, such a wealth of literature can be bewildering to a newcomer to the field. Several books have already been published on NO. However, to date, these have been either collections of papers from symposia or multi-volume edited works which, being highly academic, tend to assume specialist knowledge. Furthermore, both formats rely on contributions by many different authors resulting in variations in style and the inherent difficulty for individual contributors in knowing precisely what information has been provided elsewhere in the text. This is the first book that has attempted to cover all the different aspects of the biology of NO in one volume written by the same authors. While each chapter stands in its own right, they have all been written with a detailed knowledge of the overall content and cross-referenced where necessary. We have not assumed any prior knowledge on the part of the reader and have provided basic information as well as covering more complex issues and areas of controversy. Unusually, we have also included detailed protocols for techniques that can be used in the study of NO. Thus, readers will not only have a full scientific background to the biology of NO but will also have access to practical information enabling them to design their own experiments.

We have organized the book into several sections in order to examine research into the biology of NO from different perspectives. We think that this approach is important because it is evident that in order to fully understand the biology of NO it is necessary to cross between different fields of research, such as vascular biology, neuronal signalling and immunocytotoxicity and between different scientific disciplines, such as molecular biology, biochemistry, microscopy and pharmacology. We have also dedicated an entire section to the pathological implications of NO since it is becoming increasingly evident that NO has the potential to contribute to a wide range of different pathological processes.

The intense interest that has arisen in NO is due to the fact that it has a provided a new mechanism for cell signalling in mammalian systems. While recognizing this and focusing on it in this book, we feel that it is important to remember that NO does not function in isolation. Throughout the text we have provided examples of where NO interacts with other reactive oxygen species and with other established signalling systems such as neurotransmitters, including purines and peptides, vasoactive endothelium-derived factors, such as endothelin and prostaglandins, and with hormones. In many of these cases the nature of the interactions have yet to be fully elucidated and this will undoubtedly be an important area for research in the future.

JILL LINCOLN
CHARLES HOYLE
GEOFF BURNSTOCK

Abbreviations

A23187	Ca^{2+} ionophore
ABC method	avidin–biotin complex method
ACh	acetylcholine
ADM	antibody diluting medium
ADMA	asymmetrical dimethyl-arginine
AGEs	advanced glycation end-products
ATP	adenosine 5′-triphosphate
BH_4	tetrahydrobiopterin
BSPT	2-(2′-benzothiazolyl)-5-styryl-3-(4′-phthalhydrazidyl) tetrazolium chloride
CaMKII	Ca^{2+}/calmodulin-dependent protein kinase II
cAMP	adenosine 3′,5′-cyclic monophosphate
R-*p*-cAMPS	R-*p*-adenosine-3′,5′-cyclic phosphorothioate (inhibitor of cAMP-dependent protein kinase)
cGMP	guanosine 3′,5′-cyclic monophosphate
CGRP	calcitonin gene-related peptide
CHAPS	3-[(3-cholamidopropyl)-dimethylammonio]-1-propane-sulphonate
CNS	central nervous system
carboxy-PTIO	2-(4-carboxyphenyl)-4,4,5,5-tetraethylimidozoline-1-oxyl-3-oxide
CPR	cytochrome P_{450} reductase
DAB	3,3′-diaminobenzidine
DAHP	2,4-diamino-6-hydroxy-pyrimidine
DEAE	diethylaminoethyl
denbufylline	1,3-di-n-butyl-7-(2′-oxopropyl)xanthine
DMPP	1,1-dimethyl-4-phenylpiperazinium
DNICs	dinitrosyl-non-haem-iron complexes with thiol
DPI	diphenylene iodonium

DPN	diphosphopyridine nucleotide (former name for NAD)
DRG	dorsal root ganglion
DT-diaphorase	enzyme responsible for catalysing the oxidation of NADH or NADPH
EAE	experimental autoimmune encephalomyelitis
EC-SOD C	extracellular superoxide dismutase type C
EDNO	endothelium-derived nitric oxide
EDRF	endothelium-derived relaxing factor
EDTA	ethylenediaminetetra acetic acid
EFS	electrical field stimulation
EGTA	(ethylene glycol-bis (β-aminoethylether)) *N,N,N′,N′*-tetraacetic acid
ELISA	enzyme-linked immunosorbent assay
EM	electron microscopy
EPR	electron paramagnetic spectroscopy
e.p.s.ps	excitatory postsynaptic potentials
FAD	flavin adenine dinucleotide
FITC	fluorescein isothiocyanate
FMN	flavin mononucleotide
GABA	γ-aminobutyric acid
GAPDH	glyceraldehyde-3-phosphate dehydrogenase
GMCSF	granulocyte macrophage colony stimulating factor
GTN	glyceryltrinitrate
HEPES	4-(2-hydroxyethyl)-1-piperazine-ethanesulphonic acid
HIV	human immunodeficiency virus
5-HT	5-hydroxytryptamine
IBMX	3-isobutyl-1-methylxanthine
i.c.v.	intracerebroventricular
IDDM	insulin-dependent diabetes mellitus
IFN	interferon
Ig	immunoglobulin
IJP	inhibitory junction potentials
IL	interleukin
i.p.	intraperitoneal
IRF	interferon regulatory factor
i.t.	intrathecal
i.v.	intravenous

KT5823	8*R*,9*S*,11*S*-(−)-9-methoxy-carbamyl-8-methyl-2,3,9,10-tetrahydro-8,11-epoxy-1*H*,8*H*,11*H*-2,7b,11a-trizadibenzo-(a,g)-cycloocta-(c,d,e)-trinden-1-one (an indolocarbazole-type cGMP-dependent protein kinase inhibitor)
LDL	low-density lipoprotein
LM	light microscopy
LOS	lower oesophageal sphincter
LPS	lipopolysaccharide
LTD	long-term depression
LTP	long-term potentiation
LY 83583	6-anilino-5,8-quinolenedione
M&B22948	2-*o*-propoxyphenyl-8-azapurine-6-one (also known as zaprinast, it is an inhibitor of cGMP-dependent phosphodiesterase)
MetHb	methaemoglobin
N_2O	nitrous oxide
NADPH	nicotinamide-adenine dinucleotide phosphate, reduced
form	
L-NAME	N^G-nitro-L-arginine methyl ester
NANC	non-adrenergic, non-cholinergic
NBT	nitroblue tetrazolium
NF	nuclear factor
7-NI	7-nitroindazole
L-NIL	L-*N*-(iminoethyl)-lysine
L-NIO	L-*N*-(iminoethyl)-ornithine
NMDA	*N*-methyl D-aspartate
L-NMMA	N^G-monomethyl-L-arginine
L-NNA	L-N^G-nitroarginine
NO	nitric oxide
L-NOARG	L-nitroarginine
NOS	nitric oxide synthase
cNOS	constitutive nitric oxide synthase
eNOS (ecNOS)	endothelial nitric oxide synthase
iNOS	inducible nitric oxide synthase
nNOS (ncNOS)	neuronal nitric oxide synthase
NPY	neuropeptide Y
ODQ	1 *H*-[1,2,4]oxadiazolo[4,3-*a*]quinoxalin-1-one

6-OHDA	6-hydroxydopamine
p_aCO_2	carbon dioxide tension in arterial blood
p_aO_2	oxygen tension in arterial blood
PAF	platelet activating factor
PAP	peroxidase–antiperoxidase complex
PARS	polyadenosinediphosphoribose synthetase
PGE_2	prostaglandin E_2
p.o.	per os
PVA	polyvinyl alcohol
PVN	paraventricular nucleus
RBC	red blood cell
SCG	superior cervical ganglion
siguazodan	2-cyano-1-methyl-3-[4-(4-methyl-6-oxo-1,4,5,6-tetrahydropyridazin-3-yl)phenyl]guanidine, and was formerly known as SK&F9436
SIN-1	3-morpholinylsydnoneimine
SNAP	*S*-nitroso-*N*-acetyl-D,L-penicillamine
SNP	sodium nitroprusside
SOD	superoxide dismutase
SP	substance P
TDRF	tumour-derived recognition factor
TFP	trifluoperazine
TGF	transforming growth factor
Th cells	T helper cells
TNF	tumour necrosis factor
TPN	triphosphopyridine nucleotide (former name for NADP)
Tris	tris(hydroxymethyl)aminomethane
TRITC	tetrarhodamine isothiocyanate
TTX	tetrodotoxin
VIP	vasoactive intestinal polypeptide
W-7	*N*-(6-aminohexyl)-5-chloro-1-naphthalenesulphonamide (a calmodulin antagonist)
zaprinast	see M&B22948

Section 1

Basic principles and recent advances

1

Nitric oxide: introduction and historical background

Nitric oxide (NO) is one of the ten smallest molecules found in nature, consisting simply of one atom of nitrogen and one atom of oxygen. However, the discoveries, made in the 1980s, that the free radical NO could be synthesized by mammalian cells and could act both as a physiological messenger and as a cytotoxic agent have had a dramatic effect on biological research in the fields of both health and disease. NO has aroused such interest for a variety of reasons, as follows. NO can be synthesized by many different cell types, where it can control or influence a number of different important physiological processes; thus, its discovery has had an impact on several major and different fields of biological research. In addition, the elucidation of the mechanisms by which NO is synthesized has uncovered novel aspects of eukaryotic enzymology. The enzyme responsible for the synthesis of NO, nitric oxide synthase (NOS), has properties that appear to be unique in mammalian systems. Furthermore, NO can move freely through membranes and has thus revealed a new way in which cells can communicate with one another. Perhaps the most intriguing aspect of NO is that the same molecule can both mediate normal physiological events and be highly toxic. Attempts to reconcile such opposing effects have stimulated research into how the internal chemical environment of the cell can modify the actions of a chemical messenger. Finally, the toxicity of NO under certain conditions makes it an attractive candidate for involvement in a wide range of pathological processes.

Despite its recent discovery, the history of NO has already been reviewed many times (see Lancaster, 1992; Nathan, 1992; Snyder & Bredt, 1992 as examples). The story is fascinating not only because of the nature of NO itself but also because, in times of increasing specialization, it provides a particularly good example of how important it can be to cross boundaries both between different fields of research and between different disciplines in order to make scientific progress. In this chapter, the major findings leading to the elucidation of the biology of NO will be reviewed briefly. These will be divided into the three major fields of research that each have contributed so much to our present understanding: immunocytotoxicity, the regulation of vascular tone and neuronal signalling.

1.1 Immunocytotoxicity

Nitrites and nitrates are the stable end-products of the reaction of NO with oxygen and can release NO on acidification. Both nitrite and nitrate have been used for thousands of years in the preservation of meat. They have the effect of killing the bacterium responsible for botulism, with the additional advantage to the food industry of deepening the red colour of meat. Research was carried out to determine the mechanism by which nitrite exerts its bactericidal effect when alternative preservatives were required once the food additive sodium nitrite was shown to form carcinogenic nitrosamines at high temperatures. NO had already been used for 50 years as a scientific tool in electron paramagnetic resonance (EPR) spectro-scopy, due to its ability to bind to metals in metalloproteins. In 1983, Lancaster demonstrated by EPR spectroscopy that, on exposure to nitrite, NO bound to metalloenzymes in bacteria, which could lead to their death. At the time this was not considered to be of physiological significance in mammalian systems since mammalian cells were not considered capable of synthesizing nitrogen oxides (Lancaster, 1992). The toxicity of NO, however, was well recognized, it proving nearly fatal to Sir Humphrey Davy when he inhaled a small amount during his work on laughing gas (nitrous oxide, N_2O). Much of the research on NO at this stage was concerned with its harmful effects as an atmospheric pollutant (Lancaster, 1992; Nathan, 1992; Snyder & Bredt, 1992).

As early as 1916, Mitchell reported that humans excreted more nitrites and nitrates than could be accounted for by dietary intake alone and stated

> The problem is of peculiar theoretical interest, since the production of an oxidized nitrogenous radicle by animal tissues would be unique [*sic*] (Mitchell, Shonle & Grindley, 1916).

How such production occurred remained a mystery and, in 1978, when Tannenbaum and his colleagues reported the endogenous synthesis of nitrite and nitrate in human intestine it was still assumed that this was the result of intestinal microbial metabolism (Tannenbaum *et al.*, 1978). This assumption was shortly rejected by the same laboratory when it was shown that both germ-free and conventional rats can generate nitrate and it was concluded that nitrate biosynthesis occurred by a mammalian metabolic pathway (Green *et al.*, 1981). Subsequently, it was reported that the treatment of rats with bacterial lipopolysaccharide (LPS) resulted in a marked increase in urinary excretion of nitrate (Wagner, Young & Tannenbaum, 1983). In 1985, Stuehr and Marletta demonstrated, in mice, that macrophages were responsible for the synthesis of nitrite and nitrate in response to LPS and suggested that this process could be involved in some aspect of cytotoxicity (Stuehr & Marletta, 1985). In studies of the tumouricidal actions of cytokine-activated macrophages, Hibbs and his colleagues had demonstrated that activated macrophages could inhibit mitochondrial respiration (mediated by metal-containing enzymes) in tumour cells. In 1987, the same research group reported that macrophage cytotoxicity required the presence of L-arginine (but not D-arginine), that citrulline was a co-product of nitrite synthesis and that such synthesis could be inhibited by the arginine analogue N^G-monomethyl-L-arginine (L-NMMA) (Hibbs, Taintor & Vavrin, 1987). Meanwhile, Marletta and his colleagues demonstrated that arginine is the only amino acid essential for the synthesis of nitrite and nitrate, the nitrogen being derived from the terminal guanidino nitrogen of arginine (Iyengar, Stuehr & Marletta, 1987). However, the crucial intermediate step in the pathway from arginine to nitrite and nitrate was not discovered from studies of macrophage cytotoxicity. This step was only recognized in 1987, when research in the field of vascular biology identified a labile humoral agent released from endothelial cells as NO (see Section 1.2). In 1988, Marletta and his colleagues first reported that activated macrophages oxidize arginine to nitrite and nitrate via the production of the intermediate NO. It was additionally noted that NO synthesis required the presence of NADPH (Marletta *et al.*, 1988). Hibbs, Lancaster and Drapier all then demonstrated that the signals obtained with EPR spectroscopy from NO-producing cells were the same as those observed previously to occur in nitrite-treated cells of the botulinal bacterium (see Lancaster, 1992; Henry *et al.*, 1993). It is now known that NO rather than nitrite or nitrate is the mediator of cytoxicity. The interaction of NO with transition metals no longer represents simply a scientific tool with which to study metalloenzymes but is now recognized as one of the main mechanisms by which NO exerts its biological effects. The ability of NO to produce a deep red colour in meat by its interaction with myoglobin in muscle is no longer of interest only to the food industry: the spectral shifts following NO binding with iron-containing proteins are now used as a technique for measuring the production of NO in biological systems. The first isolation of an enzyme responsible for synthesis, NOS, was reported in 1990 but this was not achieved with either activated macrophages or endothelial cells, rather it represented an important stage in another line of research into neuronal signalling mechanisms in the brain (see Section 1.3). Shortly after, NOS was also isolated from activated macrophages (Stuehr *et al.*, 1991a) where it exists in an isoform that is fundamentally different from those present in neurones or endothelial cells, in that its synthesizing activity is not dependent on Ca^{2+}. In addition, it was demonstrated for the first time that NOS is one of very few eukaryotic enzymes containing both flavin adenine dinucleotide and flavin mononucleotide. It is now known that the expression of NOS may be induced in a variety of different cell types as part of an immune response and that NO plays an important role in the hosts natural defence mechanisms against invading organisms and pathogens.

1.2 The regulation of vascular tone

The history of NO in vascular biology starts in the second half of the nineteenth century, with the use of organic nitrates in the treatment of angina. This came about because of the collaboration of chemists and physicians committed to the importance of experimental physiology and pharmacology to medical research. The sequence of events has been described in detail in excellent reviews by Fye and Berlin (Fye, 1986; Berlin, 1987). Amyl nitrite and nitroglycerine had been synthesized by chemists who noted at the time the powerful effects of amyl nitrite on inhalation. In the 1860s a physician, Richardson, demonstrated the effects of amyl nitrite at a scientific meeting, stating that amyl nitrite

> when inhaled, produced an immediate action of the heart, increasing the action of the organ more powerfully than any other known agent

He proved his point by passing amyl nitrite around the audience for them to inhale. Richardson did not recommend amyl nitrite for use in medicine

> because of the intensity of its action (see Fye, 1986).

While his clinical observations proved to be correct, the effects of amyl nitrite have not been forgotten and in the 1980s it resurfaced as a 'recreational drug' in the form of 'poppers'. Brunton became aware of his contemporaries findings and, in 1869, provided the first comprehensive report of the potential use of amyl nitrite in the treatment of angina. In 1879, Murrell published a series of papers on the effects of the inhalation of both nitroglycerine and amyl nitrite, proposing nitroglycerine as a remedy for angina. Alfred Nobel, who made his fortune in the nineteenth century from discovering the use of nitroglycerine as an explosive, had to take nitroglycerine for angina later in his life, writing to a friend

> It sounds like the irony of fate that I should be ordered by my doctor to take nitroglycerin internally [*sic*] (see Murrell, 1879; Fye, 1986; Snyder & Bredt, 1992).

In the intervening years, other forms of organic nitrates have been synthesized and assessed clinically. In the early 1970s, the effectiveness of oral administration of nitrates was questioned when they were found to be rapidly metabolized and the parent drugs could not be detected in the circulation (Needleman, 1973). However, nitrate therapy continued to be recognized as being clinically effective in angina .

In 1871, Brunton proposed that the hypotension induced by inhalation of amyl nitrite

> is not due to the weakening of the hearts action, but to a dilatation of the vessels, and that this depends on the action of the nitrite on the walls of the vessels themselves. Whether this is due to its action on the muscular walls themselves, or the nerve-ends in them, cannot at present be said (see Fye, 1986).

Over a hundred years later, work in Diamond's and Murad's laboratories demonstrated that a variety of nitrovasodilators, including organic nitrates and indeed NO itself, activate guanylate cyclase – a response that could be inhibited by haemoglobin and myoglobin (Diamond & Blisard, 1976; Murad *et al.*, 1978). Ignarro and his colleagues reported that NO and nitroprusside caused relaxation of vascular smooth muscle and the production of cGMP by the muscle (Gruetter *et al.*, 1979). Furthermore, they demonstrated that in order to produce vasodilatation organic nitrates and nitrites need to be metabolized to NO and form nitrosothiols (Ignarro *et al.*, 1981). Ignarro recalls discussing at the time that NO would be an ideal candidate as an endogenous modulator of blood flow, but the theory was rejected because it was thought that NO could not be synthesized by mammalian cells (see Ignarro, 1992).

In 1980, the field of vascular biology was revolutionized when Furchgott and Zawadzki first reported that the relaxation response of vascular smooth muscle required the presence of an intact endothelium (Furchgott & Zawadzki, 1980). This was an accidental finding following the observation that the responses of the aorta to carbachol differed depending on the type of preparation used. A ring preparation relaxed in response to carbachol whereas helical strips, in which the endothelium was unintentionally damaged, did not. It was proposed that acetylcholine acted on receptors on the endothelium to stimulate the production of an endothelium-derived relaxing factor (EDRF). As early as 1981, Furchgott was speculating that EDRF was a labile hydroperoxide or free radical that could activate guanylate cyclase in the smooth muscle cells of arteries (Furchgott, 1981). By 1985, the similarity between EDRF and NO had been noted and the production of EDRF was known to require an influx of Ca^{2+} while the effects of EDRF could be inhibited by haemoglobin (see Furchgott, 1983; Furchgott & Vanhoutte, 1989 for reviews). Once it had been demonstrated, in studies of activated macrophage cytotoxicity, that mammalian cells could indeed synthesize nitrogen oxides (see Section 1.1), it finally became possible to consider that EDRF was in fact NO. This was first proposed independently by Furchgott and Ignarro at the same symposium in 1986. By 1987, Ignarro's and Moncada's laboratories provided the first direct chemical and pharmacological evidence that identified EDRF as NO (Ignarro *et al.*, 1987a,b; Palmer, Ferrige & Moncada, 1987). In 1988, Moncada and his colleagues demonstrated that vascular endothelial cells could synthesize NO from the terminal guanidino nitrogen of L-arginine (Palmer, Ashton & Moncada, 1988). NOS was purified from endothelial cells in 1991 and its activity, like NOS in neurones (see Section 1.3), was found to be regulated by Ca^{2+} and calmodulin (Pollock *et al.*, 1991). Moncada's laboratory demonstrated that L-NMMA, which had previously been shown to inhibit nitrite production by activated macrophages, could also inhibit endothelium-dependent responses in isolated blood vessel preparations. Furthermore, L-NMMA increased blood pressure when administered *in vivo* (see Moncada, 1992 for review). Thus important tools for the investigation of the biological effects of NO synthesis were established and it was shown that NO is an important regulator of vascular tone under normal physiological conditions.

1.3 Neuronal signalling

Two separate lines of research led to the discovery that NO can also act as a neuronal messenger, one investigating neuronal signalling in the brain and the other neuromuscular transmission in the periphery. In 1975, the isolation of a factor that mimicked the inhibitory response to electrical field stimulation (EFS) of nerves in the bovine retractor penis muscle was reported (Ambache, Killick & Zar, 1975). In a series of studies, Gillespie and his colleagues investigated this factor, which was not any of the non-adrenergic, non-cholinergic (NANC) transmitters known at the time. In 1980, they reported the presence of a stable material in extracts of both bovine retractor penis and rat anococcygeus muscles that, on acidification, elicited powerful inhibitory responses (Gillespie & Martin, 1980). Subsequently, they showed that this factor was a vasodilator, could be inhibited by haemoglobin and superoxide generators and it acted via the generation of cGMP, properties closely resembling those of EDRF (Bowman, Gillespie & Pollock, 1982; Bowman & Drummond, 1984). It is now known that the inhibitory factor was in fact nitrite which produces NO on acidification (Martin *et al.*, 1988). Once EDRF had been identified as NO and NOS inhibitors became available to demonstrate that a biological response was mediated by NO (see Section 1.2), they were able to prove that electrical stimulation of these NANC nerves was activating NOS and that NO was mediating the inhibitory response (Gillespie, Liu & Martin, 1989). Initial reluctance to describe NO as a neurotransmitter was due, in part, to the fact that NO does not conform to all the classical criteria that have been laid down for a substance to be defined as a neurotransmitter (see Eccles, 1964). Interestingly, Werman pointed out 30 years ago that the requirement for separate inactivating mechanisms such as enzyme degradation or reuptake may not be necessary for all neurotransmitters:

> There is a possibility that an inactivating mechanism is inherent in the nature of the transmitter itself. This could be accomplished by having as [a] transmitter an unstable compound with a short biological half-life (e.g. a free radical or an unstable compound) (Werman, 1966).

It is now accepted that NO can act as a neurotransmitter in the central and peripheral nervous systems (see Hoyle & Burnstock, 1995).

In 1964, Thomas and Pearse demonstrated the presence of an enzyme in subpopulations of neurones in the brain that was able to reduce tetrazolium salts in the presence of NADPH, producing a blue formazan product that could be observed by light microscopy (Thomas & Pearse, 1964). This enzyme was termed an NADPH diaphorase and, in 1988, it was investigated biochemically but the natural substrate for the enzyme was not known (Kuonen, Kemp & Roberts, 1988). In the 1970s it was demonstrated that, in the brain, stimulation of excitatory pathways results in an increase in cGMP levels (see Garthwaite, 1991, 1993). In 1982, Deguchi and Yoshioka discovered that the formation of cGMP required the presence of arginine (Deguchi & Yoshioka, 1982). In a series of studies carried out by Garthwaite in the early 1980s, it was shown that the stimulation of cGMP synthesis occurred following the activation of *N*-methyl D-aspartate (NMDA) receptors by the excitatory transmitter glutamate. Importantly, it was demonstrated that the cells on which the NMDA receptors were localized were not the same as the cells that synthesized cGMP. Thus, the presence of an intercellular messenger that could link the two events was first indicated (see Garthwaite, 1991, 1993).

In 1988, Garthwaite and his colleagues reported that NMDA receptor activation induced the release of a diffusible factor that was strikingly similar to EDRF (or NO) and that this was the intercellular messenger that mediated the activation of guanylate cyclase (Garthwaite, Charles & Chess-Williams, 1988). In 1989, it was demonstrated that the cytosol from bovine brain could synthesize nitrite and nitrate from arginine with the concomitant formation of citrulline by an enzyme that required the presence of NADPH (Schmidt *et al.*, 1989b). Bredt and Snyder were the first to isolate any isoform of NOS to homogeneity and this was achieved using rat brain. In their initial attempts to purify NOS they noted the loss of enzymatic activity during a purification step, indicating the removal of an important co-factor. Since they knew that the production of NO by endothelial cells required Ca^{2+}, they speculated that this was also the case in neurones and that calmodulin, a Ca^{2+}-binding protein, might be that co-factor (Bredt & Snyder, 1990; Snyder & Bredt, 1992). Addition of calmodulin in the presence of Ca^{2+} was found to restore enzyme activity. Thus, they uncovered the mechanism by which the production of NO by NOS in both endothelial cells and neurones can be strictly regulated by the intracellular concentration of Ca^{2+}. Shortly after it was shown that NOS also has NADPH diaphorase activity (Hope *et al.*, 1991) and the histochemical stain for NADPH diaphorase is now commonly used to localize NOS in neurones both in the central nervous system and in the periphery.

1.4 The present

For a molecule to act as an intercellular biological messenger and perform a specific role it must be able to overcome the physical barriers that exist between cells and be able to exert a specific effect in the target. This applies whether the messenger acts over long distances, as in the case of hormones released into the circulation, or is extremely localized, as in the case of chemical transmitters in nerves. The mechanisms which enable this to occur have now been established as basic concepts in biology. Thus, in neurotransmission, it is known that there is sophisticated machinery for the controlled release of transmitters from nerves. A specific response is achieved by the transmitter interacting with specific receptors on the membrane surface of the effector. In hormone research, it has been shown that receptors do not need to be localized on the extracellular membrane. For example, once oestrogen has entered the target cell, it can act on intracellular receptors. However, whatever the biological process or the localization of the receptor, the interaction of the messenger with a receptor is fundamental. The specificity of the interaction is determined by the molecular shape of both the receptor and the messenger itself. Thus, biological messengers frequently have complex structures and they have been regarded as very specialized molecules, the synthesis of which is restricted to specific cell populations involved in a particular physiological process. However, the traditional views on the characteristics of biological messengers are constantly having to be revised. More recently, widely distributed molecules such as prostanoids have also been shown to act as intercellular biological messengers. Even more strikingly, ATP, a key component of energy metabolism that exists in every living cell, is now known to mediate specific intercellular events, notably in autonomic neuromuscular transmission. It is now accepted that a molecule that is involved in *intra*cellular metabolism can also act as an *inter*cellular messenger and ubiquity does not prevent some messengers from performing highly specialized functions provided that specific receptors are present. The importance of the discovery of NO is that it has revealed a new way in which signals between cells can be achieved. NO provides the first example of an intercellular messenger for which the cell membrane does not represent a physical barrier at all and which does not interact with a specific receptor to produce a response. The ability of NO to exert specific effects, whether they be in host defence or normal physiological processes, is not governed by its molecular shape but rather by the reactions it can undergo.

The progress that has been made between the early studies described in this chapter and the present day is impressive. It has been achieved by considerable research in laboratories all over the world. Many of these findings will form the basis of the remainder of this book. As with the initial discovery of NO in biology, our current understanding of the biological roles of NO crosses different physiological systems and uses a range of experimental approaches which will be discussed separately. Clinical studies are becoming increasingly involved, as the contribution that NO makes to a variety of pathological processes is being assessed. The story of NO is not complete and problems that remain to be solved will be emphasized. The wealth of recent literature on NO is overwhelming but, as more progress is made, it is inevitable that such output will decrease. The next stage will be for NO to take its place alongside other established or novel biological mechanisms rather than being studied in isolation.

1.5 Selected references

Lancaster, J. R. (1992). Nitric oxide in cells. *American Scientist*, **80**, 248–59.

Snyder, S. H. & Bredt, D. S. (1992). Biological roles of nitric oxide. *Scientific American*, **266**, 28–35.

2

Synthesis and properties of nitric oxide

2.1 Introduction

Nitric oxide (NO) is a small molecule (mol. wt. 30 Da). The combination of one atom of nitrogen with one atom of oxygen results in the presence of an unpaired electron, thus NO is paramagnetic and is a radical. NO is less reactive than many free radicals in that it cannot react with itself. In aqueous solution, in the presence of O_2, it has a half-life of as long as 4 min. However, in biological systems, it is considerably less stable with a half-life of less than 30 s. It is uncharged and can therefore diffuse freely within and between cells across membranes. Such characteristics make it an unlikely candidate as a messenger in biological systems where the needs for specificity of effects and targeted sites of action are normally fulfilled by molecules of complex structure released by highly controlled mechanisms. NO is also highly toxic and is known to act as a mediator of cytotoxicity during host defence. It can appear difficult to reconcile the facts that NO can, on the one hand, be effective in causing cell death and, on the other hand, be involved in physiological processes such as neurotransmission and the control of vascular tone. That this can occur can only be appreciated by understanding the mechanisms involved in its synthesis and the interactions it can undergo once it has been formed. The simplicity of NO as a molecule is in marked contrast to the complexity of NOS, the enzyme responsible for its synthesis. Similarly, NO has the potential to interact with a variety of intracellular targets producing a diverse array of metabolic effects.

2.2 Properties and isoforms of NOS

Studies of activated macrophages, neurones in the brain and endothelial cells soon detected the presence of the enzyme NOS and rapidly demonstrated that it can exist in several isoforms (Bredt & Snyder, 1990; Pollock *et al.*, 1991; Stuehr *et al.*, 1991a). Some of the original terminology used to describe these isoforms is, in retrospect, inappropriate and can be misleading. It is only now that the different isoforms have been purified and cloned that some ambiguities can be resolved. In this section some of the original terms will be described leading to the classification that will be used throughout the rest of this book.

The first distinction that has been made between different isoforms is whether it is constitutive (cNOS) or inducible (iNOS). In certain populations of neurones and endothelial cells, NOS is ever present under normal conditions and is thus constitutive. In macrophages, NOS is only expressed following activation by endotoxin or cytokines. The activity of constitutive NOS is dependent on Ca^{2+} and calmodulin whereas inducible NOS has been described as both Ca^{2+}- and calmodulin-independent. In endothelial cells, NOS is found in the particulate fraction, indicating that it is membrane bound, whereas in neurones NOS is free in the cytoplasm and soluble. Thus a further distinction has been made on the basis of the cell type of origin using the terms endothelial NOS (eNOS or ecNOS) and neuronal NOS (nNOS or ncNOS) (see Förstermann *et al.*, 1991b; Nathan, 1992 for review). Following the isolation and purification of NOS, immunohistochemical studies using antibodies raised against the different isoforms revealed that inducible NOS can be present constitutively. In addition, neuronal NOS has been localized in a variety of non-neuronal cell types. Endothelial NOS has also been demonstrated in some neurones. Although some of these results may be explained on the basis of a lack of specificity of the antibody, it is now clear that the constitutive isoforms of NOS are not only expressed by neurones and endothelial cells. It should also be emphasized that it is quite common for constitutive enzymes to be induced under certain conditions and that NOS is no exception to this (Knowles & Moncada, 1994; Nathan & Xie, 1994b).

A family of three genes has now been identified to encode for NOS (Michel & Lamas, 1992; Sessa, 1994). The nature of these molecular biological studies means that it is unlikely that a major new NOS of the same family will be found, at least in mammals. It has now been proposed (Nathan & Xie, 1994b) to combine the findings on the NOS genes with an earlier numerical classification (Förstermann *et al.*, 1991b) to define the different isoforms in a way that does not specify the cells in which they may occur or whether they are induced:

Type I NOS is NOS the activity of which is dependent on elevated Ca^{2+} (above 70–100 nM, the level in resting cells) and which is of the type first identified in neurones;

Type II NOS is NOS the activity of which is *independent* of elevated Ca^{2+};

Type III NOS is the NOS the activity of which is dependent on elevated Ca^{2+} and which is of the type first identified in endothelial cells.

The primary structures of the different isoforms were only determined when NOS was purified and cloned from neurones, activated macrophages and endothelial cells (Bredt *et al.*, 1991b; Marsden *et al.*, 1992; Michel & Lamas, 1992; Nishida *et al.*, 1992; Sessa *et al.*, 1992; Xie *et al.*, 1992) (see Fig. 2.1). This has revealed the full complexity of NOS. NOS is a homodimer (Schmidt, Lohmann &

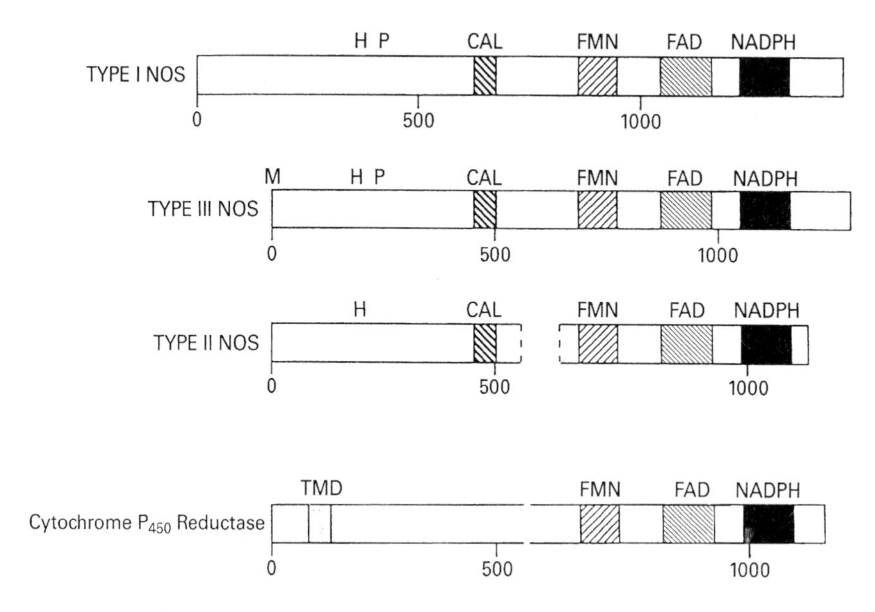

Fig. 2.1 Schematic alignment of the deduced amino acid sequences of type I, type II and type III NOS with cytochrome P_{450} reductase. Consensus sequence sites include binding sites for NADPH, FAD, FMN, calmodulin (CAL) and haem (H). The arginine-binding sites have yet to be identified unequivocally but are likely to occur in the haem domain of NOS. 0 indicates the N-terminal region. Only type III NOS contains a site for myristoylation (M) which occurs at the N-terminal. P indicates consensus sequences for phosphorylation. TMD indicates the transmembrane domain of cytochrome P_{450} reductase. Note the similarity between the C-terminal region of all three isoforms of NOS and cytochrome P_{450} reductase. (Adapted from Dawson & Snyder, 1994; *Journal of Neuroscience*, **14**, 5147–59, with permission.)

Walter, 1993). The monomers of type II and type III NOS have a molecular weight of 130 kDa and 133 kDa, respectively. The N-terminal region of type I NOS is longer, having a molecular weight of 160 kDa (Förstermann, 1994). There is an average of 50% homology between the different isoforms and 90% homology between the same isoforms in different species indicating that, at least in mammals, NOS is highly conserved (Knowles & Moncada, 1994; Nathan & Xie, 1994b; Sessa, 1994). The only other mammalian enzyme possessing substantial sequence homology with NOS is cytochrome P_{450} reductase (CPR) (Bredt *et al.*, 1991b). This is particularly evident in the C-terminal portion of NOS (60% homology) containing the binding sites for NADPH, flavin adenine dinucleotide (FAD) and flavin mononucleotide (FMN) (Fig. 2.1) (Marletta, 1993; Dawson & Snyder, 1994). In mammals, CPR serves to produce electrons from NADPH which are subsequently donated to cytochrome P_{450} drug-metabolizing enzymes. These enzymes characteristically contain the iron-protoporphyrin IX prosthetic group (haem) (Marletta,

1993). Carbon monoxide (CO) reacts with NOS to form a species that absorbs at 445 nm, indicating that NOS also contains haem (Stuehr & Ikeda-Saito, 1992). Although the binding sites for haem have yet to be indentified, they are likely to occur towards the N-terminal portion where potential cysteine residues for haem binding have been located (Marletta, 1993). The presence of a CPR domain with a haem domain in the same molecule indicates that NOS is a combination of two enzymes, one supplying the electrons and the other producing NO (Schmidt *et al.*, 1993). This is unique in mammalian systems. To date, the only other example of a 'self-sufficient' cytochrome P_{450} enzyme is in bacteria (Marletta, 1993, 1994).

All three isoforms of NOS contain calmodulin recognition sites. At first, this seemed surprising since type II activity is not dependent on Ca^{2+} whereas types I and III are activated by calmodulin via a Ca^{2+}-dependent mechanism. However, it has been shown that type II NOS also binds calmodulin, but it is so tightly bound that it is unaffected by the concentration of Ca^{2+} (Nathan, 1992). It has been demonstrated that the region responsible for the binding of calmodulin in type I NOS can adopt an α-helical structure (Zhang & Vogel, 1994). How the site differs in type II NOS to account for Ca^{2+}-independent calmodulin binding is not known and merits investigation. Tetrahydrobiopterin (BH_4) binds to all the isoforms of NOS but the binding sites for BH_4 have yet to be identified. The binding sites for arginine have also not been established but are likely to occur towards the N-terminal (Marletta, 1993; Sessa, 1994). An important question that remains to be answered is whether the arginine-binding sites differ significantly between the different isoforms. If they do, then this could provide a rationale for the development of inhibitors specific to individual types of NOS (see Chapter 10). Consensus sequences for phosphorylation sites have been identified in type I and type III NOS (Dawson & Snyder, 1994). An important difference has been observed to occur in type III NOS at the N-terminal. Only type III NOS contains a sequence indicating a myristoylation site. The process of myristoylation can enable a protein to be anchored to membranes. This may account for the fact that, in endothelial cells, NOS is associated with the particulate fraction without containing a region consistent with a transmembrane domain (Busconi & Michel, 1993) (see Chapter 3). Both type I and type II NOS are generally regarded as being cytosolic. However, type I NOS has a longer N-terminal region than the other two isoforms and, recently, evidence has been presented that this may also be involved in the subcellular localization of the enzyme. In skeletal muscle, type I NOS is localized on the sarcolemma and NOS activity is associated with the membrane fraction. This association has been shown to be due to the interaction of the N-terminal domain of type I NOS with the dystrophin complex within skeletal muscle (Brenman *et al.*, 1995).

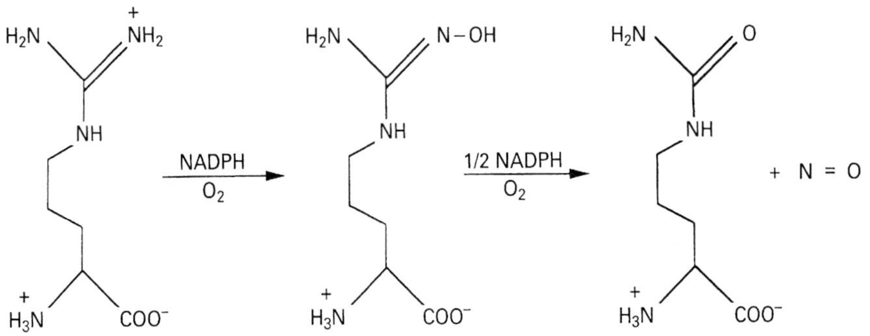

L-arginine N^{ω}-hydroxy-L-arginine L-citrulline nitric oxide

Fig. 2.2 Schematic representation of the pathway for the synthesis of NO from L-arginine. (Reproduced from Dawson & Snyder, 1994; *Journal of Neuroscience*, **14**, 5147–59, with permission.)

2.3 Synthesis of NO

Arginine is converted to NO and citrulline by NOS in a two-step process via the formation of the intermediate N^{ω}-hydroxy-L-arginine (Fig. 2.2). NO synthesis involves five electrons, three co-substrates and five co-factors or prosthetic groups (Nathan, 1992; Dawson & Snyder, 1994; Knowles & Moncada, 1994). The three co-substrates involved are arginine, molecular O_2 and NADPH. In studies of activated macrophages, it was shown that arginine is the only naturally occurring amino acid that can give rise to NO synthesis (Marletta, 1989). Isotope studies have demonstrated that the nitrogen atom in NO comes from a guanidino nitrogen in arginine and the oxygen atom comes from molecular O_2 (Marletta, 1993). The electrons required to drive this process are derived from NADPH. Thus NO synthesis cannot occur without NADPH being present as a co-substrate (Bredt & Snyder, 1992; Marletta, 1993). In order to describe the sequence of events that ultimately result in the formation of NO, and the role of the co-factors/prosthetic groups in this process, it is convenient to consider the CPR-like and the haem domains of NOS separately.

By analogy with CPR, it is thought likely that NADPH binds at the site at the C-terminal end of NOS, where it reduces FAD which binds in close proximity. The reduced form of FAD subsequently reduces FMN. Thus the flavin co-factors direct the electron flow from the C-terminal towards the N-terminal portion of NOS where the haem domain is located (Bredt & Snyder, 1992; Marletta, 1993; Knowles & Moncada, 1994). It should be noted that when NADPH is present but arginine is absent, NOS can act as an electron donor to reduce tetrazolium salts (Schmidt *et al.*, 1993). This accounts for the fact that nitroblue tetrazolium can be used to detect NADPH diaphorase activity which, under certain conditions, can be used to localize NOS in neurones (see Chapter 12). Calmodulin has no effect on the binding affinity of NOS for arginine (Matsuoka *et al.*, 1994), indicating that the activation of NOS by calmodulin does not involve it altering the active site where arginine binds. However, the oxidation of NADPH by NOS in the presence of O_2 but the absence of arginine is triggered by calmodulin (Abu-Soud & Stuehr, 1993) and results in the reduction of the haem group on NOS. The binding site for calmodulin is located between the CPR domain and the haem domain (Fig. 2.1). In NOS, calmodulin appears to play the unusual role of triggering the production and transfer of electrons from the CPR domain to the haem domain of the molecule (Abu-Soud & Stuehr, 1993; Abu-Soud, Yoho & Stuehr, 1994b; Matsuoka *et al.*, 1994). One of the consequences of the potential for uncoupling the enzyme reactions performed by NOS is that, in the absence of arginine, the activation of NOS by calmodulin can result in the generation of superoxides from NADPH and molecular O_2 (Abu-Soud & Stuehr, 1993; Dinerman, Lowenstein & Snyder, 1993).

When arginine binds in the haem domain of NOS it alters the ligand-binding properties of the haem group and increases the reduction potential of the iron (Matsuoka *et al.*, 1994). Following the passage of two electrons from the CPR domain, in the presence of molecular O_2, arginine is hydroxylated to N^{ω}-hydroxy-L-arginine (Fig. 2.2). This intermediate has been identified (Stuehr *et al.*, 1991b). The steps by which N^{ω}-hydroxy-L-arginine is converted to citrulline and NO have not been fully elucidated; however, it is clear that a further three electrons and molecular O_2 are required. CO can inhibit the conversion of N^{ω}-hydroxy-L-arginine, indicating that the haem group is also involved in the process (Marletta, 1993). BH_4 is required for the conversion of arginine to citrulline but how it regulates the reactions is the subject of some debate. It has not been established if BH_4 is metabolized during catalysis, acts as an allosteric activator of NOS or if its main role is in promoting the dimerization of NOS, which is a prerequisite for activity to occur (Marletta, 1993; Schmidt *et al.*, 1993).

2.4 Regulation of NOS expression and NOS activity

It is when the regulation of NOS expression and its activity are considered that clear distinctions can be made between the different isoforms of NOS. These are fundamental to the subsequent function of the NO that is synthesized. A summary of the properties of the different isoforms and their regulation is given in Table 2.1. The regulation of type II NOS is such that high levels (nmoles) can be produced and sustained for a long period of time. The regulation of types I and III NOS results in low levels (pmoles) of NO being produced, often only after the

Table 2.1 *Summary of the characteristics and properties of the different isoforms of NOS*

	Nitric oxide synthase isoform		
Characteristic	Type I NOS	Type II NOS	Type III NOS
Chromosomal assignment of human gene	12q24.2	17cen-q12	7q35-36
Mol. wt. of monomer	160 kDa	130 kDa	133 kDa
Subcellular localization	Largely cytosolic (except in skeletal muscle where it is membrane bound)	Cytosolic	Membrane bound
Regulation of expression	Constitutive Upregulated by sex hormones and after nerve injury	Not normally present Expression induced by cytokines/ endotoxins	Constitutive Upregulated by sex hormones and shear stress
Substrates	Arginine, O_2, NADPH	Arginine, O_2, NADPH	Arginine, O_2, NADPH
Co-factors	FAD, FMN, BH_4	FAD, FMN, BH_4	FAD, FMN, BH_4
Prosthetic groups	Haem, calmodulin	Haem, calmodulin	Haem, calmodulin
Ca^{2+}-dependency of calmodulin binding	Yes Activated by $[Ca^{2+}]_i$ above normal resting concentration of cells	No Tightly bound – activity independent of $[Ca^{2+}]_i$	Yes Activated by $[Ca^{2+}]_i$ above normal resting concentration of cells
Levels of NO produced	pmoles	nmoles	pmoles
Major function	Neuronal messenger	Immunocytotoxicity	Relaxation of vascular smooth muscle

appropriate stimulus (Moncada, Palmer & Higgs, 1991; Lowenstein, Dinerman & Snyder, 1994).

At the level of transcription, the majority of research has concentrated on type II NOS (see Förstermann *et al.*, 1991b; Moncada *et al.*, 1991; Moncada, 1992; Schmidt *et al.*, 1993; Nathan & Xie, 1994a for review). In studies of macrophages, primarily from the mouse, it has been shown that interferon-γ (IFN-γ) and bacterial lipopolysaccharide (LPS), either alone or in combination, result in the stimulation of NO production. This requires the *de novo* synthesis of type II NOS. Since this time, numerous other agents, e.g. interleukin-1 (IL-1), tumour necrosis factor (TNF) and granulocyte macrophage-colony stimulating factor, have been shown to induce NO synthesis. Some agents have no effect on their own but can act synergistically with other agents (see Chapter 6). Type II NOS expression can also be induced in many other cell types including neutrophils, epithelial cells, endothelial cells, vascular smooth muscle, hepatocytes, pancreatic islet cells and chondrocytes. The promoter/enhancer region of the type II NOS gene has been identified in mice. It has been shown to contain elements homologous to consensus sequences for the binding of transcription factors involved in the inducibility of other genes by cytokines or bacterial products (Nathan & Xie, 1994a). The detailed characterization of the same region of the human gene for type II NOS may shed some light on the fact that it has proved difficult to induce type II NOS expression in human macrophages although it can be induced in other cell types. IFN-γ can also act at the post-transcriptional level to stabilize type II NOS mRNA. Transforming growth factor (TGF-β), IL-4, IL-10 and corticosteroids have all been shown to suppress the induction of type II NOS but it is not clear whether these act at the level of the gene. TGF-β is known to destabilize type II NOS mRNA and decrease its translation (Moncada, 1992; Nathan & Xie, 1994a).

Less is known concerning the transcriptional and post-transcriptional control of type I and type III NOS. TNF-α, which induces type II NOS, destabilizes type III NOS mRNA (Nathan & Xie, 1994a). Type I NOS may be regulated at the level of alternative splicing. Transcripts lacking two exons encoding for the N-terminal end of the NOS have been found which, when expressed, produce an enzyme with NADPH diaphorase activity that is unable to convert arginine to citrulline (Ogura *et al.*, 1993; Nathan & Xie, 1994a). Type I and type III NOS expression is increased during pregnancy (Weiner *et al.*, 1994) and NOS is upregulated in sensory and motor neurones following injury (Wu & Li, 1993; Zhang *et al.*, 1993). How this occurs is not established. It is becoming increasingly evident that the expression of type I NOS in the brain and spinal cord can change markedly during embryonic and post-natal development (see Chapter 9). NO has been implicated in synaptic plasticity in the adult (see Chapter 4) and in regulating neurite outgrowth (Hess *et al.*, 1993). The regulation of gene expression for type I NOS during development of the nervous system and its role in establishing ordered neural connections is an intriguing area for future research.

Once NOS has been synthesized, its activity can then be regulated by post-translational modifications and it is at this level that NO synthesis by type I and type III NOS can be tightly controlled. The major example of this is calmodulin binding. At the normal resting level of Ca^{2+} in the cell, both type I and type III NOS are inactive. However, when Ca^{2+} levels increase, this triggers the binding of calmodulin to NOS resulting in the stimulation of catalytic activity. As little as 500 nM Ca^{2+} is required for such activation to occur (Schmidt *et al.*, 1992b). The importance of this mechanism is that it enables NO synthesis to be coupled with known physiological stimuli. In neurones, the arrival of an action potential results in an increase in free intracellular Ca^{2+}, which is sufficient to stimulate NO synthesis and this will only continue for as long as the stimulus lasts. Thus, NO can act as a transmitter during neurotransmission (see Chapters 4 and 5). A similar process could also regulate NO synthesis in fast-twitch skeletal muscle fibres which are known to express type I NOS (Kobzik *et al.*, 1994). Shear stress and receptor stimulation by a variety of agonists also raise intracellular Ca^{2+} levels in endothelial cells. Thus, synthesis of NO by type III NOS can be regulated by stimuli which are known to modify vascular tone (see Chapter 3). In marked contrast, calmodulin is tightly bound to type II NOS irrespective of the Ca^{2+} concentration (Nathan, 1992). Thus, once it is formed, type II NOS synthesizes NO continuously. The major site for regulation of NO synthesis by type II NOS is therefore at the level of transcription. All three isoforms of NOS can be phosphorylated. Phosphorylation of type I NOS by protein kinase C inhibits NO synthesis (Lowenstein & Snyder, 1992). However, studies of the actions of a variety of protein kinases have not demonstrated any clear pattern of regulation of NO synthesis by phosphorylation (Schmidt *et al.*, 1993; Nathan & Xie, 1994a).

Substrate and co-factor availability could also influence NO synthesis by all forms of NOS. The normal intracellular concentration of arginine is high ($\approx 100\,\mu M$) (Knowles & Moncada, 1994). Therefore, it appears unlikely that NO synthesis by types I and III NOS is regulated by intracellular arginine. However, it has been hypothesized that the arginine, acting as a substrate for type III NOS in endothelial cells, may come from a pool of arginine that is newly taken up and which may be rate limiting (Bogle *et al.*, 1996). It can be envisaged that, during the sustained continuous production of NO by type II NOS, arginine might be depleted. If this occurred and NADPH was still present, then NOS activity would result in superoxide rather than NO production. It has been shown that immune and inflammatory agents can, in certain cases, also induce argininosuccinate synthetase, which converts citrulline back to arginine. Thus there may be a mechanism for arginine to be recycled and maintain type II NOS activity (Nathan & Xie, 1994a). However, the synthesis of NO by type II NOS in activated macrophages can be prevented by removing arginine from the medium (Assreuy & Moncada, 1992). Naturally occurring arginine analogues that inhibit NOS activity have been reported and these could play a role in altering NO synthesis in pathological conditions (see Chapter 8). With regard to acute changes in availability of O_2 and NADPH, little is known of their effects on NOS activity. However, there is some evidence that chronic hypoxia reduces the expression of type III NOS mRNA (McQuillan *et al.*, 1994) and a reduction in the availability of NADPH in diabetes mellitus has been implicated in altered NO synthesis in this condition (see Chapter 8). There are some indications that BH_4 synthesis needs to take place for maximal NOS activity to occur. GTP cyclohydrolase I, the rate-limiting enzyme responsible for the synthesis of BH_4, can also be induced by the same agents that induce type II NOS in endothelial cells (Schoedon *et al.*, 1993; Nathan & Xie, 1994a). Finally, NO has been shown to inhibit its own synthesis by the type I (Rengasamy & Johns, 1993), type II (Assreuy *et al.*, 1993) and type III (Buga *et al.*, 1993) NOS isoforms. In the case of type I and type III NOS, it is easy to visualize how feedback inhibition could be important in restricting NO production with respect to both time and concentration. However, type II NOS is known to produce high levels of NO over a long period of time. Whether feedback inhibition of type II NOS by NO only occurs under certain experimental conditions or if the internal environment of the cell can influence whether feedback inhibition takes place has not been established (Marletta, 1994). It has been reported recently that BH_4 may protect NOS against inhibition by NO (Hyun *et al.*, 1995).

2.5 Properties and targets of NO

As a gas, NO reacts with molecular oxygen to produce nitrogen dioxide. It is incorrect to refer to NO as a 'biological messenger gas' since, under biological conditions, NO is in solution and all the reactions it undergoes are in solution. NO can react with O_2, superoxide and transition metals. The products of some of these reactions are then able to react further at nucleophilic centres, of which thiols are the major group. Thus the main target sites for NO within the cell are metal- and thiol-containing proteins and low molecular weight thiols (see Änggård, 1994; Stamler, 1994 for review). This provides a wide range of potential targets, many of which are biologically active and are involved in a diverse array of metabolic pathways. The seemingly random nature of the interactions that NO can undergo do not appear conducive to its producing a predictable response either as a physiological messenger or as a cytotoxic agent. However, there are two factors which are likely to have a major influence on the final outcome. Firstly, the reactions that can occur are likely to be dependent on the concentration of NO present. Thus, the high levels of NO produced by type II NOS will trigger a set of reactions that will not occur during the low-level NO synthesis by type I and type III NOS. Secondly, the internal environment of the cell can substantially alter the nature of the reactions that can occur. Together with the level of NO, this probably determines whether the overall effect is toxic or not. A schematic representation of

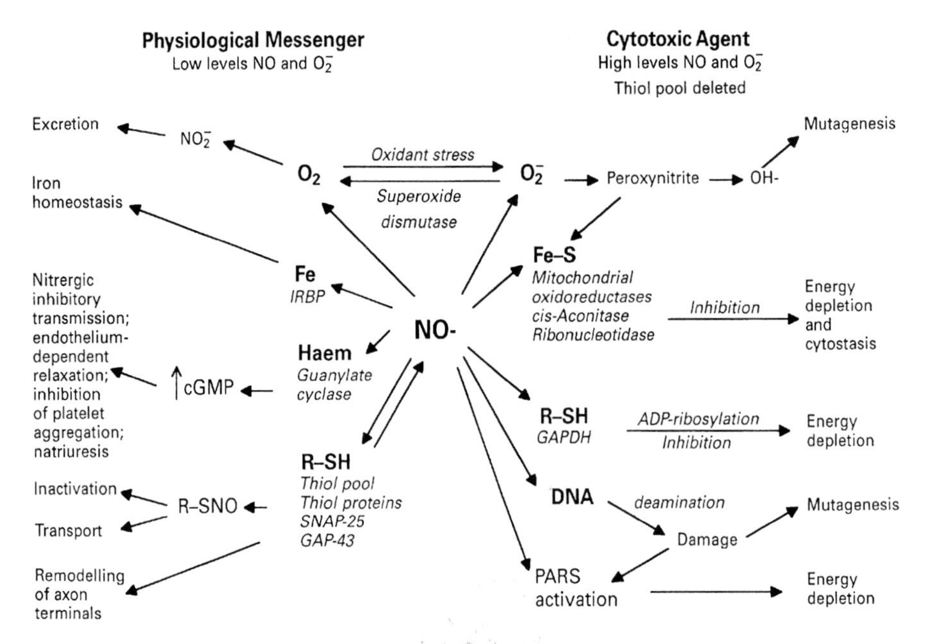

Fig. 2.3 Schematic representation of the interations that NO can undergo and the effects of NO both as a physiological messenger and a cytotoxic agent. Under normal conditions, low levels of NO are synthesized by type I and type III NOS and intracellular levels of superoxides (O_2^-) are low. NO mediates many of its physiological effects by interacting with haem on guanylate cyclase, increasing the production of cGMP. It is inactivated by reaction with molecular oxygen to form nitrite in which form it is excreted. NO may also interact with iron in proteins that are involved in iron homeostasis (e.g. iron response element-binding protein, IRBP). NO reacts with thiols (R–SH) to form nitrosothiols (R–SNO) in which form it may be transported. Nitrosylation of thiol proteins may also be involved in remodelling of axon terminals. Under conditions of oxidant stress, e.g. when high levels of NO are synthesized by type II NOS and intracellular levels of superoxides are high, the intracellular thiol pool is depleted. NO can react with superoxides to form peroxynitrite and subsequently the hydroxyl radical, which are both more toxic than NO itself. NO can inhibit enzyme activity by reacting with Fe–S groups or R–SH groups. Nitrosylation of glyceraldehyde-3-phosphate dehydrogenase (GAPDH) results in irreversible ADP-ribosylation. NO can also cause deamination of DNA resulting in damage that activates polyADP-ribose synthetase (PARS). The effects of these interactions are to cause cytostasis, energy depletion, mutagenesis and ultimately cell death. Please note that this scheme does not indicate the precise redox form in which NO undergoes each interaction.

the interactions that NO can undergo together with the cellular responses that occur as a consequence are given in Fig. 2.3. It should be emphasized that no attempt has been made to describe the precise chemical reactions taking place. Thus, some modifications to protein or thiol targets may occur via a reactive species derived from NO rather than via NO itself (Stamler, 1994). The specific role for NO in neuroprotection and neurotoxicity in the central nervous system is discussed separately in Chapter 4.

In addition to thiol proteins, there are high concentrations of low molecular weight thiols, such as glutathione and cysteine, within the cell which form a pool that can protect the cell against oxidant stress. This pool would tend to inactivate NO. It is known that NO (or its metabolites) can interact with thiols to produce nitrosothiols by a reversible process. Nitrosylation of thiols or thiol proteins has been suggested to provide a mechanism whereby NO can be transported in a stable form (Myers *et al.*, 1990; Stamler *et al.*, 1992; Änggård, 1994; Stamler, 1994). The concept of NO carriers is attractive when considering NO as a physiological messenger, since it extends the range over which NO can exert its effects beyond the distance it can diffuse as an unstable molecule. The half-life of nitrosothiols is considerably greater than that of free NO (Mülsch, 1994). The question of whether NO is released and transported as a nitrosothiol has been examined particularly with regard to NO produced in endothelial cells (see Chapter 3). There is still some controversy as to whether nitrosothiols mimic fully the physiological effects that have been attributed to NO. In addition, such a system requires that there are mechanisms by which NO can be released from its carrier or transported into the target cell. Although there are some indications as to how this may occur (Stamler *et al.*, 1992), these mechanisms have yet to be established. Cysteine residues on GAP-43 and SNAP-25 may also be targets for NO. These are two proteins that are present in axon terminals during development and are involved in axon growth and synaptogenesis. Interaction with NO results in reduced fatty acylation of the proteins. Reactions of this type could play a role in remodelling axon terminals during development and in establishing ordered patterns of neuronal connectivity (Hess *et al.*, 1993).

It is not known whether there is any selectivity with regard to the thiol groups with which NO interacts. However, under conditions of oxidant stress, the intracellular thiol pool is depleted. It may be that in these circumstances additional thiol proteins may become susceptible to NO. The high levels of NO produced by type II NOS have been shown to cause nitrosylation of the glycolytic enzyme, glyceraldehyde-3-phosphate dehydrogenase (GAPDH). Following nitrosylation, GAPDH can undergo ADP-ribosylation. This is an irreversible process resulting in inhibition of enzyme activity. It has been suggested that NO may mediate some of its cytotoxic effects by this mechanism since it would reduce the capacity of the target cell for energy production (Brüne *et al.*, 1994).

Under normal conditions, NO can react with O_2 to produce nitrite and it is by this pathway that NO is excreted (Änggård, 1994). NO also reacts readily with superoxides to form peroxynitrite which subsequently produces hydroxyl radicals (Beckman *et al.*, 1994b). It has been proposed that a major factor in determining whether or not NO is cytotoxic is the intracellular concentration of superoxides (Lipton *et al.*, 1993; Stamler, 1994). Superoxide levels within cells are normally low. However, activated macrophages and neutrophils can produce NO and superoxides at similar rates (Beckman *et al.*, 1994b). Increased levels of both these reactive species will also deplete the thiol pool. A combination of these events would provide an environment where NO would be more likely to form peroxynitrite and hydroxyl radicals and these are more toxic than NO itself (Dinerman *et al.*, 1993; Lipton *et al.*, 1993; Beckman *et al.*, 1994b). The hydroxyl radical is a powerful mutagen and peroxynitrite causes extensive protein tyrosine nitration. Cells possess protective mechanisms in the face of oxidant stress that include the upregulation of superoxide dismutase (SOD). Interestingly, cytokines can also induce SOD; therefore, it has been speculated that the cells which generate high levels of NO via type II NOS may have a natural defence in place against the toxic effects of NO. However, the targets for NO-mediated toxicity (invading microorganisms or tumour cells) could be less resistant (Stamler, 1994).

Although NO can potentially react with all transition metals, only iron, present in proteins as haem or Fe–S groups, has been investigated in detail. It has been suggested that, by reacting with a metalloregulatory protein (iron response element-binding protein), NO may provide an intracellular signal for the regulation of iron homeostasis (Stamler, 1994). In parallel with its discovery as a physiological messenger, NO was demonstrated to exert many of its physiological effects via the production of cGMP (Knowles *et al.*, 1989; Schmidt *et al.*, 1993). NO has been shown to interact with the haem group on soluble guanylate cyclase. This interaction produces a conformational change that activates the enzyme leading to increased intracellular levels of cGMP. The most potent activator of soluble guanylate cyclase is NO and the low levels of NO produced by type I and type III NOS are sufficient for enzyme activation. Not all cells contain soluble guanylate cyclase. The intracellular concentrations of Ca^{2+} required to activate NOS inhibit guanylate cyclase (Knowles *et al.*, 1989). Taken together, this enables NO to act as an intercellular messenger and would confer some specificity with respect to the cell target. The list of physiological processes that have been attributed to NO-stimulated cGMP production is growing all the time. It includes: endothelium-dependent relaxation of vascular smooth muscle and inhibition of platelet aggregation and adhesion (see Chapter 3); retrograde signalling in the brain (see Chapter 4); nitrergic inhibitory transmission in the gastrointestinal, urogenital and cardiovascular systems (see Chapter 5); pressure-induced natriuresis and tubuloglomerular feedback in the kidney (Bachmann & Mundel, 1994; Salazar & Llinás, 1996); and modulation of force development in fast-twitch skeletal muscle fibres (Kobzik *et al.*, 1994).

High levels of NO produced by activation of macrophages have been shown to attack a number of enzymes that contain Fe–S groups. Enzymes in the mitochondrial electron transport chain (NADH:ubiquinone oxidoreductase and NADH:succinate oxidoreductase) are both inhibited by NO (Stuehr & Nathan, 1989; Stamler, 1994). In addition, the activity of *cis*-aconitase, a cytoplasmic enzyme forming part of the tricarboxylic acid cycle, is reduced by NO. In this case, inhibition appears to occur via peroxynitrite rather than NO itself (Hibbs *et al.*, 1988; Stamler, 1994). All of these NO-induced changes would cause energy depletion within the target cell. Ribonucleotidase activity is also reduced by NO, resulting in cytostasis in tumour cells (Lepoivre *et al.*, 1990). It is not clear whether this occurs via an interaction with iron, thiol groups or a tyrosyl free radical on the enzyme (Stamler, 1994).

Finally, under conditions where fewer thiol groups are available for interaction (e.g. oxidant stress), nucleophilic centres on DNA are also potential targets for NO. NO has been shown to cause direct damage to DNA by deamination (Wink *et al.*, 1991). Both DNA damage and NO itself activate polyADP-ribose synthetase (Zhang *et al.*, 1994). It has been suggested that this would initiate a futile cycle in which large quantities of ATP are consumed leading to energy depletion and cell death (Zhang *et al.*, 1994). Clearly, the list of interactions that NO can undergo and the subsequent metabolic effects is likely to grow with continued research. A more complete definition of the intracellular conditions that favour individual interactions is required before the roles of NO as an intra- and intercellular messenger and as a cytotoxic agent can be fully elucidated.

2.6 Summary

NO is synthesized from arginine and O_2 by NOS, the electrons required for this process being derived from NADPH. NOS is complex, consisting of two enzymes, the one in the CPR domain supplies the electrons, while the other in the haem domain produces NO. NOS can exist in three isoforms. Type I and type III NOS are regulated at the post-translational level by calmodulin via a Ca^{2+}-dependent mechanism. Activity of type I and type III NOS produces low levels of NO for a short period of time. Type II NOS is regulated at the level of transcription, expression being induced by a variety of cytokines. Type II NOS is not regulated by Ca^{2+} and provides a continuous supply of high levels of NO. Under normal conditions, low levels of NO activate guanylate cyclase. Thus, NO mediates its effects as a physiological messenger via the production of cGMP. Interactions of NO with thiol groups may also provide a mechanism whereby NO can be transported to the target cell. In high concentrations, particularly in the presence of superoxides, NO can inhibit a variety of metabolic processes and can also cause direct damage to DNA. Such interactions result in energy depletion, cytostasis and ultimately cell death and form the basis by which NO can also be a cytotoxic agent.

2.7 Selected references

Förstermann, U. (1994). Biochemistry and molecular biology of nitric oxide synthases. *Arzneimittel-Forschung*, **44** (Suppl 3A), 402–7.

Knowles, R. G. & Moncada, S. (1994). Nitric oxide synthases in mammals. *Biochemical Journal*, **298**, 249–58.

Nathan, C. (1992). Nitric oxide as a secretory product of mammalian cells. *FASEB Journal*, **6**, 3051–64.

Nathan, C. & Xie, Q.-W. (1994). Regulation of biosynthesis of nitric oxide. *Journal of Biological Chemistry*, **269**, 13 725–8.

Stamler, J. S. (1994). Redox signaling: nitrosylation and related target interactions of nitric oxide. *Cell*, **78**, 931–6.

3

Endothelium-derived nitric oxide

3.1 Introduction

It is now known that the endothelium is not simply an inert barrier between the blood and the vascular smooth muscle. In addition to its barrier function, the endothelium is actively involved in the regulation of vascular tone and blood flow, in maintaining a balance between blood fluidity and thrombosis, in modulating control by perivascular nerves and in the regulation of cell growth within the vessel wall. To a greater or lesser extent, NO has been implicated as a mediator contributing to all of these endothelial functions. Of these, the role of NO as the endothelium-derived relaxing factor first described by Furchgott and Zawadzki has been subjected to the most intensive research (Furchgott & Zawadzki, 1980). Despite this, it is still a matter for debate as to whether EDRF is identical to NO or more closely resembles a vasoactive agent derived from NO. Considerable progress has been made with regard to elucidating the mechanisms by which NO is synthesized in endothelial cells and the regulation of such synthesis. Much of the experimental evidence for these mechanisms has been obtained from isolated blood vessel preparations or endothelial cells in culture using artificial media or physiological salt solutions. More recently, this research has been extended to consider the role of NO in the context of the chemical environment of the endothelium that exists *in vivo* and the dynamics of blood flow.

3.2 The identity of EDRF and its release

In 1987, two laboratories reported that EDRF possesses biological and chemical properties that are identical to those of NO (Ignarro *et al.*, 1987a,b; Palmer, Ferrige & Moncada, 1987). These studies investigated EDRF released following stimulation with bradykinin or acetylcholine of isolated blood vessels or endothelial cells in culture. The properties of EDRF were then compared with those of authentic NO. Both EDRF and NO were shown to cause relaxation of vascular smooth muscle denuded of endothelium and to increase cGMP levels in the muscle. Both substances were reported to be equally unstable, their biological effects being inhibited by haemoglobin and enhanced by superoxide dismutase. NO and EDRF caused similar shifts in the Soret peak for haemoglobin (a characteristic intense band at a wavelength of approximately 400 nm which occurs in the absorption spectra of porphyrins and their derivatives) and, following reflux under acidic reducing conditions, reacted with ozone to produce chemiluminescence. Finally, using diazotization and spectrophotometry, NO was detected in the effluent in amounts that were proportional to the degree of vascular smooth muscle relaxation achieved. These studies were able to demonstrate that nitrite was not responsible for the results observed. However, it was noted at the time, and has been emphasized since, that these experiments do not distinguish unequivocally whether EDRF is free NO or a labile nitroso compound.

Since this time, some evidence has accumulated that EDRF may not be identical to NO (for reviews of the evidence for and against NO see Moncada, Palmer & Higgs, 1991; Berkowitz & Ohlstein, 1992; Ignarro, 1992). Pharmacological studies have reported different patterns of relaxation with EDRF and NO, particularly in non-vascular smooth muscle. Several studies have distinguished between EDRF and NO on the basis of their chemical properties. These include the ability of certain anion exchange resins to adsorb EDRF but not NO and the differential pattern in the effects of EDRF and NO on smooth muscle cGMP levels. In addition, although superoxides can block relaxation by both EDRF and NO, the rate of inactivation is greater for NO than for EDRF (Furchgott, Jothianandan, & Ansari, 1995). Using a chemiluminescence assay that was able to distinguish between free NO and labile nitroso compounds, it was reported that the amount of NO detected in the effluent from endothelial cells, both under basal conditions and during stimulation, was not sufficient to account for the vascular smooth relaxation observed in the bioassay (Myers, Guerra & Harrison, 1989). Possibly the most convincing evidence that EDRF is not NO is the finding that free NO reacts with haemoglobin to produce nitrosyl haemoglobin, which has a characteristic electron paramagnetic resonance signal whereas EDRF does not (Greenberg, Wilcox & Rubanyi, 1990). On the basis of these and other findings it has been suggested that EDRF is more likely to be a labile nitroso precursor that can subsequently release NO to produce its biological effect.

Major candidates for the labile precursor include nitrosothiols (Myers *et al.*, 1990; Stamler *et al.*, 1992) and dinitrosyl-non-haem iron complexes with thiol ligands (DNICs) (Mülsch, 1994). However, recently, the pharmacodynamic profile of EDRF has been compared with those of several potential EDRF candidates, including NO, *S*-nitroso-cysteine, hydroxylamine and DNIC with cysteine, using a bioassay (Feelisch *et al.*, 1994). Only NO had the same profile as EDRF in terms of its half-life, sensitivity to haemoglobin and the effects of cysteine. Clearly, this controversy has yet to be resolved. It may even be the case that EDRF could

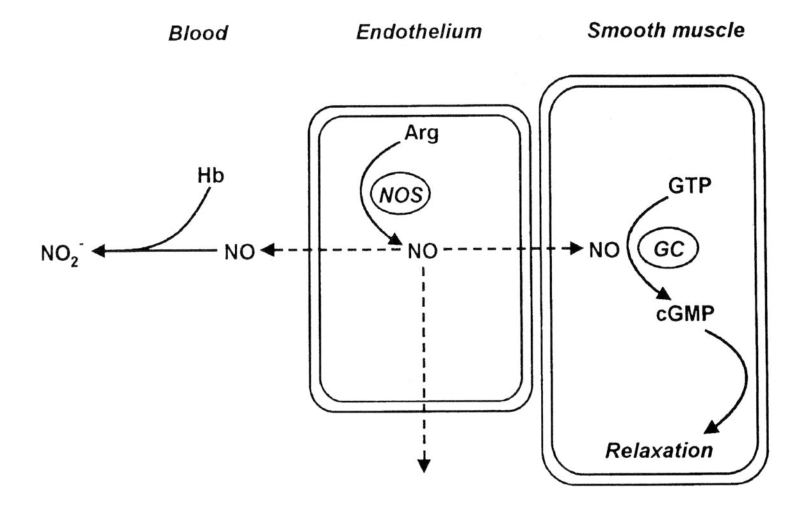

Fig. 3.1 Schematic representation of endothelium-derived-NO-mediated vasodilatation in which NO transport occurs by simple diffusion. NO is synthesized within the endothelium by type III NOS. NO is then able to move randomly in all directions by simple diffusion. NO that reaches the blood is rapidly inactivated by haemoglobin (Hb). NO that diffuses into the vascular smooth muscle activates soluble guanylate cyclase (GC) leading to increased production of cGMP and relaxation of the muscle.

represent a mixture of NO and nitroso precursors (Ignarro, 1992; Mülsch, 1994). At this stage, it is common for EDRF to be referred to as EDNO since this does not make any assumptions as to whether NO is free or not. It should be emphasized that the controversy over the identity of EDRF does not in any way diminish the importance of NO in endothelium-dependent relaxation of vascular smooth muscle. It is not disputed that NO synthesis is required for EDRF release or that EDRF exerts its effects on vascular smooth muscle by the action of NO on guanylate cyclase. However, whether NO is stored or released as a labile precursor is of particular significance when one considers the mechanisms of action, transport and function of EDNO under *in vivo* conditions.

3.3 Release and transport of EDNO

The simplest scheme for the release of EDNO (Fig. 3.1) is that free NO diffuses randomly, making the effect of NO extremely localized. It has been argued that free NO is too reactive for it to be able to reach the underlying vascular smooth muscle and produce its effect. Based on the rates of reaction of NO with molecular O_2 in solution and on the rate of diffusion of NO, a kinetic model for the movement of NO has been devised (Lancaster, 1994). On this theoretical basis it has been

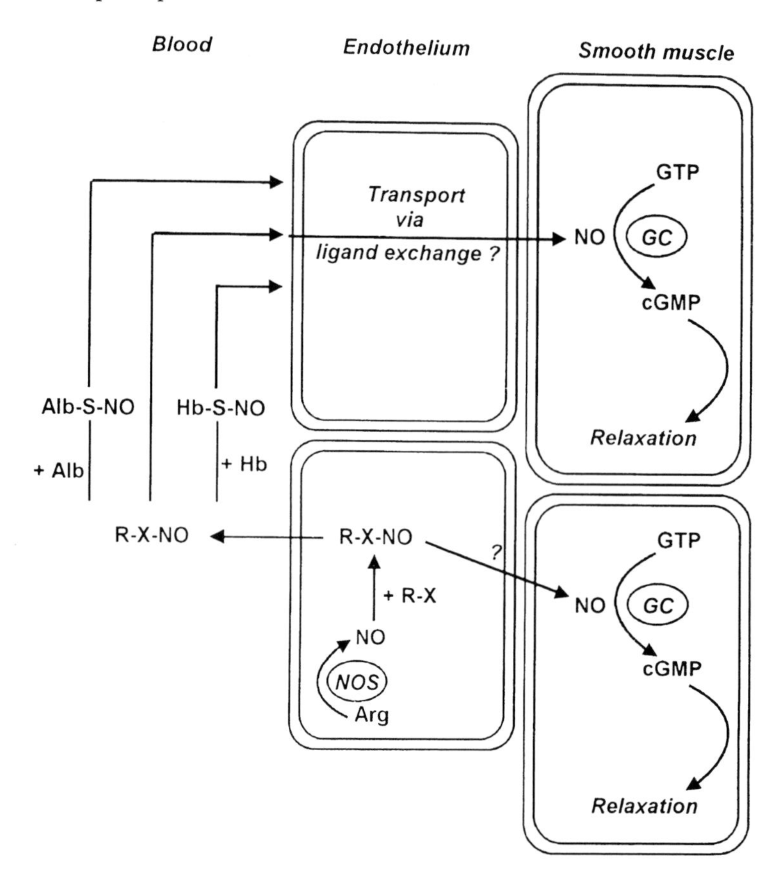

Fig. 3.2 Schematic representation of endothelium-derived-NO-mediated vasodilatation in which NO circulates in a stable form as an *S*-nitrosoprotein and is transported to the vascular smooth muscle by ligand exchange. NO is synthesized within the endothelium by type III NOS where it is converted to a more stable form (R–X–NO) by reaction with intracellular low-molecular-weight ligands such as thiols or non-haem iron groups. Once released into the circulation, R–X–NO is less susceptible to inactivation by haemoglobin (Hb). However, R–X–NO can interact with either Hb or serum albumin (Alb) to form *S*-nitrosoproteins which are still vasoactive. NO may then be transported back across the endothelium to the vascular smooth muscle by a series of exchange reactions with ligands present in the membranes and cytosol, although the precise mechanisms for this have yet to be established. Once NO reaches the vascular smooth muscle it activates soluble guanylate cyclase (GC) leading to increased production of cGMP and relaxation of the muscle.

shown that sufficient free NO could reach its target to produce a response. However, if additional factors are added to the kinetic model, the picture changes. It is well known that free NO is inactivated by reaction with the iron centres of haemoglobin. The presence of haemoglobin in the circulation would result in a concentration gradient for free NO whereby more NO would diffuse into the lumen to be inactivated than would diffuse in the abluminal direction to the underlying smooth muscle (Lancaster, 1994). This has been used as an argument that NO must be released on a carrier or react with other components in the circulation in such a way that it can retain its biological activity (Fig. 3.2).

EDRF, in contrast to free NO, has been shown to interact with sulphhydryl groups of haemoglobin to produce *S*-nitroso-haemoglobin, which is still a potent vasodilator *in vivo* (Jia, Bonaventura, & Stamler, 1995). Nitrosothiols are less susceptible to inactivation by haemoglobin (Day *et al.*, 1995), although low molecular weight nitrosothiols are still very labile. Serum albumin represents a rich source of reduced thiol in plasma. Both EDRF and NO can nitrosylate serum albumin under physiological conditions. Although less potent than low molecular weight nitrosothiols, *S*-nitroso-serum albumin is still vasoactive *in vivo* and, furthermore, it is considerably more stable (Stamler *et al.*, 1992; Keaney *et al.*, 1993). DNICs have been detected within endothelial cells but not in the culture medium following stimulation with bradykinin. Interestingly, when the medium was supplemented with serum albumin and *N*-acetyl-L-cysteine prior to stimulation, then the DNIC levels were decreased in the endothelial cells but increased in the medium (Mülsch, 1994). Thus the presence of extracellular thiols appears to promote the release of EDNO. Taken together, all of these studies indicate that EDNO as it occurs *in vivo* may not be the same chemical entity that has been investigated under *in vitro* conditions with perfusion by artificial salt solutions.

If EDNO is released into the lumen and circulates in a relatively stable form, then, for it to have biological activity, NO must be transferred back across the endothelium to act on vascular smooth muscle. The fact that *S*-nitroso-serum albumin and *S*-nitroso-cysteine can reduce arterial pressure following intravenous administration *in vivo* (Keaney *et al.*, 1993) indicates that this can occur. To date there is no experimental evidence for the presence of carriers or channels for the transport of the precursors. The reactions by which nitrosothiols and DNICs are formed are reversible. Once they have been formed they can undergo exchange reactions with other ligands such that NO can be transferred from one ligand to another. Therefore, it has been suggested that NO may be transferred across membranes and the endothelial layer by a series of interactions between nitrosothiols and/or DNICs and thiols in the membranes and cytoplasm (Stamler *et al.*, 1992; Keaney *et al.*, 1993; Mülsch, 1994) (Fig. 3.2). One of the consequences of this scheme is that the distance between the endothelial cell that produces EDNO and the vascular smooth muscle cell that responds to EDNO can be increased. In the case of *S*-nitroso-serum albumin, which is relatively stable, this effect could be considerable. In addition, not all nitroso compounds produce the same effect in different blood vessels. This raises the possibility that responsiveness may be determined by the presence of specific 'acceptor' molecules on the endothelial cell surface. However, experimental evidence for this is limited (Keaney *et al.*, 1993).

Endothelial cells can continuously synthesize EDNO under basal conditions. Following stimulation, EDNO synthesis and release are increased (see next section). A third scheme for the release of EDNO has been hypothesized (Ignarro, 1992) and is given in Fig. 3.3. It should be noted that this scheme is speculative. However, it has been included because it distinguishes between basal and stimulated EDNO release and it also suggests a possible mechanism whereby stimulated EDNO release could be *directed* towards the vascular smooth muscle. High concentrations of arginine analogues that inhibit NO synthesis do not necessarily block endothelium-dependent relaxations (Ross *et al.*, 1991). This could be argued to indicate that EDNO was not responsible for the endothelium-dependent relaxation in these experiments. However, it has been suggested that such findings could equally be explained if stimulated EDNO release was not entirely dependent on *newly synthesized* NO (Ross *et al.*, 1991; Ignarro, 1992). Based on the differential effects of inhibitors or nitroglycerine tolerance on basal tone and stimulated relaxation, it has been suggested that basally released EDNO and EDNO released on stimulation are not one and the same chemical entity (Ignarro, 1992). It has been speculated that some of the NO synthesized under basal conditions may be converted to a stable form stored within the endothelial cells. This could occur by diffusion of NO into acid lysosomes where it could react with intralysosomal thiols such as cysteine to form a nitrosothiol that is relatively stable under acid conditions. The increase in intracellular Ca^{2+} that occurs during stimulation could then both increase the synthesis of NO and stimulate the exocytosis of the stored form of EDNO. Thus, basally released EDNO would be free NO whereas EDNO released on stimulation would be a mixture of free NO and a nitrosothiol. There is preliminary evidence that inhibitors of exocytosis cause partial inhibition of endothelium-dependent relaxation without affecting basal synthesis and release of NO. Such exocytosis could occur preferentially in the abluminal direction. Following release, exposure of the nitrosothiol to increased pH would result in spontaneous decomposition to NO which could then exert its effects on the vascular smooth muscle (see Ignarro, 1992).

A considerable amount of research is required before the precise sequence of events between the synthesis of NO and its action on vascular smooth muscle can be fully elucidated. Although several hypotheses have been described here, much of the evidence for each is either indirect or preliminary in nature. It may even be

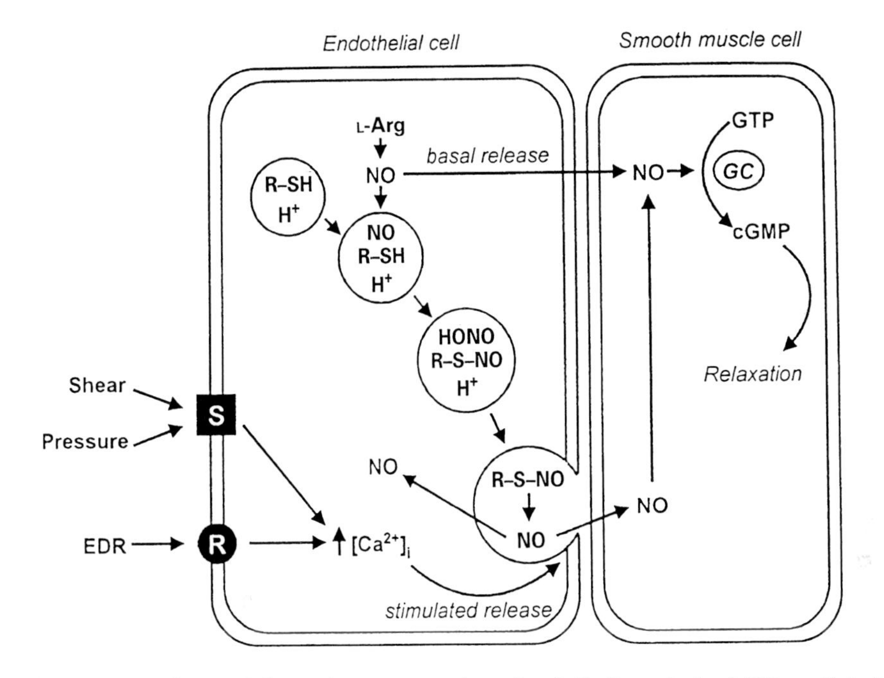

Fig. 3.3 Schematic representation of endothelium-derived-NO-mediated vasodilatation in which it has been hypothesized that NO is released in different forms depending on whether it is synthesized under basal conditions or following stimulation. Under basal conditions, NO synthesized by type III NOS in the endothelium diffuses freely in all directions. NO that reaches the vascular smooth muscle activates soluble guanylate cyclase resulting in increased production of cGMP and relaxation of the smooth muscle. This provides a constant vasodilator tone mediated by NO. Some of the basal NO diffuses within the endothelium into intracellular acid lysosomes where it interacts with low-molecular-weight thiols to form a nitrosothiol (R–S–NO) which is stable under acid conditions and can thus be stored. Following a stimulus such as shear stress (S) or agonist receptor occupation (R) an increase in intracellular Ca^{2+} triggers the exocytosis of the stored, R–S–NO. Such exocytosis could occur preferentially in the abluminal direction. Once R–S–NO reaches the extracellular space, the increase in pH will promote the release of NO which subsequently acts on soluble guanylate cyclase as above. (Reproduced from Ignarro, 1992; In *Endothelial Regulation of Vascular Tone*, eds. U. S. Ryan & G. M. Rubanyi, Marcell Dekker Inc., with permission.)

that the aim to find mechanisms for the stabilization, storage and directed release of NO from endothelial cells is misplaced. It may represent an attempt to make NO conform to traditional 'rules' developed in other systems for how intercellular physiological messengers work. In addition, the regulation of vascular tone is not the only potential function of EDNO. Other functions such as the maintenance of blood fluidity and the anti-adhesive properties of the endothelium (see Section 3.5) might be better achieved by the continuous luminal release of free NO.

3.4 Endothelial NOS and regulation of its activity

3.4.1 General characteristics

NO is synthesized in endothelial cells by the type III isoform of NOS. As with all the isoforms, arginine, NADPH and O_2 are co-substrates, FAD, FMN and BH_4 are co-factors and the prosthetic groups calmodulin and haem are required for activity (see Chapter 2). Under basal conditions, low levels of NO are synthesized and released continuously. However, increased NO synthesis can occur following a number of stimuli including activation of endothelial receptors by a variety of vasoactive agents, shear stress and low arterial O_2 tension (pO_2) (Lewis & Smith, 1992; Busse, Fleming & Hecker, 1993). Whatever the stimulus, such increased synthesis is generally thought to involve an influx of extracellular Ca^{2+} which promotes calmodulin binding and activation of NOS. Both shear stress and agonist stimulation have been shown to increase intracellular Ca^{2+} although, in the case of shear stress, the Ca^{2+} transients appear to have a shorter duration compared with the response of NOS (Lewis & Smith, 1992; Busse, Hecker & Fleming, 1994; Blatter *et al.*, 1995).

Although intracellular arginine concentrations are high and appear unlikely to be rate limiting, it has been suggested that extracellular arginine taken up by endothelial cells may form a pool used for NO synthesis and that this could regulate NOS activity. This has been speculated on the basis of studies of arginine uptake and its regulation (Bogle *et al.*, 1996). In this context it is of note that bradykinin and ATP, which both stimulate endothelial NO synthesis, also stimulate arginine uptake into endothelial cells (Bogle *et al.*, 1991). The synthesis of NO following stimulation of endothelial cells in culture by bradykinin or the Ca^{2+} ionophore, A23187, can be reduced by inhibiting BH_4 synthesis (K. Schmidt *et al.*, 1992). In addition, during inhibition of BH_4 synthesis, stimulation of endothelial cells by A23187 results in the production of hydrogen peroxide that can be prevented by the NOS inhibitor N^G-nitro-L-arginine methyl ester (L-NAME) (Cosentino & Katusic, 1995). Since NOS can synthesize hydrogen peroxide in the absence of arginine but the presence of NADPH and O_2 (see Chapter 2), this may indicate that BH_4 acts to promote arginine binding to type III NOS. Type III NOS can be phosphorylated by both protein kinase A and protein kinase C but only the latter appears to affect NO synthesis, causing inhibition (Hirata *et al.*, 1995). Inhibition of protein kinase C has also been shown to increase the expression of type III NOS mRNA in endothelial cells (Ohara *et al.*, 1995). Thus NO synthesis by type III NOS can be regulated by a variety of mechanisms, most of which are probably common to type I NOS. However, there is one important factor in the regulation of NO synthesis by endothelial cells that is unique. Under *in vivo* conditions, the luminal surface of the endothelium is constantly subjected to blood flow and it is becoming increasingly evident that there is a close relationship between shear stress and the synthesis and release of EDNO.

3.4.2 Regulation by shear stress

Shear stress is caused by the viscous drag of the blood against the surface of the endothelium. Shear stress can be increased both by an increase in flow and by a decrease in vessel diameter under constant-flow conditions. Increased shear stress results in an increase in the synthesis and release of EDNO (Buga *et al.*, 1991; Busse *et al.*, 1993, 1994; Korenaga *et al.*, 1994; Kuchan & Frangos, 1994; Noris *et al.*, 1995); furthermore, shear stress has been shown to potentiate agonist-stimulated EDNO release (Busse *et al.*, 1994). Some of these agonists are present within the endothelial cell and can be released by shear stress. Their subsequent activation of endothelial receptors could also provide an indirect mechanism whereby shear stress stimulates EDNO synthesis (see Section 3.5.1).

Studies have reported that shear stress-induced EDNO release is dependent on extracellular Ca^{2+} (Buga *et al.*, 1991; Korenaga *et al.*, 1994). However, other workers have reported that continuous EDNO release during sustained shear stress is not dependent on Ca^{2+} or calmodulin (Busse *et al.*, 1993, 1994; Kuchan & Frangos, 1994). This is also likely to be the case for the continuous basal release of EDNO since sustained increases in intracellular Ca^{2+} are generally regarded as harmful to cells. Synthesis of NO by type II NOS, which is Ca^{2+} *independent*, has been excluded as an explanation for this finding (Kuchan & Frangos, 1994). Thus the activation of type III NOS in endothelial cells appears to involve a mechanism in addition to Ca^{2+}-induced activation of calmodulin binding. In this context, the intracellular localization of type III NOS takes on particular significance. It is known that the endothelium can carry out mechanotransduction, i.e. it can 'sense' changes in physical force at the luminal surface and convert this to a biochemical or physiological response within the cell (Davies, 1993). A 'shear stress sensor' has not yet been identified, but chemical modification of the glycocalyx covering the luminal surface has been shown to prevent shear-stress-induced EDNO release while having no effect on agonist-stimulated release (Busse *et al.*, 1993). Mechanical signals can initiate a variety of intracellular second messenger systems to produce a response. However, it is also possible that mechanical force can produce conformational changes in proteins that are membrane bound (Davies, 1993; Busse *et al.*, 1993, 1994; Hecker *et al.*, 1994).

Microscopical studies of endothelial cells have reported that NADPH diaphorase activity and NOS-immunoreactivity can be found in the cytoplasm but also in association with membranes, particularly those of the endoplasmic reticulum and Golgi complex (Tomimoto *et al.*, 1994; Loesch & Burnstock, 1995). This has been confirmed by biochemical studies that have shown that the majority of NOS activity in endothelial cells is membrane bound (Tracey *et al.*, 1993; Hecker *et al.*, 1994). Unlike the other isoforms of NOS, type III NOS can incorporate myristic acid via an amide linkage at the N-terminal (Liu & Sessa, 1994). Mutation of the myristoylation site converts type III NOS from a membrane bound to a cytoplasmic enzyme (Busconi & Michel, 1993; Sessa, Barber & Lynch, 1993). It has been speculated that the process of myristoylation and membrane binding may enable NOS to be localized close to an arginine pool or to sites of local release of intracellular Ca^{2+} stores (Sessa *et al.*, 1993). Cytoplasmic NOS in endothelial cells has a lower specific activity than membrane-bound NOS (Hecker *et al.*, 1994). Furthermore, translocation of NOS from the membrane to the cytoplasm, either by disruption of the Golgi complex (Stanboli & Morin, 1994) or by a mutation of the myristoylation site (Sessa *et al.*, 1995), results in a reduction of the capacity to synthesize NO. Once it has been translocated to the cytoplasm, type III NOS can be phosphorylated (Robinson, Busconi & Michel, 1995). Thus the membrane binding of type III NOS appears to be an important factor in determining its activity. It is also possible that its location on the membrane may provide a mechanism whereby mechanical forces can regulate NOS activity by conformational changes without the need for raising intracellular Ca^{2+} concentrations (Busse *et al.*, 1993, 1994).

Blood flow may not simply act as an instant trigger to increase NOS activity; there is evidence that chronic subjection to flow (as occurs *in vivo*) can regulate the expression of type III NOS. NO synthesis and NOS content are significantly higher in freshly isolated endothelial cells than in endothelial cells from the same source that have been cultured under static conditions (Hecker *et al.*, 1994). Prolonged exposure of endothelial cells in culture to flow results in the upregulation of type III NOS mRNA and EDNO synthesis (Noris *et al.*, 1995).

3.4.3 Regulation by other stimuli

Other long-term physiological stimuli also appear to regulate NO synthesis by endothelial cells. Exercise training has been shown to increase endothelium-dependent vasodilatation mediated by NO in skeletal muscle arterioles (Koller *et al.*, 1995) and to increase type III NOS mRNA expression in the coronary circulation (Sessa *et al.*, 1994). Whether this effect is due to the haemodynamic changes that occur during adaptation to exercise is not known. While acute hypoxia does not inhibit NO synthesis, indeeed, it can cause endothelium-dependent vasodilatation in certain vascular beds, chronic hypoxia results in a reduction in both the expression and stability of type III NOS mRNA in endothelial cells in culture (McQuillan *et al.*, 1994). Finally, it is becoming increasingly evident that both NO synthesis and the expression of type III NOS can be regulated by sex hormones. EDNO release is significantly greater in isolated aortae from female, compared with male, rats and rabbits (Hayashi *et al.*, 1992). Type III NOS mRNA expression is upregulated during pregnancy (Goetz *et al.*, 1994; Weiner *et al.*, 1994). Oestrogen treatment, both *in vivo* and *in vitro*, results in increased NOS activity and NOS mRNA expression (Weiner *et al.*, 1994; Hishikawa *et al.*, 1995). This may account for the clinical observation that the incidence of atherosclerosis is significantly lower in premenopausal women than in men (Hayashi *et al.*, 1992). Investigation of the gene encoding for type III NOS is likely to be important in determining the mechanisms that can regulate NOS expression in endothelial cells. An initial study has already reported the presence of numerous putative transcription factor binding sites including several that are responsive to oestrogen or fluid shear stress (Venema *et al.*, 1994).

3.5 Physiological roles of EDNO

3.5.1 Regulation of vascular tone

When considering the role of EDNO in the control of vascular tone, it is important to remember that the endothelium is one part of a dual control system, the other part consisting of nervous control by perivascular sympathetic, parasympathetic and sensory nerves (see Lincoln & Burnstock, 1990; Ralevic & Burnstock, 1993 for review). In certain vessels, NO can also be synthesized by a subpopulation of perivascular nerves (see Chapter 5). Furthermore, within the endothelium, EDNO is only one of a variety of substances that are vasoactive (dilators or constrictors) of which prostacyclins and endothelin are major examples (see Lüscher, 1993 for review). At its simplest, the constant basal release of EDNO provides a vasodilator tone acting against sympathetic vasoconstriction. Administration of NOS inhibitors *in vivo* causes vasoconstriction in a variety of vascular beds and produces a hypertensive response (Moncada & Higgs, 1993). Removal of the endothelium or NOS inhibition results in increased vasoconstrictor responses to sympathetic nerve stimulation in isolated blood vessel preparations (Ralevic & Burnstock, 1993). Conversely, sympathetic activity can regulate the physiological release of EDNO. Removal of sympathetic tone by pithing or ganglionic blockade in rats attenuates the pressor effect of NOS inhibition. This attenuation can be reversed by induction of vasoconstriction with phenylephrine (Vargas, Ignarro & Chaudhuri, 1990). It seems likely that the relationship between vasodilatation induced by EDNO release and vasoconstriction induced by sympathetic nerves is mediated by changes in shear stress. EDNO can also act to oppose the potent vasoconstrictor effects of endothelin, which is also synthesized by the endothelium. Endothelin acts on ET_A receptors on vascular smooth muscle to produce a long-lasting contraction. NO has been shown to displace endothelin from these receptors and thus terminate the response (Goligorsky *et al.*, 1994). In addition, endothelial cells can express ET_B receptors which, when activated by endothelin, produce vasodilatation through the synthesis of NO rather than vasoconstriction (Tsukahara *et al.*, 1994).

It is now known that a wide variety of agents can act on endothelial receptors to produce vasodilatation by the stimulation of NO synthesis (see Furchgott & Vanhoutte, 1989; Ralevic & Burnstock, 1993 for review). Some agents, such as ATP and 5-hydroxytryptamine (5-HT), can be released from aggregating platelets and the endothelium may act to protect against inappropriate vasoconstriction in response to these agents under these circumstances. Others, such as bradykinin, histamine and substance P (SP), may be released locally during inflammation and thus contribute to the vasodilatation that occurs in the inflammatory process. However, the source of acetylcholine (ACh), in particular, which also causes endothelium-dependent relaxation, has proved problematic (Furchgott & Vanhoutte, 1989). Many of the agents that can act on endothelial receptors have also been localized in perivascular nerves. However, it is considered unlikely that neurotransmitters, released on nerve stimulation, could diffuse through the vessel wall (particularly in the larger vessels), without degradation, to act on receptors on the luminal surface of the endothelium. Evidence has accumulated that the endothelium itself has the capacity to synthesize and/or take up and store a variety of agents, including SP, ACh, 5-HT, histamine, vasopressin, angiotensin II and ATP, that can act on endothelial receptors to stimulate NO synthesis. In addition, many of these agents have been shown to be released from isolated endothelial cells or vascular beds following stimuli such as hypoxia or increased shear stress, both of which can cause vasodilatation (see Lincoln, Ralevic & Burnstock, 1991b; Ralevic & Burnstock, 1993 for review). In the case of shear stress, it appears that several endothelium-dependent mechanisms described here could act in concert to modify vascular tone. Shear stress can directly activate NO synthesis by some form of mechnotransduction and potentiate the response to agonist stimulation. NO synthesis can be further increased by stimulation of endothelial receptors by agents that are themselves released from the endothelium by shear stress. Finally, some agents such as ATP or bradykinin can increase the uptake of arginine, a substrate for NO synthesis. It should be noted that the general mechanisms described here for the regulation of vascular tone by EDNO make no attempt to distinguish between different vessels or vascular beds. It is known that there is considerable heterogeneity in the vasculature that can be demonstrated by variations in the expression of endothelial receptors, in the susceptibility to shear stress, in the pattern of innervation and in the ability to autoregulate. It remains to be determined how regional differences in the regulation and expression of type III NOS and EDNO synthesis contribute to such heterogeneity.

3.5.2 Regulation of blood fluidity and thrombus formation

In addition to the regulation of vascular tone, the endothelium plays a key role in maintaining blood fluidity and in preventing thrombus formation and EDNO is now known to contribute to such functions by inhibiting platelet aggregation and adhesion (Lüscher, 1993; Radomski & Moncada, 1993). Under normal conditions, platelets do not aggregate. Aggregation occurs following activation with agents such as thrombin, collagen and ADP but can also occur in the absence of these agonists during increased shear stress (Ruggeri, 1993). Synthetic nitrosothiols can inhibit platelet aggregation both *in vivo* and *in vitro* and this occurs via the activation of soluble guanylate cyclase (Radomski *et al.*, 1992; De Belder *et al.*, 1994). Similarly, authentic NO and EDNO, released from endothelial cells in culture following stimulation, increase cGMP levels within platelets and inhibit aggregation (Radomski, Palmer & Moncada, 1987b). Furthermore, stimulation of EDNO release *in vivo* inhibits platelet aggregation (Radomski & Moncada, 1993). Platelets themselves contain NOS and have the capacity to synthesize NO from arginine via a Ca^{2+}-dependent mechanism. However, the amounts of NO produced by platelets is small and it is thought likely that NO derived from the endothelium makes a major contribution to the prevention of platelet aggregation (Radomski & Moncada, 1993). The fact that ATP and 5-HT released from platelets during aggregation can stimulate type III NOS via endothelial receptors indicates that there may be protective mechanisms for the local stimulation of EDNO release as the situation demands (Lüscher, 1993). Prostacyclins, also synthesized by the endothelium, are powerful inhibitors of platelet aggregation and there is evidence that the anti-aggregatory effects of EDNO and prostacyclins can be synergistic (Radomski & Moncada, 1993). EDNO can also modify the adhesive properties of the luminal surface of the endothelium. Both authentic NO and EDNO released by stimulation with bradykinin inhibit platelet adhesion to endothelial cells in culture (Radomski, Palmer & Moncada, 1987a). More recently, it has been shown that inhibition of NO synthesis by endothelial cells in culture increases platelet adhesion both under static conditions and during perfusion with whole blood. This indicates that the basal release of EDNO is sufficient to produce the effect (De Graaf *et al.*, 1992; Radomski *et al.*, 1993).

3.5.3 Regulation of cell permeability and cell growth

With regard to the potential role of EDNO in the regulation of endothelial cell permeability and cell growth in the vascular wall, there is more conflict in the literature. In studies using cultures of endothelial cells derived from large vessels (umbilical vein, aorta, pulmonary artery), thrombin- and bradykinin-induced increases in permeability have been shown to be attenuated by EDNO (Oliver, 1992; Westendorp *et al.*, 1994; Draijer *et al.*, 1995). Similarly, NOS inhibition by L-NAME *in vivo* increases vascular permeability in the coronary circulation (Filep, Földes-Filep & Sirois, 1993). Thus, in these vessels, EDNO appears to reduce endothelial cell permeability. L-NAME has also been reported to cause an increase in permeability in the microvasculature of the small intestine (Kubes & Granger, 1992). However, other studies of microvessels (hamster cheek pouch, coronary venules, brain capillary endothelial cells) have found the opposite effect, in that NO donors increased permeability and NOS inhibitors attenuated the increase in permeability induced by histamine or bradykinin (Rubin *et al.*, 1991; Mayhan, 1992; Yuan *et al.*, 1993).

It has been reported that NO donors stimulate endothelial cell growth and migration *in vitro*. Further, SP-induced angiogenesis can be potentiated by NO donors and attenuated by NOS inhibitors (Ziche *et al.*, 1994). However, other studies have reported that NO donors inhibit endothelial cell proliferation and angiogenesis while NOS inhibitors stimulate angiogenesis (Pipili-Synetos *et al.*, 1994; Yang *et al.*, 1994). There is more consistency with regard to the effects of NO on vascular smooth muscle proliferation. An intact and quiescent endothelium appears to suppress vascular smooth muscle proliferation. Removal of the endothelium results in intimal thickening, which is largely the result of proliferation of smooth mucle cells in the intimal region (McNamara *et al.*, 1993; Soyombo, Thurston & Newby, 1993). Regression of intimal hyperplasia is associated with regrowth of the endothelium. Interestingly, the expression of type III NOS mRNA and protein and the activity of NOS have all been shown to be increased in endothelial cells that are actively dividing (Arnal *et al.*, 1994). Investigation of the effects of NO donors and cGMP analogues have consistently concluded that NO inhibits mitogenesis and proliferation of vascular smooth muscle cells (Garg & Hassid, 1989; Soyombo *et al.*, 1993; Peiró *et al.*, 1995). However, it has been emphasized that EDNO is unlikely to be the only factor synthesized by endothelial cells to produce this effect. Thus, in addition to playing a major role in the regulation of vascular tone, particularly in response to local changes in the environment, EDNO may contribute to other functions of the endothelium. These include the maintenance of blood fluidity, anti-thrombogenecity and barrier function and the control of vascular wall remodelling. In view of this it is not surprising that changes in the endothelial production of NO have already been implicated in the pathogenesis of a variety of cardiovascular disorders (see Chapter 8).

3.6 Summary

EDNO is synthesized in endothelial cells by the type III isoform of NOS. Type III NOS binds to membranes by a process involving myristoylation of the N-terminal end of the enzyme. The localization of type III NOS on the membrane appears to play a key role in enabling EDNO synthesis to be regulated by mechanical forces such as flow and increased shear stress. EDNO synthesis can also be stimulated by raising intracellular Ca^{2+} following activation of endothelial receptors by a variety of vasoactive agents. It is still a matter for debate whether EDNO is released as a free NO radical or whether it is stored and released in a more stable form following interactions with thiol or non-haem iron ligands. Under basal conditions of flow, EDNO provides a constant vasodilator tone acting against sympathetic vasoconstriction. Following changes in the environment, such as increased shear stress, hypoxia or platelet aggregation, EDNO synthesis can be stimulated to provide a mechanism for the local regulation of vascular tone and blood flow. In addition, EDNO can inhibit platelet aggregation and adhesion thus contributing to the maintenance of blood fluidity and the anti-adhesive properties of the endothelium. EDNO may also act, together with other factors produced by the endothelium, to influence endothelial permeability and vascular smooth muscle cell proliferation.

3.7 Selected references

Busse, R., Hecker, M. & Fleming, I. (1994). Control of nitric oxide and prostacyclin synthesis in endothelial cells. *Arzneimittel-Forschung*, **44** (Suppl 3A), 392–6.

Feelisch, M., te Poel, M., Zamora, R., Deussen, A. & Moncada, S. (1994). Understanding the controversy over the identity of EDRF. *Nature*, **368**, 62–5.

Lüscher, T. F. (1993). Platelet-vessel wall interaction: role of nitric oxide, prostaglandins and endothelins. *Baillières Clinical Haematology*, **6**, 609–27.

Mülsch, A. (1994). Nitrogen monoxide transport mechanisms. *Arzneimittel-Forschung*, **44** (Suppl 3A), 408–11.

Radomski, M. W. & Moncada, S. (1993). Regulation of vascular homeostasis by nitric oxide. *Thrombosis and Haemostasis*, **70**, 36–41.

4

Nitric oxide in the central nervous system

4.1 Introduction

In the central nervous system (CNS), it is known that increased activity in excitatory pathways causes an increase in cGMP levels through the action of the transmitter, glutamate. Since it was shown that cGMP levels rise in cells other than those that are stimulated by glutamate receptor agonists, it is evident that another intercellular factor is involved in this process. Garthwaite, Charles & Chess-Williams (1988) first proposed that endothelium-derived relaxing factor (which came to be known as NO) is this intercellular messenger in the brain (see also East & Garthwaite, 1991; Garthwaite, 1991, 1993; Garthwaite & Boulton, 1995). It is now established that several populations of neurones in the CNS have the capacity to synthesize NO (see Appendix I).

Within the CNS NO has several functions. The three types of NOS are found within the brain: type I (nNOS) in neurones, type II (iNOS) in glial cells (astrocytes) and type III (ecNOS) within endothelial cells and neurones. Most is known about the neuronal NOS, and it is discussed in this chapter. The NOS in endothelial cells of cerebral blood vessels is mostly related to vascular function, and differs little from peripheral vascular endothelial NOS (see Chapters 3 and 5), but some is related to hippocampal neurones involved in long-term potentiation (LTP) (see section 4.5). Type II NOS in brain tissues is confined to glial cells and has not been studied intensively. The main activities of neuronal NOS in the CNS can be broadly divided into four categories: neurotoxicity, neuroprotection, synaptic plasticity, which includes LTP and long-term depression, and modulatory, including involvement in pathways that affect behaviours such as learning or expression of pain. Within each of these categories there are conflicting reports about the involvement of NO, and it is necessary to pay close attention to experimental models and conditions.

4.2 Control of NOS activity

The mechanisms for NO synthesis by NOS are described in Chapter 2. In terms of control of NOS activity in neurones in the CNS the most important determinant is the free cytosolic Ca^{2+} concentration. Constitutive and inducible NOS can readily be distinguished by their requirement for Ca^{2+} for activity. Calmodulin, a Ca^{2+}-binding protein, binds reversibly to constitutive NOS. Intracellular Ca^{2+} concentrations above or equal to 400 nM are required for calmodulin to bind to NOS, and for the enzyme to become fully active (Knowles *et al.*, 1989; Nathan, 1992; Schmidt *et al.*, 1992b; Vincent, 1994). Constitutive NOS activity can therefore be inhibited by Ca^{2+} chelators and calmodulin antagonists. However, calmodulin binds irreversibly to inducible NOS, thus the activity of inducible NOS is independent of the intracellular Ca^{2+} concentration, and is unaffected by Ca^{2+} chelators or calmodulin antagonists (Nathan, 1992). This difference is important in terms of the regulation of NO synthesis and its potential function. At normal resting intracellular Ca^{2+} concentrations, neuronal NOS is inactive (Knowles *et al.*, 1989; Förstermann *et al.*, 1991a; Schmidt *et al.*, 1992b). During action potential discharge voltage-dependent Ca^{2+} channels in the neurolemma open allowing Ca^{2+} to enter the neurone and stimulate release of Ca^{2+} from intracellular stores, elevating the cytosolic Ca^{2+} concentration to sufficient levels to allow the Ca^{2+} to bind to calmodulin. In this activated state Ca^{2+}-calmodulin binds to and activates the NOS (Sheng *et al.*, 1992). Effectively calmodulin acts as a Ca^{2+}-sensitive switch that turns NOS on and off. When the cytosolic levels of Ca^{2+} fall the Ca^{2+} dissociates from the calmodulin, which, in turn, dissociates from the NOS. Thus, in nerves, NO production is tightly controlled by Ca^{2+}, which is strictly regulated by action potential discharge, and the concentrations produced are relatively low. However, in the CNS Ca^{2+} can also enter the cell through voltage-dependent channels activated by glutamate, i.e. NMDA-receptors, and this mechanism may be important in physiological and pathophysiological events (described in the following sections).

Type I NOS can also be phosphorylated and this may form an additional mechanism for regulating its activity (Bredt, Ferris & Snyder, 1992; Schmidt, 1992; Schmidt *et al.*, 1992b; Schmidt, Lohman & Walter, 1993; Lowenstein, Dinerman & Snyder, 1994). It has been suggested that long-term increases in intracellular Ca^{2+} concentrations may downregulate NO formation due to phosphorylation by Ca^{2+}/calmodulin-dependent protein kinase II, which inhibits NOS activity (Schmidt *et al.*, 1992b, 1993). The NOS molecule contains some serine residues towards its N-terminal, and phosphorylation of these serines affects the NOS activity. Each different serine can be phosphorylated specifically by a different kinase, either cAMP-dependent protein kinase (Brüne & Lapetina, 1991; Bredt *et al.*, 1992), protein kinase C (Nakane *et al.*, 1991; Bredt *et al.*, 1992) or Ca^{2+}/calmodulin-dependent protein kinase II (Nakane *et al.*, 1991; Bredt *et al.*, 1992; Schmidt *et al.*, 1992b). The effect of phosphorylation is unclear and, depending upon the experimental conditions, it can result in enhanced, inhibited or unaltered production of NO (see Schuman & Madison, 1994a).

There is also evidence that NOS is under control by negative feedback. Using pyridine haemochrome spectral analysis, electron paramagnetic resonance and Raman spectroscopy, it has been shown that NOS has marked similarities to ferroprotoporphyrin-IX-containing enzymes such as cytochrome P_{450} (Klatt, Schmidt & Mayer, 1992; Stuehr & Ikeda-Saito, 1992; Wang *et al.*, 1993a). The stoichiometry between the haem content and the enzyme is 1:1 (Klatt *et al.*, 1992; Stuehr & Ikeda-Saito, 1992). The haem group seems to be a catalytic centre (Stuehr & Ikeda-Saito, 1992) and it is probably involved in the nucleotide electron transfer sequence. Carbon monoxide and NO itself can interact with this haem group, and consequently inhibit NO production (Klatt *et al.*, 1992). Thus, NO produced by NOS can feed back onto the NOS and inhibit further production of NO.

Mechanisms controlling NO activity may extend to the mechanisms that control its major effector, soluble guanylate cyclase, which synthesizes cGMP. Elevation of intracellular Ca^{2+} levels induces the degradation of cGMP by activating a Ca^{2+}/calmodulin-dependent cGMP phosphodiesterase (Mayer *et al.*, 1992). The activation of this phosphodiesterase occurs at Ca^{2+} concentrations that stimulate NOS activity. Thus, it would seem that cGMP levels are kept low in the cells that produce NO while NO is being produced.

4.3 Neurotoxicity

The neural damage that results from a cerebrovascular accident is primarily due to the release of glutamate and consequent activation of NMDA-receptors. This has been demonstrated in animal models of stroke in which the size of infarct due to arterial occlusion is reduced by treatment with NMDA-receptor antagonists. During some pathological conditions, such as ischaemic hypoxia or hypoxic hypoxia, glutamate is release from neurones in a Ca^{2+}-independent fashion. Non-vesicular Ca^{2+}-independent glutamate release is probably mediated via a reverse uptake mechanism (Szatkowski, Barbour & Attwell, 1990; Attwell, Barbour & Szatkowski, 1993). Normally the glutamate carrier in the neurolemma is driven chemiosmotically with Na^+ symport, and K^+ and OH^- antiport: with a stoichiometry of $2Na^+$:glutamate$^-$:K^+:OH^- it is electrogenic. It is thought that this carrier runs backwards during anoxia (Hirsch & Gibson, 1984; Kauppinen, McMahon & Nicholls, 1988). Because anoxia drastically inhibits cellular oxidative metabolism, and ATP production, plasmalemmal Na^+ pumps fail, with a resultant elevation of extracellular K^+ and intracellular Na^+, decrease of intracellular K^+ and extracellular Na^+ levels, and a consequent depolarization. These conditions are favourable for driving the glutamate transporter backwards, and will result in an estimated extracellular glutamate concentration of more that 250 μM (Attwell *et al.*, 1993), which is sufficient to cause neurone death via activation of

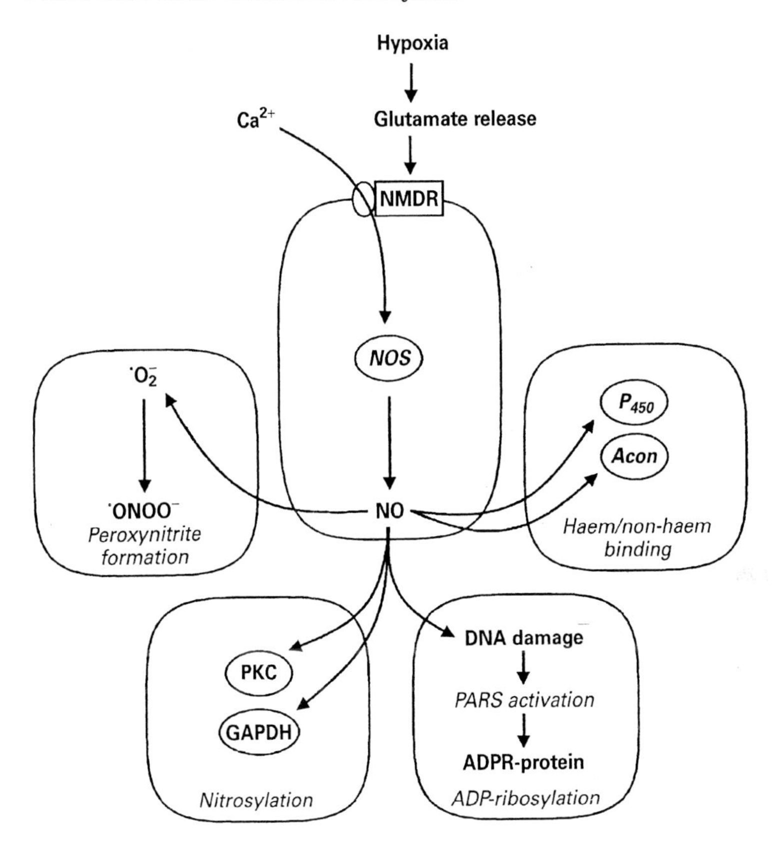

Fig. 4.1 Some possible mechanisms for neurotoxic effects of NO induced by glutamate. Hypoxia stimulates massive release of glutamate, which acts on NMDA-receptors (NMDR) to induce Ca^{2+} entry into the neurone, and subsequent calmodulin-dependent activation of NOS. The NO produced can interact with superoxide anions, resulting in peroxynitrite formation, which has cytotoxic activity. NO produced can interact with superoxide anions, resulting in peroxynitrite formation, which has cytotoxic activity. NO can nitrosylate several enzymes, including phosphokinase C (PKC), with various consequences, and glyceraldehyde-3-phosphate dehydrogenase (GAPDH), which results in inhibition of glycolysis. NO can cause DNA mutation and strand breaking, which in turn stimulates polyADP-ribose synthase (PARS) activation, ADP-ribosylation of proteins (ADPR-protein), and ultimately causing massive energy depletion and cell death (see text). Another site of action is iron in haem or non-haem (Fe–S) complexes that are associated with enzymes such as cytochrome P_{450} (P_{450}) or aconitases (Acon), which have deleterious effects, including inhibition of glycolysis. In nitrosylation and DNA-damage mechanisms, NO may be acting indirectly, in a different redox form, rather than as the free radical.

NMDA-receptors and subsequent excessive, cytotoxic influx of Ca^{2+} into the postsynaptic cell, if it remains so elevated for a period of minutes.

Several enzymatic processes contribute to the neurotoxicity induced by glutamate receptors, and they all seem to be Ca^{2+} dependent (Brorson, Manzolillo & Miller, 1994; Brorson, Marcuccilli & Miller, 1995). One of these is the overproduction of NO, some of the effects of which are summarized in Fig. 4.1. In experimental models the effects of inhibitors are very varied, in some instances they can prevent nearly all the neurotoxic damage, and in some cases they can markedly exacerbate it (Dawson, 1994). The different results obtained by different groups is probably attributable to the different experimental models and conditions, and more particularly whether or not peroxynitrite can be produced in the experimental system (Lipton & Stamler, 1994).

Elevation in cGMP induced by stimulating NMDA-receptors in rat and mouse cerebellar slices is prevented by inhibiting NO synthesis with N^G-monomethyl L-arginine (L-NMMA) or L-N^G-nitroarginine (L-NNA), or by complexing nascent NO with haemoglobin (Bredt & Snyder, 1989; Garthwaite *et al.*, 1989; East & Garthwaite, 1990; Wood *et al.*, 1990; Southam & Garthwaite, 1991; Kollegger, McBean & Tipton, 1993). The significance of the antagonism by haemoglobin is that the NO must leave the cell from which it is produced in order to stimulate guanylate cyclase in another cell, because the haemoglobin itself remains extracellular. Perhaps the best evidence for this comes from a microdialysis study in which a tube is implanted into the rat cerebellum and, after recovery, the extracellular fluid that dialyses across the wall of the tube in the conscious ambulant animal can be assayed (Luo, Knezevich & Vincent, 1993). When NMDA is applied via the microdialysis tube nitrite and nitrate increases can be measured in the dialysate. These increases, and the basal levels of nitrite and nitrate are suppressed by either NMDA-receptor antagonists or inhibition of NOS (Luo *et al.*, 1993), indicating that in addition to NMDA-receptor activation being transduced to NO production, there is a tonic level of glutamate-stimulated NO production in these animals. A direct demonstration of the production of NO in the rat brain *in vivo* in response to bilateral carotid artery occlusion has been performed using electron paramagnetic resonance (Tominaga *et al.*, 1993). With this technique, spin-trapping agents form adducts with NO radicals, and the electron-spin signal is measured spectroscopically. After induction of ischaemia there is a strong signal, but pretreatment with L-NAME prevents its appearance (Tominaga *et al.*, 1993). Direct evidence for the relationship between NMDA-receptors and NOS-containing neurones comes from studies of rat brain, in which NOS is visualized immunohistochemically simultaneously with the mRNA, using *in situ* hybridization, for NMDA-receptors (NR1 mRNA) (Price, Mayer & Beitz, 1993). In the cortex, striatum and mesencephalon, neurones containing NOS contain higher levels of NR1 mRNA than do non-NOS-containing neurones. Also, within any region there is a variability in the NR1 mRNA content of NOS-containing cells, with some NOS-positive cells not expressing any NMDA-receptor mRNA (Price *et al.*, 1993).

In cultures of the rat brain, glutamate induces elevation of cGMP levels, this effect is mimicked by NMDA, and is blocked by the NMDA-receptor antagonist MK-801 and by L-NMMA or L-NNA (Demerle Pallardy *et al.*, 1991). However, when the cultures were made anoxic although MK-801 could protect against cell death the NOS inhibitors could not (Demerle Pallardy *et al.*, 1991), suggesting that glutamate neurotoxicity does not involve NO-induced cGMP formation. In support of this finding, *in vivo* experiments have shown that chronic pretreatment of rats with NOS inhibitors (i.p. injections twice daily for 4 days) has no effect on lesions induced by focal hippocampal NMDA injections (Lerner Natoli *et al.*, 1992). In contrast, in a slightly different whole-animal model, when the NOS antagonist is applied after occlusion of the middle cerebral artery, to induce an ischaemic infarct and focal cerebral oedema, the deleterious effects are substantially attenuated (Nagafuji *et al.*, 1992; Carreau *et al.*, 1994). Occlusion of the middle cerebral artery in rats causes a transient increase in cGMP levels, nitrite levels and NOS activity in the cortex (Kader *et al.*, 1993). Pretreatment with L-NNA abolishes the rise in nitrite and cGMP, indicating overall that cerebral ischaemia induces a transient increase in NO production and activation of guanylate cyclase (Kader *et al.*, 1993). Similarly, L-NNA affords protection of neonatal rats against glutamate-dependent neuronal damage induced by ischaemic hypoxia (Trifiletti, 1992). In further contrast, bilateral occlusion of the carotid artery in the gerbil for 5 min causes cerebral ischaemia with a selective destruction of hippocampal CA1 neurones. Treatment with L-NNA (50 $mg\,kg^{-1}$) 4 h prior to ischaemia significantly extended the hippocampal and cerebral damage (Weissman *et al.*, 1992). Thus in this model inhibition of NO generation exacerbates neurotoxicity, perhaps indicating that NO is actually having a cytoprotective role in this instance. It has also been suggested that there might actually be a narrow therapeutic window for inhibition of NOS in the gerbil, because an optimal dose of L-NNA (3 $mg\,kg^{-1}$) affords greater protection of CA1 neurones against ischaemia due to carotid artery occlusion than does a higher or lower dose (10 $mg\,kg^{-1}$ or 0.3 $mg\,kg^{-1}$, respectively) (Nagafuji *et al.*, 1993).

In models where NO has been implicated in neurotoxicity, it may also stimulate ADP-ribosylation of proteins (Wallis *et al.*, 1993; Brüne *et al.*, 1994; Zhang *et al.*, 1994). In rat hippocampal slices, NO is profoundly neurotoxic to CA1 neurones, and this can be prevented by treatment with nicotinamide. Since nicotinamide inhibits ADP-ribosylation, this suggests that ADP-ribosylation is involved in NO-induced neuronal injury (Wallis *et al.*, 1993). The mechanism by which ADP-ribosylation is involved in cell death is quite interesting, because ultimately there is a massive depletion of cellular energy, initially triggered by NO causing DNA mutations and strand breaking (Bredt & Snyder, 1994; Zhang & Snyder, 1995). The enzyme that is responsible for ADP-ribosylation is poly-adenosinediphosphoribose synthetase (PARS). PARS is a nuclear enzyme, and is activated by DNA damage, being involved in DNA repair mechanisms. It utilizes NAD, removing the adenosine diphosphoribose moiety thus yielding nicotinamide, to catalyse the attachment of 50–100 ADP-ribose moieties to nuclear proteins (mostly histones), and to PARS itself. The leads to the depletion of NAD. In the reconversion of a molecule of nicotinamide back to NAD a total of four ATP molecules is required, and there is a consequential drop in the energy charge of the cell. If the extent of DNA damage is small, activation of PARS facilitates DNA repair, but if the damage to DNA is extensive this energy-depleting cycle results in cell death. Other enzymes can become ADP-ribosylated, one in particular that has been identified that undergoes NO-stimulated ADP-ribosylation is glyceraldehyde-3-phosphate dehydrogenase, resulting in inhibition of glycolysis (Zhang & Snyder, 1992, 1993).

A mechanism for the neurotoxic effects of glutamate-induced NO production has been suggested that involves hydroxyl radicals. Using a microdialysis technique that allows sampling of extracellular fluid in specific brain regions, it has been found that activation of NMDA-receptors increases hydroxyl radical production in the striatum (Hammer, Parker & Bennett, 1993). Furthermore, antagonists of NOS and of protein kinase C prevent this increase, implying that neuronal death might involve NOS- and protein-kinase C-dependent synthesis of hydroxyl radicals (Hammer *et al.*, 1993).

4.4 Neuroprotection

Although NO production is implicated in neurotoxicity, it can also have cytoprotective effects. Whether or not NO is protective under physiological conditions is not known, and whether or not it confers cytoprotection *in vitro* depends upon the experimental conditions. In general the free radical of NO can take part in a cytoprotective mechanism, but if superoxides are present, which result in the production of peroxynitrite, then neurotoxic effects are more likely (Lipton & Stamler, 1994).

The mechanism of cytoprotection afforded by NO centres on the redox state of a particular site of the NMDA-receptor channel (Korenaga *et al.*, 1994; Lipton & Stamler, 1994). At this site two cysteine residues appose one another and can form a disulphide bridge. When this site is reduced by dithiothreitol the Ca^{2+} current flow increases, and when it is oxidized by 5,5′-dithio-bis(2-nitrobenzoic acid) the current flow decreases (Lei *et al.*, 1992). NO can oxidize the free cysteine sulphydryl groups, to form *S*-nitrosothiols, accelerating disulphide bridge formation resulting in a decrease in Ca^{2+} flow through the channel (Lipton, Singel & Stamler, 1994; Lipton & Stamler, 1994).

NO may also be cytoprotective because it can inhibit the binding of glutamate to its receptor. In preparations of rat brain synaptic membranes NO causes a haemoglobin-sensitive, concentration-dependent antagonism of glutamate (Fujimori & Pan Hou, 1991). NO and NO donors can reduce the Ca^{2+} flux through NMDA-receptor channels in a manner unrelated to oxidation of the redox site, in cultured rat forebrain neurones (Hoyt *et al.*, 1992). This effect is likely to be mediated by an action at the level of the NMDA-receptor channel because it is not mimicked by application of the cGMP analogue 8-bromo-cGMP (Hoyt *et al.*, 1992). Decreasing the efficacy of released glutamate is another example of negative feedback control of NO synthesis, with NO inhibiting its own stimulus.

As already mentioned, bilateral occlusion of the carotid artery in the gerbil for 5 min causes cerebral ischaemia with a selective destruction of hippocampal CA1 neurones, and inhibition of NOS by L-NNA (i.p.) prior to ischaemia exacerbates the damage (Weissman *et al.*, 1992), indicating that NO is playing a cytoprotective role. Bilateral carotid artery occlusion (5 min) in gerbils also raises the level of activity of NOS in forebrain, midbrain and hindbrain (Caldwell *et al.*, 1994). Four days after surgery histological neuronal death is found particularly in the hippocampal CA1 layer. The animals also become hyperactive for up to 3 days following surgery. In nNOS– mice (i.e. mice in which the gene encoding type I NOS was disrupted) infarct volumes and histological damage caused by occluding the middle cerebral artery are significantly smaller than they are in control mice (Huang *et al.*, 1994). Furthermore, inhibition of residual NOS activity (type III) in nNOS– mice by L-NNA causes a significantly larger infarct volume. Thus, it appears that type I NOS is responsible for production of the NO that exacerbates damage induced by ischaemia, and that NO from type III NOS is responsible for protective activity. This probably reflects the differential localization of these two isoforms. Some of the type III NOS is in vascular endothelial cells as well as neurones, and it could be the activity of the endothelial NOS that is cytoprotective (Huang *et al.*, 1994).

It is not fully understood whether or not cytoprotection is pathophysiological. In Alzheimer's disease there is a sparing of NOS-containing neurones in brain regions where neuronal degeneration is severe (Mufson & Brandabur, 1994).

4.5 Long-term potentiation

Long-term potentiation is an enduring enhancement in the efficiency of synaptic transmission, or an activity-dependent increase in synaptic strength, lasting from hours to weeks or even longer, and is a property of many excitatory synapses in the CNS. It was first described by Bliss and Lømo (1973), who reported that when granule cells were stimulated at high frequency for several seconds and then tested with single shocks, the excitatory postsynaptic potential (e.p.s.p) of the synapse increased. These findings were interesting in that they appeared to be in agreement with Hebb's original statement (Hebb, 1949), which postulated that if a nerve terminal A stimulates cell B then over a period of time, due to a growth process or metabolic change taking place in either cell, A's efficiency in firing B is increased. It is thought that this type of synaptic plasticity might be a synaptic correlate of learning and memory. The process is most pronounced in higher brain centres involved in cognitive function, particularly in the cerebral cortex and hippocampus. LTP is often demonstrated in the hippocampus, between the Schaffer collateral synapse of cornu ammonis (CA1) and CA3, CA1 pyramidal cells being postsynaptic and CA3 being presynaptic. *In vitro*, in brain slice preparations e.p.s.ps can be recorded from postsynaptic pyramidal cells, using an intracellular microelectrode. LTP occurs when the excitatory synapses are stimulated such that the depolarization of postsynaptic neurones is coincident with the release of the neurotransmitter (glutamate) from the presynaptic nerve terminals (Schuman & Madison, 1994a,b). After tetanic stimuli (20–100 pulses at 100 Hz) have been applied to the presynaptic nerve, the e.p.s.ps recorded in response to a single stimulus markedly increase. The mechanism of induction of LTP is not fully understood, but in its simplest terms it involves glutamate and excitatory amino acid receptors. The normal e.p.s.p is mediated by glutamate released from the presynaptic nerve acting on AMPA-receptors, resulting in the opening of Na^+ ion channels, and depolarization of the postsynaptic cell. However, during the tetanic stimulation NMDA-receptors are also activated, which results in Ca^{2+} entering the postsynaptic cell. This Ca^{2+} influx brings about changes in the activity of protein kinases, proteases, phospholipases and NOS in the postsynaptic cell (Manilow, Schulman & Tsien, 1989; Williams, Errington & Bliss, 1989; Schuman & Madison, 1991). Ca^{2+}/calmodulin-dependent protein kinase II (CaMKII) and protein tyrosine kinases are particularly involved in LTP.

There are three main sites that have been implicated in the expression of LTP: the presynaptic nerve, the postsynaptic receptor and its related transduction mechanisms, and the postsynaptic dendritic spines. From the nerve terminal there may be increased glutamate release following the establishment of LTP (Bekkers & Stevens, 1990; Manilow & Tsien, 1990), then there may be increased ion flux through the glutamate-receptor ion channels, or an increase in the number of receptors, or there may be remodelling of the nerve terminal and the postsynaptic dendritic spines such that a greater area of synaptic contact is made, enabling greater postsynaptic depolarization for a given stimulus in the presynaptic terminal. Because the induction and maintenance of LTP involve both pre- and postsynaptic sites, it is thought that there must be a retrograde signal that passes from the postsynaptic cell to the presynaptic cell. It is thought that NO may provide the retrograde link that appears to be necessary for co-ordinating the enhancement of both pre- and postsynaptic mechanisms that contribute to LTP.

When NO was found to be synthesized and released from cultured cerebellar granule cells, and result in cGMP formation when NMDA-receptors were activated, it was put forward that NO could be this retrograde interneuronal messenger involved in LTP (Garthwaite *et al.*, 1989; East & Garthwaite, 1991). The basic evidence for the involvement of NO in the establishment of LTP comes from studies of hippocampal slices or dissociated cell cultures, recording membrane potentials of CA1 cells, or recording the population potential extracellularly. If inhibitors of NOS are present when the tetanic stimulation is applied, LTP fails to develop properly (Bohme *et al.*, 1991; O'Dell *et al.*, 1991; Schuman & Madison, 1991; Haley, Wilcox & Chapman, 1992b). Furthermore, in all these cases the inhibition of NOS was overcome by application of L-arginine. Haemoglobin, which remains extracellular, prevents the LTP from developing (O'Dell *et al.*, 1991; Schuman & Madison, 1991; Haley *et al.*, 1992b), indicating that the NO must diffuse out of the cell in which it is formed and enter another cell in order to have its action. It seems likely that the CA1 pyramidal cell is the source of NO, because intracellular injection of these cells with NOS inhibitors prevents the LTP from developing (O'Dell *et al.*, 1991; Schuman & Madison, 1991). The conditions used for inducing LTP are of paramount importance, particularly the working temperature and whether or not weak or strong tetanic stimuli are applied. For example, at 31 °C, LTP induced by weak tetanic stimuli can be blocked by L-NNA, whereas that induced by a strong stimulus cannot (Haley, Malen & Chapman, 1993).

Hebb's original rule, which stated that only two cells, A and B, were involved in the increase in efficiency, has been contested on the grounds of the nature of NO. Because it can diffuse through membranes over long distances it can, potentially, affect synapses, resulting in a non-Hebbian, heterosynaptic, potentiation (Schuman & Madison, 1994). This phenomenon has been demonstrated in culture: when LTP is induced in a single hippocampal CA1 neurone LTP is also expressed in adjacent postsynaptic cells within 500 μm (Bonhoeffer, Staiger & Aersten, 1989; Kossel, Bonhoeffer & Bolz, 1990; Schuman & Madison, 1991).

Neither type I NOS nor NADPH-diaphorase activity is usually found in hippocampal pyramidal CA1 neurones (Vincent, 1994), and this casts doubt upon the CA1 neurones being the source of NO during induction of LTP. However, it has become clear that the CA1 pyramidal cells contain type III NOS rather than type I (Dinerman *et al.*, 1994; O'Dell *et al.*, 1994). The reason for the lack of NADPH-diaphorase activity is puzzling, but using glutaraldehyde rather than formaldehyde as a fixative allows such activity to be found in these hippocampal CA1 neurones (Dinerman *et al.*, 1994). Some authors have found a very small population of type-I-NOS-containing and NADPH-diaphorase-active hippocampal pyramidal CA1 neurones (Valtschanoff *et al.*, 1993a) and, by using relatively gentle fixation methods, type I NOS has been visualized immunohistochemically in substantial populations of these neurones (Wendland *et al.*, 1994). Furthermore, although the mRNA for type I NOS could not be localized in CA1 pyramidal cells by conventional *in situ* hybridization techniques (see Vincent, 1994), using single-cell polymerase chain reaction methods to amplify type I NOS mRNA the amino acid sequence of the product of the amplified mRNA has been found to be identical to that of the neuronal isoform of NOS (i.e. type I NOS) (Chiang *et al.*, 1994).

Gene targeting has confirmed that type III NOS is involved in LTP. In hippocampal slices from nNOS– mice, which therefore cannot synthesize NO from type I NOS, LTP could be induced in the normal way, and its induction is sensitive to inhibition of NOS by analogues of L-arginine (O'Dell *et al.*, 1994). The nNOS– mice, just like wild-type mice, express the endothelial form of NOS (type III) in hippocampal CA1 neurones, and thus type III NOS rather than type I NOS could be responsible for the neuronal generation of NO during LTP (O'Dell *et al.*, 1994).

Whether or not guanylate cyclase is the effector of NO in LTP induction has not been established. Although NO donors can produce LTP (Bonhoeffer *et al.*, 1989; Bohme *et al.*, 1991), the membrane-permeable analogue of cGMP, dibutyryl cGMP (db-cGMP), does not (Haley *et al.*, 1992b). However, it has also been shown that application of db-cGMP attenuates the effect of L-NAME in inhibiting LTP establishment (Haley *et al.*, 1992b), which would argue for a role in favour of cGMP. Given that elevation of intracellular Ca^{2+} levels induces the degradation of cGMP by activating a Ca^{2+}/calmodulin-dependent cGMP phosphodiesterase (Mayer *et al.* , 1992), keeping cGMP levels low in the cells that produce NO, if the guanylate cyclase/cGMP system does have a role in LTP it would be restricted to presynaptic mechanisms.

ADP-ribosylation has also been suggested to be involved in the presynaptic aspect of LTP induction (Brüne & Lapetina, 1989). Inhibitors of cytosolic ADP-ribosyltransferase can prevent LTP when applied extracellularly, but not when applied intracellularly in the postsynaptic neurone (Schuman *et al.*, 1992).

4.6 Long-term depression

Long-term depression (LTD) is a possible model for learning motor movements in the cerebellum. It is a long-lasting depression of parallel fibre synapses following repeated excitation by climbing fibres of Purkinje cells. It can be observed in many higher regions of the brain, including the hippocampus, visual cortex and cerebellum (Ito, 1989).

In the cerebellum the proposed mechanism of LTD involves the synapses of parallel fibres and climbing fibres, from granule cells and the inferior olive, respectively, with Purkinje cells. In both types of synapse the neurotransmitter is glutamate, acting via AMPA-receptors. It is thought that continuous climbing fibre activation elicits e.p.s.ps that lead to a decrease in synaptic strength at the synapse between the parallel fibres and the Purkinje cells. The reduction in synaptic strength is due to a decrease in sensitivity of the postsynaptic AMPA-receptors for glutamate (Hirano, 1991; Linden & Connor, 1991). The depolarization evoked by the parallel fibres results in the activation of Ca^{2+} channels and influx of Ca^{2+} via metabotropic glutamate receptors (Ito & Karachot, 1990). The increase in Ca^{2+} is a result of the activation of voltage-sensitive Ca^{2+} channels, and also the activation of phospholipase C via the metabotropic receptor. Elevated Ca^{2+} levels activate protein kinases, and this seems to be essential in the development of LTD (Crepel & Jaillard, 1990); activated protein kinase C phosphorylates subunits of the AMPA-receptor, thereby reducing its sensitivity (Linden & Connor, 1991, 1993). The elevated Ca^{2+} levels also stimulate NOS, resulting in cGMP synthesis (Ito & Karachot, 1990; Linden & Connor, 1993), and subsequent activation of protein kinase G, which, in turn, contributes to the receptor desensitization (Ito & Karachot, 1990). Inhibition of protein kinase C or protein kinase G prevents induction of LTD (Hartell, 1994).

Induction of LTD is prevented by haemoglobin, NOS inhibitors, methylene blue and sodium nitroprusside, implying that it involves the generation of NO (Crepel & Jaillard, 1990; Ito & Karachot, 1990; Shibuki & Okada, 1991). When climbing fibres are stimulated NO is released from cerebellar slices, and LTD can be induced by applying NO or 8-bromo-cGMP instead of stimulating the presynaptic elements (Shibuki & Okada, 1991; Daniel *et al.*, 1993). If cGMP or NO donors are injected directly into Purkinje cells LTD is induced (Daniel *et al.*, 1993; Hartell, 1994), and this provides evidence that the soluble guanylate cyclase in the Purkinje cell itself is the site of action for evoked NO, synthesized during the stimulus for LTD. The cerebellar granule cells contain mRNA for soluble guanylate cyclase (Matsuoka *et al.*, 1992).

The site of NO production has not been identified with any certainty. In mature animals the cerebellar Purkinje cells do not appear to contain either NADPH-diaphorase activity or type I NOS (Brüning, 1993; Yan, Jen & Garey, 1993; Chen & Aston Jones, 1994; Schilling, Schmidt & Baader, 1994), although they do in neonatal animals (Brüning, 1993; Yan *et al.*, 1993). Whether or not they are like hippocampal CA1 neurones, and contain type III NOS and fixation-sensitive NADPH-diaphorase (see Section 4.5), does not appear to have been reported.

However, in the rat cerebellum NOS is found in granule cells (Bredt, Hwang & Snyder, 1990; Schilling *et. al.*, 1994), but in humans type I NOS is found in cerebellar Purkinje cells in addition to basket and granule cells (Egberongbe *et al.*, 1994).

In the 'nervous mouse', a mutant deficient in cerebellar Purkinje cells, NMDA evokes large increases in cerebellar cGMP, implying that non-Purkinje cells produce substantial levels of NOS, and cGMP (Wood, Emmett & Wood, 1994). However, in similar mutant mice cerebellar NOS activity is reduced by approximately 50%, implying that the Purkinje cells do indeed contain NOS (Ikeda *et al.*, 1993). It is worth pointing out that in this study NOS activity was measured rather than NOS presence, and under the conditions used the activity of non-type I NOS would also be measured.

4.7 Behavioural responses involving NO

Treatment of rats with L-NNA (i.p.), at doses that block LTP in hippocampal slices, impairs spatial learning in a radial arm maze and olfactory memory, but has no effect on shock-avoidance learning (Bohme *et al.*, 1993). Another study has shown that inhibition of NOS impairs behavioural performance during task acquisition, but has no effect once spatial learning has been accomplished (Bannerman *et al.*, 1994; Yamada *et al.*, 1995). Also, L-NAME decreases locomotor activity in habituation tasks (Yamada *et al.*, 1995). Impairment of learning skills could involve catecholaminergic and serotonergic pathways, since the same doses of L-NAME that inhibited learning decreased the accumulation of 5-HT metabolites and increased noradrenaline metabolites (Yamada *et al.*, 1995). Amnesia can be induced in chicks by injecting L-NNA into the intermediate hyperstriatum ventrale after they have been trained in passive avoidance tasks: the effects of L-NAME are obviated by concomitant injection of L-arginine (Holscher & Rose, 1992, 1993).

Injection of the NO precursor L-arginine into the lateral cerebral ventricle of rats able to drink *ad libitum* has no significant effect, but in rats deprived of water for 24 h it inhibits re-hydrating drinking (Calapai *et al.*, 1992). This effect is antagonized by L-NAME, which has no effect in the absence of the L-arginine. Thus NO is implicated as an inhibitor of drinking when thirst is stimulated by water deprivation. Drinking induced by intracerebroventricular (i.c.v.) application of angiotensin II is also inhibited by L-arginine (5–10 μg, i.c.v), which also inhibits drinking in thirsting rats when injected specifically into the preoptic area (Calapai *et al.*, 1992).

In cats, respiratory pattern generation is affected by intrapontine injection of L-NNA. The period of inspiration increases, implying that endogenous NO plays an inhibitory role in normal pneumotaxic mechanisms (Ling *et al.*, 1992).

Penile erection and yawning can be induced in rats by injecting the hypothalamic paraventricular nucleus (PVN) with oxytocin (Melis, Stancampiano & Argiolas, 1994). These responses are dose-dependently reduced and eventually abolished by injection of L-NAME (5–20 μg) at the same site, or by methylene blue (100–400 μg, i.c.v.) (Melis *et al.*, 1994). The NO donor, nitroglycerine (33–99 μg, i.c.v. or 1.6–6.6 μg, into the PVN) also induces erection and yawning, and these responses are again blocked by methylene blue (200–400 μg, i.c.v.) (Melis *et al.*, 1995b). Selective 5-HT_{1c} receptor antagonists injected subcutaneously into anaesthetized rats induce penile erection and yawning. Both L-NAME and L-NMMA (50–500 μg i.c.v.) prevent these two behavioural responses (Melis *et al.*, 1995a). L-Arginine reverses the effects of the NOS inhibitors, and D-NMMA is without effect. Inhibition of guanylate cyclase by methylene blue or LY 83583 (i.c.v) also prevents the yawning and erection evoked by the 5-HT_{1c} receptor agonists (Melis, Stancampiano & Argiolas, 1995a). All these results indicate that oxytocinergic and serotonergic pathways may involve generation of NO, or stimulation of central nitrergic neurones, in these behavioural responses (Melis *et al.*, 1994, 1995a,b).

That NO is involved in behavioural mechanisms has also been demonstrated using NOS-knockout nNOS– mice (Nelson *et al.*, 1995). In this genetic knockout the animals are ostensibly normal and when housed singly they show no overt behavioural disorders (Huang *et al.*, 1993, 1994), but they are exceptionally aggressive towards other males, and repeatedly attempt to mate non-receptive females (Nelson *et al.*, 1995). This type of study is difficult to interpret, because the lack of a gene and its product may become compensated for in unknown ways during development. So although deletion of this NOS gene has marked behavioural effects it cannot be categorically stated that central nitrergic pathways are normally concerned in controlling these behaviours.

4.8 Perception of pain

NO may be involved in the perception of pain at many levels of nociceptive neural pathways. In the periphery, primary afferent neurones and dorsal root ganglia may contain NOS, and in the brainstem and thalamus several sensory structures also contain NOS (see Appendix I). Much of the work on the involvement of NO in pain pathways has been carried out using behavioural models of expression of pain using systemic application of agents via intraperitoneal (i.p.) or per os (p.o.) routes, and the more precise supraspinal administration (intrathecal, i.t.) or central administration (i.c.v.). From several studies of allodynia and hyperalgesia or hyperaesthesia it would seem that nociceptive reflexes involve a glutamatergic NMDA-receptor-mediated pathway that involves the production of NO, and this leads to enhancement of the processing, or spinal facilitation, of the afferent input that is ultimately conveyed to the cortex, and subsequently manifest in behavioural responses (see Meller & Gebhart, 1993, 1994a,b; Kawabata & Takagi, 1994). Allodynia is the condition where a non-noxious stimulus is perceived as being noxious, whereas hyperalgesia is the increased sensitivity to an already noxious stimulus.

In mice, subplantar injection of formalin causes a biphasic behavioural response of paw licking that is indicative of pain sensation. The first phase, due to stimulation of afferent C-fibres, occurs within 30 s, and the second phase, which is due to the development of inflammatory responses, occurs some 15–30 min later. L-NAME (1–75 mg kg^{-1}, i.p.) causes a dose-dependent long-lasting (at least 24 h) antinociception (Moore *et al.*, 1991). L-NAME administered i.c.v. or p.o. (0.1–100 μg per mouse or 75–150 mg kg^{-1}, respectively) is also antinociceptive. When administered by these routes the second phase responses can be abolished while the first phase responses tend to be at best reduced by up to 72% (100 μg, i.c.v) or not at all (150 mg kg^{-1}, p.o.) (Moore *et al.*, 1991). Similar results have been obtained in rats, where the antinociceptive effects of arginine analogue NOS inhibitors also show stereospecificity (Malmberg & Yaksh, 1993). Systemic application of the NOS inhibitor 7-nitroindazole (10–50 mg kg^{-1}, i.p.) to mice produces antinociception in the second phase of formalin-induced hind-paw licking, which is reversed by L-arginine (50 mg kg^{-1}, i.p.) (Moore *et al.*, 1993a). At the time of nociceptive activity, which is maximal at 18–30 min post injection, brain NOS activity is markedly reduced by 7-nitroindazole. These results indicate that NO is involved in CNS pain perception. Intrathecal application of L-NNA in rats inhibits the flinching behaviour associated with subplantar formalin injection. L-Arginine, but not D-arginine, reverses the effects of L-NNA, indicating that NO is involved in nociceptive pathways, particularly in the spinal cord (Yamamoto, Shimoyama & Mizuguchi, 1993a). In electrophysiological experiments, where the activity of single dorsal horn neurones is monitored (wide dynamic range second-order sensory neurones in L1–L3 that respond to noxious stimuli such as pinch or heat, and non-noxious stimuli such as touch and light pressure), topical application of either L-arginine or L-NAME reduces the effects of electrically stimulating the presynaptic Aδ-fibre or C-fibre inputs at their peripheral endings in the skin of the toes of the hind-paw (Haley, Dickenson & Schachter, 1992a). Systemic application of L-NAME, or application to the peripheral nerve terminals, has no significant effect; however, when formalin is applied in the middle of the defined receptive field of the primary sensory neurone in the toe skin, there is a biphasic response of firing of the dorsal horn neurone that correlates with the behavioural response. Topical application of L-NAME or L-arginine to the spinal cord inhibits the firing in both phases, and subcutaneous injection of L-NAME 10 min prior to formalin reduces only the second response (Haley *et al.*, 1992a).

Subplantar injection of carageenan into the rat paw induces an inflammatory response and concomitant thermal and mechanical hyperalgesia. L-NAME (2–200 nmol, i.t.) dose-dependently attenuates the thermal hyperalgesia, but not the mechanical hyperalgesia or the inflammation (Meller *et al.*, 1994). Furthermore, application of L-NAME before carageenan delays the onset of the thermal hyperalgesia but not the mechanical hyperalgesia (Meller *et al.*, 1994). Thus, in this model, which also involves NMDA-receptor activation, NO production is involved in both the development and maintenance of thermal hyperalgesia. Thermal hyperalgesia induced by NMDA (i.t.) in mice is also attenuated by L-NAME (Kitto, Haley & Wilcox, 1992). The fact that L-NAME alone has no effect on the radiant heat tail-flick latency indicates that NO is only involved in a modulation of this pain pathway (Kitto *et al.*, 1992). Another model of pain in which stimulation of NMDA-receptors appears to induce NO formation is the constricted sciatic nerve. Constriction injury of the rat sciatic nerve by loose ligatures induces thermal and mechanical hyperaesthesia: onset of the thermal hyperaesthesia can be delayed by pretreatment with L-NNA, L-NAME or methylene blue (Meller *et al.*, 1992; Yamamoto & Shimoyama, 1995).

Further evidence for the separate function of the thermal and mechanical nociceptive mechanisms comes from experiments in which i.t. administration of L-arginine (10 pmol to 10 nmol) produces a dose-dependent increase in the threshold of the mechanical nociceptive tail-withdrawal reflex in rats, while having no significant effect on the thermal nociceptive threshold (Zhuo, Meller & Gebhart, 1993). Atropine (i.t.) decreases the mechanical but not the thermal nociceptive threshold, and this effect can be reversed by L-arginine (Zhuo *et al.*, 1993). Methylene blue and L-NAME (i.t.) markedly enhance the nociceptive actions of atropine, but otherwise have no effect. Thus, NO is involved in the spinal synaptic plasticity and interacts with cholinergic mechanisms in mechanical nociceptive transmission (Zhuo *et al.*, 1993).

The differential involvement of NO in the thermal and mechanical nociceptive pathways is interesting. It is apparent that the thermal route requires NMDA-receptor activation that primarily induces NO synthesis, whereas the mechanical route requires non-NMDA glutamate receptors that primarily induce cyclo-oxygenase activity and subsequent arachidonic acid metabolism (Malmberg & Yaksh, 1993; Meller & Gebhart, 1994a,b).

Morphine is well known as an analgesic agent, and when applied i.t. its antinociceptive effects can be quantified in the simple radiant heat tail-flick test. L-NNA, haemoglobin or methylene blue (2 μg, 120 μg and 5 μg, i.t., respectively), all inhibitors of the NO/guanylate cyclase system, potentiate the antinociceptive activity of morphine, while L-arginine and 3-morpholinylsydnoneimine (SIN-1) (20 μg and 5 μg, respectively) attenuate it (Xu & Tseng, 1995). L-NAME (100 μg, i.c.v.) and methylene blue (10 μg, i.c.v.) both have limited antinociceptive effects on their own (Shibuta *et al.*, 1995). In marked contrast, the antinociceptive effects of β-endorphin are attenuated by L-NNA, haemoglobin and methylene blue (Xu & Tseng, 1995).

Intrathecal injection of L-arginine (5 μg) to mice induces allodynia, assessed by stroking the flank with a paintbrush (Minami *et al.*, 1995). Thus, L-arginine treatment converts a non-noxious mechanical stimulus into a noxious one. This effect of L-arginine is inhibited by L-NAME or methylene blue, suggesting that NO is involved in a spinal facilitation that occurs during non-noxious mechanosensory transmission. Prostaglandin E_2 (PGE_2, i.t.) also induces allodynia that is blocked by NMDA antagonists, and this is also sensitive to L-NAME and methylene blue (Minami *et al.*, 1995).

In a model of visceral pain, diclofenac, administered i.p., i.c.v or i.t., has antinociceptive effects that appear to involve serotonergic pathways (Bjorkman, 1995). Pain behaviour of scratching, biting and licking can be evoked by application of NMDA, and this can be prevented by pretreating with diclofenac, paracetamol or ibuprofen. The antinociceptive effects of these three anti-inflammatory agents can be reversed by L-arginine, but not D-arginine, implying that NO or nitrergic transmission is involved in central pain pathways (Bjorkman, 1995).

In summary, inhibition of NOS has antinociceptive effects in animal models when pain has been produced by chemically stimulating peripheral nerve terminals (formalin or cargeenan), or in models of thermal hyperalgesia, visceral pain or mechanical allodynia. Conversely, inhibition of NOS may exacerbate pain in models of mechanical hyperalgesia.

4.9 Vascular control

Although large cerebral arteries are innervated by NOS-containing neurones that are derived from parasympathetic and sensory ganglia, small vessels may be innervated by non-autonomic NOS-containing nerves (Iadecola *et al.*, 1993a; Nozaki *et al.*, 1993). A close association between NOS-containing central neurones and cerebral arterioles and microvessels has been demonstrated in the rat brain by NADPH-diaphorase histochemistry and NOS-immunohistochemistry at the light and electron microscopic levels (Poeggel *et al.*, 1992; Regidor, Edvinsson & Divac, 1993). This raises the possibility that NO may be involved in a mechanism whereby central NOS-containing neurones can control their own blood supply. The cerebrovasodilatation response to stimulation of the cerebellar fastigial nucleus *in vivo* has been shown to be attenuated by L-NAME. In contrast, L-NAME has no effect on the cerebrovasodilatation induced by stimulating the pontine reticular formation (Iadecola, Zhang & Xu, 1993b). The source of the NO mediating the regional response in cerebral blood flow has not been established conclusively as it is not possible to remove endothelial cells in such experiments. However, it is tempting to speculate that it might be neuronal NO released from central neurones rather than autonomic neurones. Neural and endothelial control of vascular tone are discussed in detail in Chapters 3 and 5, and blood vessels in which NO has been identified as a neurotransmitter are listed in Appendix II.

4.10 Summary

The main activities of neuronal NOS in the CNS can be broadly divided into four categories: neurotoxicity, neuroprotection, synaptic plasticity, which includes long-term potentiation and long-term depression, and modulatory, including involvement in pathways that affect behaviours such as learning or expression of pain. Within each of these categories there are conflicting reports about the involvement of NO, and it is necessary to pay close attention to experimental models and conditions. The neurotoxic effects due to over-production of NO are primarily induced by activation of NMDA receptors by glutamate, the resultant entry of calcium into the neurone stimulating calmodulin-dependent synthesis of NO by NOS. The NO produced can interact with superoxide anions, resulting in peroxynitrite formation, which has cytotoxic activity. NO can nitrosylate several enzymes, form iron-complexes that alter associated enzyme activity, and can cause DNA mutation and strand breaking, which in turn stimulates PARS activation. All these mechanisms lead to energy depletion. Cytoprotective effects are due to interactions of NO with the NMDA receptor, inhibiting calcium entry and inhibiting glutamate binding. There is evidence that NO is a retrograde signal molecule involved in LTP and LTD particularly in the hippocampus and cerebellum, respectively. NO as a neurotransmitter is implicated in many functions of the CNS, as inhibition of its synthesis can disrupt learning skills and cognitive functions, and induce amnesia, in a variety of animal models. It has also been implicated in sensory perception because inhibition of NOS has antinociceptive effects in animals when pain has been produced by chemically stimulating peripheral nerve terminals (with formalin or carageenan), and in models of thermal hyperalgesia, visceral pain or mechanical allodynia. Conversely, inhibition of NOS may exacerbate pain in models of mechanical hyperalgesia.

4.11 Selected references

Bredt, D. S., Ferris, C. D. & Snyder, S. H. (1992). Nitric oxide synthase regulatory sites: phosphorylation by cyclic AMP dependent protein kinases, protein kinase C, calcium/calmodulin protein kinase; identification of flavin and calmodulin binding sites. *Journal of Biological Chemistry*, **267**, 10976–81.

Dawson, D. A. (1994). Nitric oxide and focal cerebral ischemia: multiplicity of actions and diverse outcome. *Cerebrovascular and Brain Metabolism Reviews*, **6**, 299–324.

Garthwaite, J. & Boulton, C. L. (1995). Nitric oxide signaling in the central nervous system. *Annual Reviews of Physiology*, **57**, 683–706.

Schuman, E. M. & Madison, D. V. (1994). Nitric oxide and synaptic function. *Annual Reviews of Neuroscience*, **17**, 153–83.

5

Nitric oxide in the peripheral nervous system

5.1 Introduction

For a neural system the set of rules that a substance must obey for it to be considered a neurotransmitter are encompassed by the transmitter criteria. The criteria that have become widely accepted, and minor variations of which are most often quoted (McLennan, 1970; Mountcastle & Sastre, 1980; Burnstock, 1981), are based on those set out by Paton in 1958, and a few years later by Eccles (1964): (1) the presynaptic neurone synthesizes and stores the transmitter; (2) the transmitter is released in a Ca^{2+}-dependent manner; (3) there should be a mechanism for terminating the activity of the transmitter, either by enzymatic degradation or by cellular uptake; (4) local exogenous application of the transmitter substance should mimic its endogenous activity; (5) agents that block or potentiate the endogenous activity of the transmitter should also affect its exogenous application in the same way. Recently these criteria have been re-evaluated (Hoyle & Burnstock, 1995) in the light of many studies that have shown that a substance can still be called a neurotransmitter even though it does not conform to this original set of rules. The revised criteria, which all apply to NO, appear to be more flexible, but nevertheless provide a rigid framework within which to define a transmitter substance.

1. A neurotransmitter must be released from a nerve.

   By inference, there must be a mechanism to justify the presence of the transmitter within the nerve. Usually this would be taken to mean that the apparatus for its synthesis must be present, but otherwise the nerve must possess a specific uptake mechanism for the substance.
2. The transmitter substance must act on cells close to the site of release.

   That is to say that the transmitter must be able to reach its target by simple diffusion, not carried in the bloodstream, or passaged through other cells.
3. The transmitter must have a mechanism of inactivation.

   The inference here is that a transmitter must be removed from its site of action. Receptor-mediated transmitters must be removed from the synaptic cleft by degradation to inactive products or by uptake into local cells. Non-receptor-mediated transmitters must be inactivated by intracellular or extracellular mechanisms.

4. Application of a substance under near-physiological conditions must mimic the process of transmission.
5. Agents that potentiate or inhibit the action of the applied transmitter substance must affect the process of transmission in the same way.

Two approaches have helped to establish NO as a neurotransmitter in the peripheral nervous system, namely histochemical and functional studies. Using immunocytochemical methods to localize NOS, or histochemical methods to reveal NADPH-diaphorase activity (see Chapters 12 and 16), neurones with the potential to synthesize and release NO have been visualized (see Appendix I). Functional studies, including assays for NO release, NOS activity and guanylate cyclase activity, whether cytochemical, biochemical or pharmacological, have also proven to be invaluable in determining a transmitter role for NO. Autonomically innervated organs and tissues in which NO has been identified as a transmitter are listed in Appendix II. In many cases, especially in the enteric nervous system, blood vessels and airways, it appears that NO is probably acting as a co-transmitter alongside vasoactive intestinal polypeptide (VIP) or ATP and, in addition to functioning as a neuromuscular transmitter, it may also function as a neuromodulator, inhibiting other transmission processes.

5.2 Gastrointestinal tract

5.2.1. Oesophagus and stomach

Neurones that can synthesize NO are well represented in the enteric nervous system, throughout the gastrointestinal tract. Many NOS-immunoreactive and NADPH-diaphorase-positive neurones innervate the smooth muscle of the oesophagus. In the stomach, ganglion cells containing NOS-immunoreactivity or NADPH-diaphorase activity are found in the myenteric and submucous plexuses, with nerve fibres innervating all the muscle layers. Nerve cell bodies of the myenteric plexus that contain NOS-immunoreactivity also contain NADPH-diaphorase reactivity, but not all the NOS-positive nerve fibres also contain NADPH-diaphorase (Belai *et al.*, 1992). The majority of NADPH-diaphorase-positive fibres are intrinsic but although the vagus nerve innervating the stomach contains diaphorase-positive fibres, which are either preganglionic parasympathetic nerves or primary sensory neurones, vagal ligation does not result in a significant loss of intramural innervation (Forster & Southam, 1993). Treatment of rats with capsaicin, which destroys primary afferent neurones, also has no significant effect of the density of intramural innervation (Forster & Southam, 1993).

In the opossum, cat, dog and human lower oesophageal sphincter (LOS) *in vitro*, neurogenic NANC relaxant responses are inhibited by L-NNA or L-NAME (Christinck *et al.*, 1991; De Man *et al.*, 1991; Murray *et al.*, 1991; Tottrup, Svane & Forman, 1991b; McKirdy *et al.*, 1992; Oliveira *et al.*, 1992; Tottrup *et al.*, 1993; Kortezova *et al.*, 1994; Mizhorkova *et al.*, 1994; Preiksaitis, Tremblay & Diamant, 1994; Chakder, Rosenthal & Rattan, 1995; Conklin *et al.*, 1995). The relaxant response in the opossum can also be inhibited by recombinant human haemoglobin (Chakder *et al.*, 1995; Conklin *et al.*, 1995). Relaxation of the LOS can be evoked *in vivo* by vagal stimulation or reflexly by gastric distension or by stimulating swallowing, and in opossum, dog and rat this is inhibited by L-NNA or L-NAME (Tottrup, Knudsen & Gregersen, 1991a; Yamato, Saha & Goyal, 1992; Boulant *et al.*, 1994; Kawahara *et al.*, 1994). Systemic application of recombinant human haemoglobin to human volunteers stimulates oesophageal peristalsis and inhibits LOS relaxation (Murray *et al.*, 1995). Thus, NO is an important mediator of inhibitory neuromuscular transmission in the LOS. The oesophagus itself has been comparatively less well studied, but in both opossum and human isolated circular muscle L-NNA inhibits poststimulation rebound excitation (Murray *et al.*, 1991; Preiksaitis *et al.*, 1994). However, it does not appear that NO is an excitatory transmitter in this muscle because application of authentic NO does not cause a contraction (Murray *et al.*, 1991).

The relaxation of the smooth muscle evoked by stimulating NANC nerves in isolated fundic preparations is mimicked by application of NO, and is inhibited, but not abolished, by L-NMMA or L-NNA, and the inhibition can be reversed or prevented by L-arginine but not D-arginine (Li & Rand, 1990; Boeckxstaens *et al.*, 1991a, 1992; D'Amato, Curro & Montuschi, 1992a,b; Lefebvre, Baert & Barbier, 1992a; Lefebvre, 1993; McLaren, Li & Rand, 1993). NANC relaxation of the rat stomach *in vivo* can be elicited by vagal stimulation in a manner that is sensitive to systemic application of L-arginine-reversible NOS inhibitors (Lefebvre, Hasrat & Gobert, 1994). In the guinea-pig stomach, receptive relaxation, an intrinsic reflex causing relaxation of the fundus and corpus in response to distension such as occurs during filling, is blocked by L-NAME, L-NMMA or methylene blue, thus indicating the involvement of a nitrergic pathway (Desai, Sessa & Vane, 1991a; Desai *et al.*, 1991b). In animals pretreated with atropine and guanethidine, vagal stimulation evokes a gastric relaxation that is abolished by hexamethonium, methylene blue or L-NMMA, i.e. vagal stimulation activates postganglionic nitrergic nerves (Desai *et al.*, 1991a,b, 1994). However, induced receptive relaxation is blocked by L-NMMA, but not by hexamethonium, indicating that there are also inhibitory nitrergic nerves in this reflex arc that are not reliant upon nicotinic stimulation. Thus, there are two separate neural paths that utilize NO to bring about accommodation (Desai *et al.*, 1991a,b, 1994). Reflex descending inhibition in the rat large intestine, which in many ways is analogous to gastric receptive relaxation, also involves a nitrergic efferent pathway (Hata *et al.*, 1990).

Inhibitory transmission in the rat stomach is not solely due to NO, and a VIP-ergic component may also be involved. VIP causes a relaxation in isolated rat gastric fundus that is usually unaffected by inhibiting NOS (Li & Rand, 1990; Boeckxstaens *et al.*, 1991a, 1992; McLaren *et al.*, 1993). The effect of VIP can be abolished by preincubation with trypsin or chymotrypsin (enzymes that can degrade VIP) (Boeckxstaens *et al.*, 1992; D'Amato *et al.*, 1992b; McLaren *et al.*, 1993), or by antisera raised against VIP (α-VIP) (Li & Rand, 1990). These treatments also reduce the effect of NANC inhibitory nerve stimulation by electrical field stimulation (EFS) or nicotine (Li & Rand, 1990; Boeckxstaens *et al.*, 1992; D'Amato *et al.*, 1992b). In combination with L-NMMA or L-NNA, either trypsin, chymotrypsin or α-VIP virtually abolishes the neurogenic relaxation (Li & Rand, 1990; Boeckxstaens *et al.*, 1992; McLaren *et al.*, 1993). Whether or not this represents a true co-transmission remains to be determined, although co-existence of VIP and NOS has been found (Aimi *et al.*, 1993). The predominance of a nitrergic or VIP-ergic response may be frequency dependent: low stimulation frequencies evoke relaxations that are more sensitive to inhibition of NOS than those evoked by higher frequencies of stimulation; higher frequencies of stimulation produce relaxations that are more sensitive to trypsin (Boeckxstaens *et al.*, 1992). Co-existence of NOS and VIP, and dual inhibitory transmission by NO and VIP has been demonstrated to occur in the ferret and guinea-pig stomach (Desai *et al.*, 1991a; Grundy, Gharib Naseri & Hutson, 1993; Desai *et al.*, 1994; Schemann, Schaaf & Mader, 1995) and may be the case in cats too (Barbier & Lefebvre, 1993).

In isolated preparations of the guinea-pig gastric fundus, L-NNA only reduces, and does not abolish, responses to NANC inhibitory nerve stimulation (Desai *et al.*, 1991a,b; Lefebvre *et al.*, 1992a; Desai *et al.*, 1994). EFS of guinea-pig gastric smooth muscle evokes VIP release as well as NO production (Grider *et al.*, 1992). Although some authors find that L-NNA has no effect against exogenous VIP (Boeckxstaens *et al.*, 1992; D'Amato *et al.*, 1992b; McLaren *et al.*, 1993), some find that at a concentration at which it abolishes NO production, it decreases VIP-induced relaxations by 58% to 88%, and decreases VIP release by more than 50% (Grider *et al.*, 1992). Furthermore, VIP produces relaxation of isolated smooth muscle cells, and also NO production. Again, L-NNA abolishes NO production, and inhibits the relaxant response to VIP (Grider *et al.*, 1992). These results imply that VIP evokes NO production from smooth muscle, and that NO acts prejunctionally to promote VIP release. These studies are supported by an examination of second messenger systems: VIP induces accumulation of cAMP and cGMP in

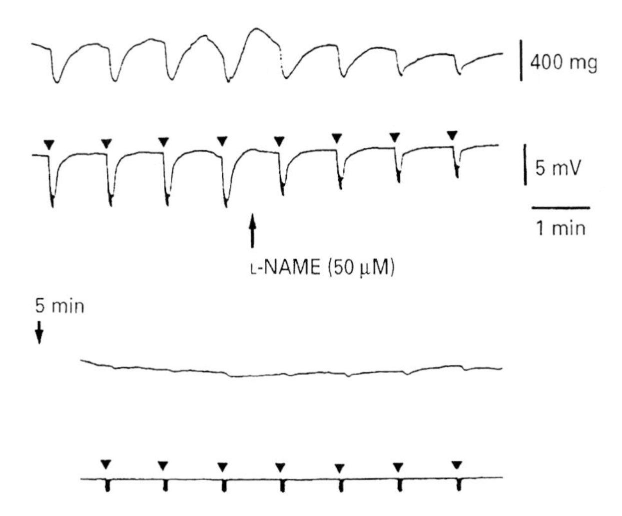

Fig. 5.1 Inhibition of inhibitory neuromuscular transmission in the human stomach by L-NAME. Sucrose-gap recording of mechanical (upper trace) and simultaneous electrical (lower trace) activity of isolated smooth muscle from human gastric cardia. In the presence of atropine (0.3 μM) and guanethidine (3.4 μM) electrical field stimulation (4 Hz, 0.03 ms, 5 s, supramaximal V) every minute (chevrons) evoked a transient relaxation and hyperpolarization. When L-NAME (50 μM) was added to the superfusing solution (at arrow, ↑) the mechanical and electrical responses were both rapidly inhibited. Less than 10 min after application they were abolished. The lower pair of traces is a continuation of the upper pair of traces, following a 5 min break (at arrow, ↓). (Reproduced from Lincoln, Hoyle & Burnstock, 1995.)

isolated gastric smooth muscle cells, while NO or sodium nitroprusside (SNP) induce elevation of only cGMP levels (Jin *et al.*, 1993). In the presence of L-NNA, VIP causes only an increase in cAMP levels. In addition, the combination of R-*p*-cAMPS and KT5823, which are, respectively, inhibitors of cAMP- and cGMP-dependent protein kinases, abolishes the relaxant effects of VIP, while each alone only attenuates the relaxation (Jin *et al.*, 1993).

In contrast to most laboratory animals, in isolated preparations of smooth muscle from human stomach, especially in the cardia circular muscle (the region of the stomach immediately below the LOS), mechanical relaxations and smooth muscle membrane hyperpolarizations in response to EFS are abolished by L-NAME (Lincoln, Hoyle & Burnstock, 1995) (Fig. 5.1). Thus it is likely that NO is the sole inhibitory transmitter, and that VIP is not involved in the neurogenic relaxant responses.

Neurally released NO may also function as an inhibitory neuromodulator in the rat stomach. Inhibition of NOS by L-NAME causes potentiation of cholinergic contractions evoked by EFS in isolated fundic smooth muscle (Lefebvre, De Vriese & Smits, 1992b). However, L-NAME does not interact directly with the muscarinic receptors since the responses to exogenous methacholine are unaffected, thus implying that NO acts prejunctionally to inhibit the release of acetylcholine (Lefebvre *et al.*, 1992b). Evidence of a similar prejunctional inhibition of cholinergic transmission has been obtained from isolated rabbit gastric corpus smooth muscle, where inhibition of NOS by L-NNA potentiates EFS- or nicotine-evoked cholinergic excitation (Baccari, Bertini & Calamai, 1993; Baccari, Calamai & Staderini, 1994). Also, L-NNA has no effect on exogenous acetylcholine or methacholine, and nor does sodium nitroprusside (SNP) (Baccari *et al.*, 1993), again indicative of an action of NO at a prejunctional rather than postjunctional level.

The pyloric region of the stomach, or pyloric sphincter, whose main function is to control gastric emptying, is particularly well endowed with NOS-containing nerves and a relaxant nitrergic mechanism (Allescher *et al.*, 1992b; Bayguinov & Sanders, 1993; Allescher & Daniel, 1994; Soediono & Burnstock, 1994). For example, in the canine pylorus, a combination of NOS inhibition and oxyhaemoglobin abolishes the inhibitory junction potential (IJP), which represents inhibitory neuromuscular transmission at an electrophysiological level. The hyperpolarization of the IJP is mimicked by 8-bromo-cGMP, and is potentiated after inhibition of cGMP-phosphodiesterase (Bayguinov & Sanders, 1993). Interestingly, in genetically engineered mice that lack the neuronal NOS gene (nNOS– mice) the most prominent abnormality, apart from the lack of NOS in peripheral nerves, is pyloric hypertrophy and gastric distension (Huang *et al.*, 1993). Patients, usually infants, suffering from pyloric stenosis also have pyloric hypertrophy and lack innervation of this region by NOS-containing nerves (Vanderwinden *et al.*, 1992; Kobayashi, O'Briain & Puri, 1995).

5.2.2 Small intestine

The distribution of NOS-containing or NADPH-diaphorase-containing neurones in the small intestine is very similar from species to species, with a substantial subpopulation of cell bodies being found in ganglia of the myenteric plexus and markedly fewer in ganglia of the submucosal plexus, with nerve fibres ramifying throughout all the muscular and neural layers. In the guinea-pig small intestine all the NO-synthesizing neurones appear to project anally (Costa *et al.*, 1992). At an ultrastructural level, immunoreactivity for type I NOS is found in nerves forming close appositions to muscle cells or other nerve cells (Llewellyn Smith *et al.*, 1992). NADPH-diaphorase activity and NOS-immunoreactivity may be found in intestinal neurones that also contain VIP-immunoreactivity (Costa *et al.*, 1992; Aimi *et al.*, 1993; Berezin *et al.*, 1994), but it is not co-localized with acetylcholinesterase (Aimi *et al.*, 1993) or substance P (SP) (Costa *et al.*, 1992; Berezin *et al.*, 1994). The ganglia of the sphincter of Oddi (choledochalduodenal sphincter) in the guinea-pig contains two main populations of NOS-positive neurones, those that also contain VIP and those that also contain neuropeptide Y (NPY); NOS is not co-localized with SP, calcitonin gene-related peptide (CGRP) or enkephalins (Wells, Talmage & Mawe, 1995).

In dog duodenum and ileum neurogenic relaxations, evoked by electrical stimulation or agents such as ATP, γ-aminobutyric acid (GABA) or 5-HT, are blocked by oxyhaemoglobin or NOS inhibitors (Toda, Baba & Okamura, 1990a; Boeckxstaens *et al.*, 1991b; Bogers *et al.*, 1991; Toda, Tanobe & Baba, 1991a; Toda *et al.*, 1992). Similarly, in the rat duodenum longitudinal muscle, NANC inhibition evoked by nicotine is blocked by L-NNA (Irie *et al.*, 1991). In the human ileum circular muscle, relaxant responses of NANC inhibitory transmission can be inhibited by up to 65% by L-NNA (Maggi *et al.*, 1991), implying that another transmitter is responsible for the residual 35%.

Neuromodulation by NO is a pronounced phenomenon in the small intestine. For example, in guinea-pig small intestine, exogenous NO decreases the nerve-stimulation-evoked release of both SP and ACh (Gustafsson *et al.*, 1990; Wiklund *et al.*, 1993b) and, in dog duodenal longitudinal muscle, inhibition of NOS results in potentiation of cholinergic transmission (Toda *et al.*, 1991a). Almost paradoxically, in the guinea-pig ileum NO can also have excitatory actions, stimulating the release of both ACh and SP from intramural neurones (Gustafsson *et al.*, 1990; Wiklund, Wiklund & Gustafsson, 1993a). Excitatory effects of NO have been found to occur in rat small intestine longitudinal muscle as well. NANC nerve stimulation evokes a contractile response, as does SNP (Barthó *et al.*, 1992), while L-NNA abolishes the NANC contraction, and L-arginine reverses this effect (Barthó *et al.*, 1992). Thus, it would appear that in this organ, under the prevalent conditions, nitrergic transmission is excitatory.

Nitrergic pathways may provide a tonic inhibitory influence over excitation in the small intestine. Inhibition of NOS in cat, rat and guinea-pig increases intestinal compliance, or causes increases in intraluminal pressure and contractile activity (Calignano *et al.*, 1992; Gustafsson & Delbro, 1993; Waterman & Costa, 1994; Waterman, Costa & Tonini, 1994), potentiating ascending excitation (Allescher *et al.*, 1992a).

The sphincter of Oddi controls the outflow of the hepatobiliary duct into the duodenum. Pressure recordings *in vivo* in prairie dogs, rabbits, guinea-pigs and cats show that it tends to have a basal pressure or motility that increases when L-NAME or L-NNA is applied systemically (Kaufman *et al.*, 1993; Mourelle *et al.*, 1993; Thune *et al.*, 1995). L-Arginine can reverse these actions. *In vitro*, the rabbit, guinea-pig, opossum and possum sphincters of Oddi relax in response to EFS when cholinergic and adrenergic transmission are prevented. NOS inhibitors block these relaxations (reversed by L-arginine), and may unmask NANC contractile responses (Allescher *et al.*, 1993; Baker *et al.*, 1993; Mourelle *et al.*, 1993; Pauletzki *et al.*, 1993). Thus, it appears that nitrergic transmission is inhibitory, and provides an ongoing tonic level of control.

5.2.3 Large intestine

The canine colon and ileocolonic junction, in which there is a rich innervation by NOS- or NADPH-diaphorase-containing neurones, have been particularly well studied. Here NANC inhibitory neuromuscular transmission has all the characteristics of being mediated predominantly via NO, being antagonized by oxyhaemoglobin, L-NAME, L-NNA, mimicked by exogenous NO or *S*-nitrosocysteine, and being potentiated by the cGMP phosphodiesterase inhibitor M&B22948 (also known as zaprinast, Dalziel *et al.*, 1991; Thornbury *et al.*, 1991; Ward *et al.*, 1992a,b; Shuttleworth, Sanders & Keef, 1993a; Shuttleworth *et al.*, 1993b). The fast IJP in the canine colon is abolished by inhibition of NOS (Ward, McKeen & Sanders, 1992c). In addition, rebound excitation seen after NANC inhibition in the colon is also attenuated by inhibition of NOS, and by oxyhaemoglobin, and is mimicked by NO (Ward *et al.*, 1992a). The same is true in the cat colon (Venkova & Krier, 1994). Target cells for NO produced by EFS can be identified using antibodies that recognize cGMP. Using this technique it has been found that NO causes a large increase in cGMP in smooth muscle cells, and also in myenteric neurones (Shuttleworth *et al.*, 1993b). Interestingly, 94% of the myenteric neurones that become cGMP positive are NADPH-diaphorase negative (Shuttleworth *et al.*, 1993b).

In the human colon, although L-NNA can reduce the amplitude of responses to NANC inhibitory transmission (Burleigh, 1992; Tam & Hillier, 1992; Boeckxstaens *et al.*, 1993), the fast IJP is unaffected by L-NNA or L-NAME (Keef *et al.*, 1993), nor is it affected by oxyhaemoglobin (Keef *et al.*, 1993). However, inhibition of NOS causes a reduction of the sustained hyperpolarization that follows the rapid IJP when trains of pulses are applied (Keef *et al.*, 1993). VIP is co-localized extensively with NOS in the human colon (Keranen *et al.*, 1995), but its role as an inhibitory transmitter remains to be substantiated. Certainly it does not mediate the IJP (Hoyle *et al.*, 1990).

In the guinea-pig large intestine, where VIP may be co-localized with NOS (Costa *et al.*, 1992; Berezin *et al.*, 1994), little or no interaction between VIP and NO has been found. Unlike gastric smooth muscle, the taenia coli does not generate NO in response to VIP (Grider *et al.*, 1992; Jin *et al.*, 1993), although NO is produced by EFS both here and in intertaenial longitudinal muscle (Shuttleworth, Murphy & Furness, 1991; Wiklund *et al.*, 1993b). Several effects of NO are only manifest in the guinea-pig taenia coli when the cholinergic system is intact, i.e. in the absence of atropine (Ward *et al.*, 1996). Under these conditions a neurogenic relaxation and hyperpolarization can be recorded that are sensitive to haemoglobin and L-NNA. This situation is curious because the actions of an inhibitory transmitter, NO, seem to be dependent upon an excitatory transmitter, acetylcholine.

The internal anal sphincter has been extensively investigated, particularly in the opossum. Like other gastrointestinal sphincters (LOS, pylorus, choledochal-duodenal sphincter or sphincter of Oddi, ileocaecal sphincter or ileocolonic junction) it receives an inhibitory nitrergic innervation. In isolated preparations, relaxant responses to EFS are abolished by L-NNA, the effects of which may be reversed by L-arginine but not D-arginine (Rattan & Chakder, 1992; Chakder & Rattan, 1993b). Release of NO as a result of EFS can be determined using a chemiluminescence assay (Chakder & Rattan, 1993a), and responses to EFS can be inhibited by hydroquinone (Chakder & Rattan, 1992) and recombinant human haemoglobin (Rattan, Rosenthal & Chakder, 1995). *In vivo*, rectal distension evokes relaxation of the internal anal sphincter via a descending inhibition, or rectoanal reflex. This reflex is blocked by L-NNA, which also blocks relaxations evoked by sacral nerve stimulation (Rattan, Sarkar & Chakder, 1992). When the hypogastric nerve is stimulated the opossum internal anal sphincter responds with an initial relaxation followed by a phase of contraction; both elements are attenuated by L-NNA (Rattan & Thatikunta, 1993). It is also likely that inhibitory transmission involves VIP since it co-exists with NOS (Lynn *et al.*, 1995). Both NO and VIP, as well as EFS, induce elevation of cGMP and cAMP levels in the smooth muscle (Chakder & Rattan, 1993b). Relaxant responses to VIP may be severely inhibited by L-NNA or hydroquinone (Chakder & Rattan, 1992), implying that its responses involve NO. However, others have found no significant effect of L-NNA on responses to VIP (Tottrup, Glavind & Svane, 1992).

The human internal anal sphincter is innervated by NOS-containing nerves, and neurogenic relaxations in isolated preparations are abolished by L-NNA or oxyhaemoglobin (Burleigh, 1992; O'Kelly, Brading & Mortensen, 1993). *In vivo*, topical application of glyceryltrinitrate (GTN, nitroglycerin) to the anus causes a reduction in anal pressure (Loder *et al.*, 1994). This observation led to the suggestion that GTN can be used to treat conditions such as anal fissures, haemorrhoids or proctalgia, which benefit from a reduced anal pressure (Loder *et al.*, 1994; Gorfine, 1995).

5.3 Urogenital tract

5.3.1 Bladder and urethra

In the mouse and rat urinary bladder, small intramural and admural ganglia and the pelvic ganglia contain nerve cell bodies that have NOS-immunoreactivity or NADPH-diaphorase reactivity, and fibres are seen projecting through the muscular wall. In the urethra, NO has been shown to be an important contributor to neurogenic NANC relaxation in many species including pig (Bridgewater, MacNeil & Brading, 1993; Persson *et al.*, 1993), rabbit (Dokita *et al.*, 1991; Zygmunt *et al.*, 1993; Lee, Wein & Levin, 1994), sheep (Garcia Pascual *et al.*, 1991; Thornbury, Hollywood & McHale, 1992; Garcia Pascual & Triguero, 1994), rat (Persson *et al.*, 1992; Bennett *et al.*, 1995), dog (Hashimoto, Kigoshi & Muramatsu, 1993) and human (Ehren *et al.*, 1994; Leone *et al.*, 1994b). The nerves from which the NO is released are not sympathetic, because neither guanethidine nor 6-hydroxydopamine (6-OHDA) has an effect on nitrergic relaxations, nor does antidromic stimulation of primary afferent neurones cause nitrergic relaxation (Garcia Pascual *et al.*, 1991; Thornbury *et al.*, 1992; Hashimoto *et al.*, 1993). Retrograde labelling experiments show that very few primary afferent fibres in the bladder and urethra contain NOS (Vizzard *et al.*, 1994). Thus the nitrergic nerves are most likely to be intramural postganglionic parasympathetic nerves.

In the isolated rabbit urethra, when the tone has been raised by a pharmacological manipulation, EFS can evoke NANC neurogenic relaxant responses. These responses are antagonized by the NOS inhibitors (Andersson *et al.*, 1991, 1992; Dokita *et al.*, 1991; Lee *et al.*, 1994). L-Arginine, but not D-arginine, potentiates the neurogenic relaxations, and attenuates the effect of NOS inhibitors (Andersson *et al.*, 1991). Methylene blue blocks responses to EFS as well as to applied NO (as acidified sodium nitrite solution) (Andersson *et al.*, 1991; Dokita *et al.*, 1991), and the selective cGMP phosphodiesterase antagonist M&B22948 potentiates the relaxant response to EFS, which is again indicative of mediation by NO (Dokita *et al.*, 1991).

Similar results have been obtained from isolated preparations from sheep bladder neck and urethra, in which inhibitors of NOS reduce or abolish the relaxations evoked by EFS (Garcia Pascual *et al.*, 1991; Thornbury *et al.*, 1992; Garcia Pascual & Triguero, 1994). In these tissues, *S*-nitroso-L-cysteine mimics the response produced by nerve stimulation; it has been suggested that *S*-nitroso-L-cysteine may be the released neurochemical rather than NO itself (Thornbury *et al.*, 1992; Garcia Pascual & Triguero, 1994). Likewise, in the pig urethra, L-NNA abolishes relaxant responses due to EFS, and this effect is reversed by L-arginine (Bridgewater *et al.*, 1993).

In the dog isolated urethra, EFS also produces relaxation of precontracted preparations. This relaxant response has two distinct components, with a rapid transient relaxation followed by a secondary, slower and more sustained relaxation (Hashimoto *et al.*, 1993). The primary, but not the secondary, phase is abolished by L-NNA or L-NMMA: L-arginine but not D-arginine reverses the effect of these NOS inhibitors (Hashimoto *et al.*, 1993). Thus the nitrergic component is responsible for the rapid transient relaxation, but not the slower, more sustained secondary component. In other animals a non-nitrergic component may be present, but without such a marked temporal separation (Persson & Andersson, 1992; Persson *et al.*, 1992).

The rat has been used to examine the presence of nitrergic mechanisms in micturition *in vivo*. Methylene blue has two major effects: firstly, it induces an increase in the rhythmic contractile activity of the bladder that occurs during filling; secondly, it reduces the relaxation of the urethra that occurs during voiding. Thus a nitrergic mechanism is implicated in both stages of the micturition cycle (Nishizawa *et al.*, 1992). L-NAME increases the rhythmic bladder contractions (Persson & Andersson, 1992; Persson *et al.*, 1992), resulting in a decreased micturition volume, and blocks the reflex relaxation of the urethra (Persson *et al.*, 1992; Bennett *et al.*, 1995).

5.3.2 Male reproductive tract

In the male reproductive tract NO-synthesizing neurones are probably postganglionic parasympathetic nerves. Histochemical studies have revealed NOS-positive nerve fibres in rat and human corpus cavernosum, and in neuronal cell bodies in the rat pelvic ganglion that are identified as projecting to the erectile tissues of the penis. Immunoreactivity for NOS, which disappears after bilateral transection of the cavernous nerves, has been localized in nerve fibres in erectile tissue of the rat (Burnett *et al.*, 1992), and NOS-immunoreactive fibres are also found in the smooth muscle and submucosa of rat vas deferens, epididymis and ejaculatory duct, and human prostate gland.

The smooth muscle of the erectile tissue is vascular in nature, forming a trabecular network in the corpus cavernosum and corpus spongiosum of the penis. The smooth muscle is lined throughout with endothelial cells. The erectile process is a parasympathetic event that involves relaxation of the penile arteries and the erectile tissue, resulting in filling of the cavernous spaces. The first experiments investigating the role of NO in the relaxation of cavernous smooth muscle showed that acetylcholine causes relaxation via an endothelium-dependent mechanism (Saenz de Tejada, Goldstein & Krane, 1988) that is inhibited by NOS inhibitors (Kim *et al.*, 1991). However, transmural electrical stimulation of cavernous smooth muscle strips denuded of endothelium also results in relaxation (Kim *et al.*, 1991); this is NANC in nature and there is now considerable evidence that NO is a major mediator. L-NMMA and L-NNA, but not D-NNA, inhibit neurogenic relaxation in human and rabbit cavernous tissue (Ignarro *et al.*, 1990; Kim *et al.*, 1991; Pickard, Powell & Zar, 1991; Holmquist, Hedlund & Andersson, 1992; Knispel, Goessl & Beckmann, 1992; Rajfer *et al.*, 1992). Inhibition by L-NNA is effectively reversed by L-arginine but not D-arginine. Authentic NO causes relaxation of cavernous smooth muscle as do the NO donors SIN-1 and *S*-nitroso-*N*-acetyl-D,L-penicillamine (SNAP) (Kim *et al.*, 1991; Holmquist *et al.*, 1992; Rajfer *et al.*, 1992). Furthermore, it has been demonstrated that methylene blue and haemoglobin inhibit NANC inhibitory transmission in this tissue (Ignarro *et al.*, 1990; Kim *et al.*, 1991; Pickard *et al.*, 1991; Holmquist *et al.*, 1992; Rajfer *et al.*, 1992). Preventing the breakdown of cGMP with an inhibitor of cGMP phosphodiesterase also enhances the relaxant responses to NANC nerve stimulation and to NO derived from SNAP (Rajfer *et al.*, 1992).

NANC nerve-mediated penile erection has also been investigated *in vivo* in dogs and rats by measuring intracavernous pressure following pelvic nerve or sacral spinal cord stimulation (Burnett *et al.*, 1992; Finberg, Levy & Vardi, 1993; Trigo-Rocha *et al.*, 1993). These studies have shown that the increase in intracavernous pressure induced by nerve stimulation is inhibited by L-NNA, L-NMMA and L-NAME but not D-NMMA. This inhibition is partially prevented by prior administration of L-arginine and partially reversed by L-arginine while D-arginine is without effect in either case. L-NAME has no effect on penile tumescence induced by administration of papaverine. Intracavernous injection of SNAP induces tumescence similar to that obtained by nerve stimulation. In addition, nerve-mediated erection is inhibited by methylene blue and enhanced by a cGMP phosphodiesterase inhibitor.

5.3.3 Anococcygeus and retractor penis muscles

Several groups have demonstrated that inhibitory transmission in the rat and mouse anococcygeus is attenuated by the addition of NOS inhibitors (Gibson *et al.*, 1989; Gillespie, Liu & Martin, 1989; Li & Rand, 1989; Ramagopal & Leighton, 1989). Relaxations are restored by the addition of L-arginine and, in general, L-NNA is a more potent inhibitor than L-NMMA. Similar results have also been reported in relation to the rabbit anococcygeus (Sneddon & Graham, 1992; Graham & Sneddon, 1993). Field stimulation of raised-tone preparations in the presence of atropine and guanethidine results in NANC relaxations at very low frequencies, with 1 or 2 Hz producing almost complete relaxation. L-NNA and L-NAME significantly reduce NANC inhibitory responses, and they can be partially reversed by L-arginine but not D-arginine. Haemoglobin also inhibits NANC nerve responses, and SNP mimics the effects of NANC nerve stimulation. In addition, IJPs recorded from smooth muscle cells by intracellular microelectrodes during NANC nerve stimulation are reduced in amplitude by L-NNA. There is still some question as to whether the transmitter is free NO or an NO-donating compound (Gibson *et al.*, 1995). The reason for this is that drugs such as hydroquinone and hydroxocobalamin have been shown to produce qualitatively different effects on NO-mediated and on NANC nerve-mediated relaxations in the anococcygeus muscle (Hobbs, Tucker & Gibson, 1991; Rajanayagam, Li & Rand, 1993). Since cysteine, a naturally occurring thiol, enhances NO-induced relaxations but has no effect on nitrergic nerve-mediated relaxations, it has been suggested that the mediator released during nitrergic transmission may be a compound such as a nitrosothiol rather than free NO (Rand & Li, 1992). Possibly the anococcygeus is similar to the bovine retractor penis, in that there are very high levels of endogenous superoxide dismutase (SOD), which could account for the lack of effect of free-radical generators against neurogenic responses (Hobbs *et al.*, 1991; Gibson *et al.*, 1995).

Apart from a few features, pharmacological studies of the bovine retractor penis muscle have revealed that inhibitory transmission resembles that found in the anococcygeus (Rand, 1992). One exception to this is that the arginine analogue L-NMMA has no effect on NANC inhibitory transmission in the bovine retractor penis whereas it partially blocks relaxation in the rat anococcygeus. However, the more potent NOS inhibitor, L-NNA, was equally effective in abolishing NANC relaxations in both preparations (Martin, Gillespie & Gibson, 1993). The reason for this is unclear, but the authors have emphasized that L-NMMA should not be used as the only arginine analogue in pharmacological studies attempting to demonstrate nitrergic transmission. A second difference between these two preparations is that the superoxide generator pyrogallol has no effect on NANC relaxations in the bovine retractor penis but it inhibits those in the rat anococcygeus (Gillespie & Sheng, 1990). It has been suggested that the bovine retractor penis has a higher level of SOD compared to the rat anococcygeus, thus providing a greater protective effect against superoxide generators (Sheng, 1994). When SOD is inhibited in the bovine retractor penis by diethyldithiocarbamate, superoxide generators do have their expected action: pyrogallol and hydroquinone (also a generator of superoxide anions) do inhibit the responses to EFS (Martin, McAllister & Paisley, 1994; Paisley & Martin, 1996) (Fig. 5.2). NOS has been partially purified from both the rat anococcygeus and the bovine retractor penis muscles (Mitchell *et al.*, 1991; Sheng *et al.*, 1992) where it has been shown to have the same characteristics as the neuronal isoform of the enzyme.

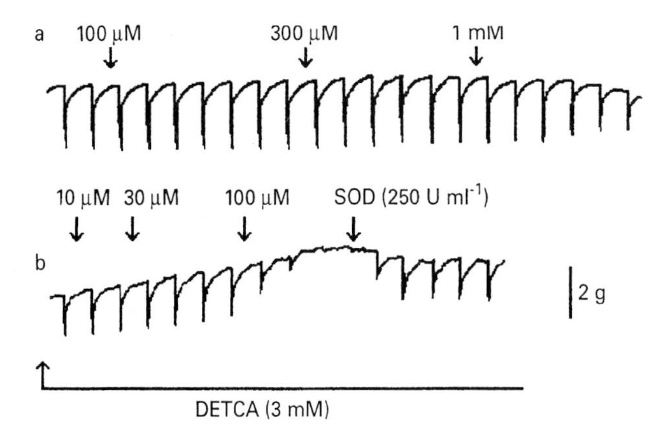

Fig. 5.2 Isolated bovine retractor penis muscle. Effects of hydroquinone and inhibition of superoxide dismutase (SOD) by diethyldithiocarbamate (DETCA) on inhibitory neuromuscular transmission. The tone of the muscle was raised by guanethidine (30 μM), which also abolishes sympathetic transmission. Every 4 min electrical field stimulation (EFS) was applied (4 Hz for 10 s), which caused transient relaxations (downward deflections). (a) Hydroquinone (applied as indicated at the arrows, ↓) had minor effects until 1 mM. (b) In the presence of DETCA (3 mM) hydroquinone inhibited the responses to EFS, abolishing them at 100 μM. Superoxide dismutase (250 U ml^{-1}) reversed the inhibition due to hydroquinone. [Reproduced (modified) with permission from Paisley & Martin, 1996.]

5.3.4 Female reproductive tract

Retrograde labelling studies in rats have also shown that preganglionic parasympathetic neurones that supply the uterus may contain NOS (Papka *et al.*, 1995a), while postganglionic neurones in the pelvic ganglia supply uterine smooth muscle, as do sensory neurones originating from thoracic, lumbar and sacral dorsal root ganglia (Papka *et al.*, 1995b). The human vagina is only lightly innervated by NO-immunoreactive neurones, most of which are associated with capillaries in the epithelial papillae, where it has been suggested that they play a role in controlling vascular permeability (Hoyle *et al.*, 1996).

In human uterus NO appears to be responsible for a tonic inhibition of spontaneous contractile activity via a cGMP pathway (Buhimschi *et al.*, 1995). L-NAME enhances the spontaneous contractile activity and L-arginine diminishes it; furthermore, in samples of uterus obtained from pregnant women the responsiveness to an NO donor (diethylamine/NO) was greatest before the onset of labour, and least during labour (Buhimschi *et al.*, 1995). Evidence for increased biosynthesis of NO during pregnancy has been provided by increased urinary excretion of nitrate and cGMP, which could be inhibited by *in vivo* administration of L-NAME (Conrad *et al.*, 1993). This has been suggested to contribute to maternal vasodilatation during pregnancy. The source of the NO synthesis has not been established. High levels of inducible NOS activity have been reported in decidual preparations from the pregnant rabbit uterus. Interestingly, NOS activity drops dramatically on the last day of pregnancy (Sladek *et al.*, 1993). These results indicate that during pregnancy, when the uterus needs to maintain a relatively relaxed state, nitrergic mechanisms may be involved, and that at term a decrease in responsiveness to NO may be involved in the initiation of labour. This is backed up by studies of rats, in which it has been shown that there is decreased neuronal NADPH-diaphorase activity in nerves innervating the uterus at the time of parturition (Natuzzi *et al.*, 1993). The neuropeptide calcitonin gene-related peptide (CGRP), which is found extensively in nerves innervating the uterus, causes relaxation that is dependent upon NOS activity, being blocked by L-NMMA (Shew *et al.*, 1993).

5.4 Cardiovascular system

NADPH-diaphorase activity or NOS-immunoreactivity has been localized in perivascular nerves supplying a number of different blood vessels from a variety of species (see Appendix I). Although type I NOS is generally thought to be the isoform responsible for the generation of NO in neurones, in the rat basilar artery a large population of axons contains type III NOS in addition to a similar sized population (approximately 27%) that contains type I NOS (Loesch & Burnstock, 1996). As in the gastrointestinal tract, NOS has been shown, in many cases, to be co-localized with VIP in vascular neurones (Kummer *et al.*, 1992). However, it is not entirely clear if these neurones are sympathetic, parasympathetic or sensory. The origins of NOS-immunoreactive perivascular nerves supplying cerebral arteries of the rat have been investigated. NOS-immunoreactivity is present in 70–80% of neuronal cell bodies in the sphenopalatine ganglia, 20% of which also contain VIP. Sectioning of the postganglionic parasympathetic fibres originating from these ganglia resulted in a 75% reduction in NOS-immunoreactive nerve fibres in the circle of Willis. NOS-immunoreactive cell bodies are not found in the sympathetic superior cervical ganglia. A few NOS-immunoreactive cell bodies are found in the sensory trigeminal ganglion and sectioning of the trigeminal nerves results in a 25% decrease in NOS-immunoreactive perivascular nerve fibres (Bogers *et al.*, 1994). That the origin of NOS-containing nerve fibres in the large cerebral arteries is the sphenopalatine ganglia has been confirmed in another study (Iadecola *et al.*, 1993a).

Several studies by Toda and colleagues have demonstrated that nerve-mediated relaxation of dog, monkey and human cerebral arteries is inhibited by L-NNA or L-NMMA (Toda, Minami & Okamura, 1990b; Toda & Okamura, 1990a,b; Toda, 1993) and the inhibition can be prevented and reversed by L-arginine but not D-arginine. Removal of the endothelium does not alter these responses. In a superfusion bioassay cascade experiment, the release of NO or a NO carrier from porcine cerebral arteries denuded of endothelium during transmural stimulation of NANC nerves has been demonstrated (Chen & Lee, 1993). In endothelium-denuded sheep cerebral arteries, L-NMMA has been shown to inhibit both neurogenic and VIP-induced relaxation, indicating an interaction between NO and VIP in this vessel (Gaw *et al.*, 1991). Neurogenic relaxation in bovine cerebral arteries is inhibited by L-NNA, tetrodotoxin, haemoglobin and methylene blue. In addition, the neurogenic relaxation is attenuated by acetylcholine and physostigmine and potentiated by atropine; thus, it may be concluded that NO is the transmitter mediating neurogenic relaxation and that its release is inhibited by acetylcholine (Ayajiki, Okamura & Toda, 1993). As in the pulmonary artery and the saphenous vein, NOS inhibition potentiated the contractile response to sympathetic nerve stimulation in the superficial temporal artery of the dog and monkey (Toda, Yoshida & Okamura, 1991b).

In guinea-pig pulmonary arteries denuded of endothelium, it has been shown that EFS in the presence of adrenergic and cholinergic blockade induces a neurogenic frequency-dependent relaxation (Liu *et al.*, 1992a,b). L-NMMA inhibits the relaxant response, an effect that is completely reversed by L-arginine but not D-arginine. The relaxation is also inhibited by methylene blue and pyrogallol, the effect of the latter being reversed by SOD. A specific inhibitor of cGMP phosphodiesterase potentiates the response while an inhibitor of cAMP phosphodiesterase has no effect. The cGMP content of the arteries increases upon nerve stimulation, a response that is also inhibited by L-NMMA. Chemical sympathectomy does not affect neurogenic stimulation (Liu *et al.*, 1992b).

The rabbit portal vein has two muscle layers and the longitudinal muscle has a non-sympathetic NANC innervation mediating relaxation that is independent of the endothelium (Brizzolara, Crowe & Burnstock, 1993). L-NAME partially inhibits the relaxation, an effect that could be reversed by the addition of L-arginine. A combination of suramin and L-NAME abolished neurogenic relaxation indicating that both ATP and NO are inhibitory transmitters in this vessel.

In the guinea-pig uterine artery, nerve stimulation in the presence of guanethidine produces monophasic relaxation at low frequencies and short (less than 20) trains of pulses. At high frequencies with trains of 50 or more pulses a biphasic relaxant response was observed (Morris, 1993). L-NAME inhibits the faster phase of the response while trypsin inhibits the slower phase. It was suggested that NO and VIP may be co-transmitters in non-adrenergic inhibitory perivascular nerves in this vessel. Pronounced vasodilatation of the maternal vasculature is known to occur during normal pregnancy.

The heart also contains NOS-positive neurones. In a dissociated cell culture preparation from the newborn guinea-pig heart, NADPH-diaphorase activity and NOS-immunoreactivity have been shown to be co-localized in a subpopulation of intrinsic neurones (Hassall *et al.*, 1992). Neuronal NADPH-diaphorase activity has also been demonstrated in ganglia located within the interatrial septum, in association with the superior and inferior venae cavae and at the points of entry of the pulmonary veins. Positive nerve fibres appear throughout the atrium in association with neuronal cell bodies and atrial myocytes in the rat and guinea-pig heart (Klimaschewski *et al.*, 1992; Tanaka *et al.*, 1993). There is little information on the possible role of neuronal NO in the nervous control of the heart, but it has been suggested that NO generation is an obligatory process in cholinergic inhibition of L-type calcium currents in cardiac pacemaker tissue from the rabbit (Han, Shimoni & Giles, 1994).

5.5 Respiratory tract

In the airways the density of NOS-containing fibres increases progressively from the top of the trachea to the bronchi. In the primary bronchi the innervation appears to be greatest, and it progressively diminishes as the bronchiole diameter decreases. At all levels of the respiratory tree nerve fibres are found in the smooth muscle and submucosa, and fine nerve bundles or single fibres appear to innervate the respiratory epithelium. The NOS-containing fibres in the airways may be extrinsic in origin (Fischer *et al.*, 1993), but paratracheal ganglia contain a substantial subpopulation of nerve cells that are NADPH-diaphorase positive (Hassall, Saffrey & Burnstock, 1993). As in blood vessels and the enteric nervous system, NOS is extensively co-localized with VIP in the innervation of the airways (Dey, Mayer & Said, 1993; Shimosegawa & Toyota, 1994).

Human airway smooth muscle contracted by methacholine or histamine relaxes in response to nerve stimulation. This relaxation, in contrast to guinea-pig airway smooth muscle, does not appear to involve VIP but, like pig and guinea-pig tissue, it does appear to involve a component of NO that is sensitive to L-NMMA, L-NAME or L-NNA (Li & Rand, 1991; Belvisi *et al.*, 1992; Ellis & Undem, 1992; Kannan & Johnson, 1992; Bai & Bramley, 1993; Belvisi *et al.*, 1993; Ward *et al.*, 1995). In the guinea-pig the parabronchial and paratracheal ganglia do not contain NOS-positive or NADPH-diaphorase-positive nerve cell bodies; the source of nitrergic neurones innervating the airways is extrinsic, originating in sympathetic and sensory ganglia (Fischer *et al.*, 1993). However, other studies have indicated that the guinea-pig paratracheal neurones do contain a large subpopulation of nerve cells that express NADPH-diaphorase activity (Hassall *et al.*, 1993; Shimosegawa & Toyota, 1994). In the airways, parasympathetic cholinergic innervation mediates constriction. This cholinergic excitation is modulated by endogenous neurogenic NO, which inhibits the extent of constriction of human bronchus and rat trachea (Sekizawa *et al.*, 1993; Ward *et al.*, 1993). The source of NO has been examined in greater detail using a guinea-pig trachea preparation, in which preganglionic stimulation of the vagus nerve and postganglionic stimulation of the intramural nerves can be separated. L-NAME has no effect against the vagal constrictions, but does potentiate the responses to intramural stimulation (Watson, Maclagan & Barnes, 1993), implying that the nitrergic nerves that modulate parasympathetic constriction are not carried in the vagus, and are therefore neither sensory nerves nor vagal postganglionic nerves themselves, leaving sympathetic and local neurones as a possible source.

In guinea-pig trachea the responses to NANC nerve stimulation and NO donors are potentiated by pretreatment with the phosphodiesterase IV inhibitors, rolipram and denbufylline, as well as the phosphodiesterase V inhibitor, zaprinast (M&B22948), while the phosphodiesterase III inhibitor, siguazodan is without effect (Ellis & Conanan, 1995). Thus the second messenger systems involve cGMP-dependent cAMP formation in addition to cGMP. Similar results have been obtained from isolated preparations of human bronchus (Fernandes, Ellis & Undem, 1994).

5.6 Preganglionic neurones and autonomic ganglia

To date there is no functional evidence for mammalian preganglionic autonomic neurones utilizing NO as a neurotransmitter or neuromodulator. Nevertheless, there is histological evidence that would be consistent with this possibility. For example, in the spinal cord, type I NOS-immunoreactivity has been located in sympathetic and parasympathetic cell bodies lying in the intermediolateral columns in thoracic, lumbar and sacral segments. Furthermore, NOS-containing nerve terminals can be identified in peripheral ganglia, including the superior cervical, superior and inferior mesenteric, coeliac and pelvic ganglia, and retrograde labelling methods have verified that these terminals arise from the spinal cord, although some of the projections to the coeliac and inferior mesenteric ganglia arise from myenteric ganglia in the large intestine (Furness & Anderson, 1994; Anderson *et al.*, 1995). Denervation of sympathetic ganglia, by transecting the preganglionic sympathetic nerve, results in a substantial loss of NOS-positive terminals from the ganglia (Ceccatelli *et al.*, 1994; Dun *et al.*, 1995). Few nerve cell bodies in sympathetic prevertebral or paravertebral ganglia contain NOS-immunoreactivity (Anderson *et al.*, 1993; Ceccatelli *et al.*, 1994) but, in contrast, most cell bodies in parasympathetic ganglia, e.g. the sphenopalatine and submandibular ganglia, do (Ceccatelli *et al.*, 1994).

Central control of sympathetic activity may well involve nitrergic mechanisms, as shown by intrathecal injection of L-NAME causing a decrease in rabbit renal nerve action potential discharge, while application of L-arginine markedly stimulates this discharge (Hakim *et al.*, 1995). Functional evidence for a role of NO in sympathetic ganglia has been obtained from studies of the chick ciliary ganglion. Here ganglionic transmission is enhanced by SNP and 8-bromo-cGMP, and long-term potentiation is inhibited by L-NAME (Scott & Bennett, 1993). In rat superior cervical ganglion (SCG) neurones, which probably are postsynaptic to preganglionic NOS-containing nerves, application of NO donors enhances Ca^{2+} currents. This is mimicked by intracellular application of cGMP, 8-bromo-cGMP and the phopshodiesterase inhibitor M&B22948, and blocked by methylene blue (Chen & Schofield, 1995).

5.7 Primary sensory neurones

Dorsal root ganglia (DRG) contain NOS-positive cell bodies at all levels of the spinal cord in rats and humans, as do the nodose ganglia in rats and guinea-pigs. Within the spinal cord the central projections of the primary afferents enter the dorsal horn, and nerve cell bodies in the superficial laminae (laminae I–IV, but especially laminae II and III) contain NOS-immunoreactivity. In the brainstem and related structures NOS-immunoreactivity is found in various sensory nuclei and sensory pathways and the petrosal ganglion. Vagal afferents project through the SCG (Dun *et al.*, 1995), and project to the airways (Fischer *et al.*, 1993). Primary afferents in the rat uterus originate in thoracic, lumbar and sacral DRG (Papka *et al.*, 1995b), those in the adrenal gland originate from T1 to L1 (Heym *et al.*, 1995).

Stimulation of cutaneous primary afferents in the rat hind-paw leads to increased expression of c-Jun and c-Fos, transcription factors of immediate early-onset genes, in NOS-positive neurones in laminae II and III of the dorsal horn of the spinal cord (Herdegen *et al.*, 1994). Peripheral axotomy causes an increase in expression of NOS and mRNA for NOS in DRG, and in the trigeminal and nodose ganglia (Hökfelt *et al.*, 1994). Similarly, sciatic nerve axotomy results in a rapid upregulation of NOS mRNA expression in lumbar DRG (Verge *et al.*, 1992).

In the rat iris, transmural stimulation evokes constriction of the irideal arterioles due to activation of adrenergic nerves (Hill & Gould, 1995). If stimuli are applied at frequent intervals the responses wane rapidly. This inhibition is prevented by treatment with capsaicin, and mimicked by an analogue of SP, Sar^9, and CGRP, which implies that sensory nerves are involved. It is also prevented by application of L-NAME, which, furthermore, inhibits the effects of Sar^9 and CGRP. Thus, NO mediates the effects of the sensory neuropeptides and modulates sympathetic vasoconstriction (Hill & Gould, 1995). The effects of CGRP appear to be linked to NO in several systems. As mentioned above, relaxant effects of CGRP in the rat uterus are NO dependent (Shew *et al.*, 1993), and in rat gastric submucosal arterioles hyperaemia-stimulated CGRP-dependent vasodilatation is mediated in part by neurogenic NO (Chen & Guth, 1995; Holzer *et al.*, 1995). In rabbit skin, the capsaicin-evoked release of CGRP from primary afferent neurones is dependent upon NO, although the dilatation evoked by CGRP is not (Brain *et al.*, 1993; Hughes & Brain, 1994).

In an *in vitro* model of peripheral pain transmission, bradykinin applied to primary afferents in the rat tail evokes depolarization of lumbar dorsal roots, which desensitize rapidly (Rueff *et al.*, 1994). L-Arginine enhances this desensitization while methylene blue and 7-nitroindazole both prevent it, implying that NO and a cGMP pathway are its mediators (Rueff *et al.*, 1994). Systemic treatment of rats with L-NAME inhibits nociceptor flexor reflexes (Verge *et al.*, 1992), an effect that is more marked after sciatic axotomy, indicating that NO is involved in the afferent limb of lumbar pain reflexes (Verge *et al.*, 1992).

NOS-immunoreactivity is localized in carotid body microganglion cells and fibres associated with parenchymal type I cells, glomus cells and blood vessels (probably projecting from the petrosal ganglion and from ganglion cells disposed along the carotid sinus nerve) (Wang *et al.*, 1993b; Hohler, Mayer & Kummer, 1994; Villar *et al.*, 1994). Carotid body chemosensory tissue synthesizes NO, as determined by the production of citrulline, at a rate that is dependent upon the pO_2 (Prabhakar *et al.*, 1993). Chemosensory activity is enhanced concentration dependently by L-NNA (Prabhakar *et al.*, 1993), indicating that endogenous NO inhibits chemosensation. Thus, the pO_2 can affect chemosensitivity via NO production.

Whether or not spinal motoneurones with their cell bodies in the ventral horn contain NOS appears to depend upon species. In the mouse spinal cord no such motoneurones can be labelled with NADPH-diaphorase activity. In the rat and cat, NOS-immunoreactive or NADPH-diaphorase-positive neurones are found in distal cervical, lumbar or sacral segments, where they may be sparse (Valtschanoff, Weinberg & Rustioni, 1992; Vizzard, Erdman & de Groat, 1993; Saito *et al.*, 1994; Vizzard *et al.*, 1994; Pullen & Humphreys, 1995). This may be similar to the human condition (Terenghi *et al.*, 1993). In the cat sacral spinal cord there is a large discrepancy between the localizations of NOS-immunoreactive and NADPH-diaphorase-positive neurones, and in Onuf's nucleus, ventrolateral and ventromedial regions the correspondence varies from approximately 44–66%, with 7–11% containing neither, 9–18% containing NOS-immunoreactivity only, and 18–23% containing NADPH-diaphorase activity only (Pullen & Humphreys, 1995). A notably poor correspondence between NOS-immunoreactivity and NADPH-diaphorase activity has also been noted in sacral autonomic visceral pathways in the rat (Vizzard *et al.*, 1994). The function of NOS-containing motoneurones remains to be established.

5.8 Summary

NO is generally accepted as a neurotransmitter in the peripheral nervous system, even though it does not satisfy the original transmitter criteria, which have now been re-worked. In the gastrointestinal tract it is principally an inhibitory neuromuscular transmitter, responsible for reflex receptive relaxation in the stomach and descending inhibition in the intestine, and for providing a tonic inhibition of the intestine. It is not the sole NANC inhibitory transmitter, in many cases, if not always, it co-transmits with either or both VIP and ATP. There are examples of NO causing contraction of intestinal muscle, as in rat ileal longitudinal muscle, but this seems to be exceptional. In addition to acting as a neurotransmitter, NO acts as a neuromodulator, inhibiting or potentiating the release of ACh and SP. NO is also an inhibitory neuromuscular transmitter in many other places: for example, in the lower urinary tract, especially in the bladder neck and urethra; in the male reproductive tract where it causes relaxation of penile cavernosal tissues and retractor penis muscles; in the female reproductive tract where it causes uterine relaxation. In addition to its being released from endothelial cells, NO is also a mediator of neurogenic vasodilatation in many blood vessels. In blood vessels NO can modulate sympathetic transmission, reducing its vasoconstrictor efficacy. Preponderantly studied in the enteric and postganglionic parasympathetic divisions of the peripheral nervous system, NOS is also found in primary afferent neurones, and may be involved in mechanisms of neurogenic inflammation and pain perception and, to a lesser extent, in motoneurones, where its function is essentially unknown. Peripheral sympathetic and parasympathetic ganglia contain preganglionic neurones that have the potential to release NO, but as yet there is little functional evidence for a particular role.

5.9 Selected references

Hoyle, C. H. V. & Burnstock, G. (1995). Criteria for defining enteric neurotransmitters. In *Handbook of Methods in Gastrointestinal Pharmacology*, ed. T. S. Gaginella, pp. 123–40. Boca Raton, FL: CRC Press.

Lincoln, J., Hoyle, C. H. V. & Burnstock, G. (1995). Transmission: nitric oxide. In

Autonomic Neuroeffector Mechanisms, eds. G. Burnstock & C. H. V. Hoyle, pp. 509–39. Reading: Harwood Academic.

Sneddon, P. & Graham, A. (1992). Role of nitric oxide in the autonomic innervation of smooth muscle. *Journal of Autonomic Pharmacology*, **12**, 445–56.

6

Nitric oxide in the immune system

6.1 Introduction

The first indication that mammalian cells could synthesize nitrogen oxides was provided by studies of the cytotoxic effects of activated macrophages. Since this time a considerable amount of research has established that NO plays a significant role in host defence acting as a diffusible toxic mediator that can attack invading organisms and tumours, some of which are too large to be phagocytosed. In contrast to the physiological effects of NO in the regulation of vascular tone and in nerve transmission, the immunocytotoxic actions of NO require that NO should only be synthesized once an immune response is triggered and that high levels of NO should be produced over a sustained period. In view of this, it is not surprising that the synthesis of NO during immune responses is carried out by an isoform of NOS, type II NOS, that is fundamentally different from type I NOS in neurones and type III NOS in endothelial cells. Type II NOS is not present in most cells under normal conditions. However, following the appropriate stimulus, the expression of type II NOS can rapidly be induced. Once this occurs, NO synthesis is not dependent on transient increases in intracellular Ca^{2+} and can thus be maintained for a prolonged time to sustain its effects. Despite this, it would be a mistake to assume that NO synthesis by type II NOS is not strictly controlled. While the toxicity of NO is useful in host defence, it also represents a potential danger to the host and it is becoming evident that there are also mechanisms to downregulate type II NOS expression. The importance of such mechanisms is illustrated by the high mortality rate associated with septic shock where high levels of NO synthesis appear to continue unopposed. The immune system is highly sophisticated and complex. It is able to respond in a variety of different ways depending on the pathogen involved and to recruit a variety of different cell types to produce the appropriate response. Although in its infancy, research is also emerging that indicates that, in addition to its toxic effects on an invading organism, NO may also play a role in immunoregulation.

Table 6.1 *Cell types expressing type II NOS*

A. Cells of the immune system
Macrophages/monocytes [blood*, bone marrow, lung, peritoneum* (a)]
Kupffer cells
Mesangial cells*
Microglia* (b)
Splenocytes (c)
Neutrophils*
T-lymphocytes including:
Natural killer cells* (d)
T helper 1 cells (not T helper 2 cells) (e)
B. Non-immune cell types
Epithelial cells (gastric mucosa, intestinal, kidney, lung*)
Endothelial cells (aorta, brain)
Vascular smooth muscle cells*
Cardiac myocytes
Chondrocytes*
Osteoblasts*
Fibroblasts
Keratinocytes*
Hepatocytes*
Pancreatic β cells (f)
Astrocytes*
Neuronal cells

Cell types that have been shown to express type II NOS after activation by lipopolysaccharide and/or cytokines. *Indicates that type II NOS has been reported in human cells. Apart from where it has been indicated in parentheses, refer to Kröncke *et al.*, 1995 for review. (a) Weinberg *et al.*, 1995; (b) Colasanti *et al.*, 1995; (c) Yang *et al.*, 1995; (d) Xiao, Eneroth & Qureshi, 1995; (e) Barnes & Liew, 1995; (f) Corbett *et al.*, 1992c.

6.2 Cell types that express type II NOS

The cell types that have been demonstrated to express type II NOS or to synthesize NO after induction are summarized in Table 6.1. As might be expected, a number of different cell types within the immune system are capable of expressing type II NOS. More surprising, is the fact that many non-immune cell types throughout the body can also be induced to express type II NOS. These include cells in the cardiovascular and nervous systems, cells involved in cartilage, bone and skin formation, epithelial cells and pancreatic cells secreting insulin (see Table 6.1). This raises the possibility that non-immune cells may be recruited to contribute to immunocytoxicity at a local level. Induction of type II NOS in endothelial cells *in vitro* has been shown to cause lysis of tumour cells. Furthermore, endothelial cells in the liver have been reported to suppress lymphoma metastasis via the production of NO by type II NOS (Rocha *et al.*, 1995). Inappropriate expression of type II NOS in these non-immune cell types can also occur in a variety of immune and inflammatory disorders and this is thought to contribute to the tissue damage that can occur as a consequence (see Chapter 7). Type II NOS expression has also been reported to occur in a variety of tumour cell lines and in human ovarian, uterine and breast tumour tissue (Henry *et al.*, 1993; Nussler & Billiar, 1993; Bani *et al.*, 1995; Thomsen *et al.*, 1995).

Within the immune system, monocytes/macrophages from the blood, bone marrow, lung, peritoneum, liver (Kupffer cells), kidney (mesangial cells) and central nervous system (microglia) can all be activated to express type II NOS. Type II NOS can also be induced in splenocytes, neutrophils and T-lymphocytes including the subpopulation that can act as natural killer cells. T helper (Th) cells are another subpopulation of T-lymphocytes that play an important role in determining the type of immune response that is brought into effect. Two subsets of Th cells, Th1 and Th2, have been classified in mice based on the pattern of different cytokines they secrete (Roitt, 1994). Th2 cells tend to suppress the inflammatory state of macrophages but are active in the production of antibodies by B-lymphocytes. In contrast Th1 cells stimulate macrophages to release inflammatory mediators (Barnes & Liew, 1995; Modolell *et al.*, 1995). Type II NOS can be induced in Th1 cells but not Th2 cells (Barnes & Liew, 1995).

There is one question that has been the subject of considerable debate concerning whether human macrophages are able to express type II NOS as part of an immune response (see Denis, 1994; Kröncke, Fehsel & Kolb-Bachofen, 1995). It has proved difficult to demonstrate high levels of NO synthesis after stimulation of human macrophages *in vitro* with agents that readily induce NO synthesis in rodent macrophages. One of the reasons for this may be that human cell cultures have generally been derived from peripheral blood monocytes which are likely to differ physiologically from the tissue-derived macrophages that are more commonly studied in animals (James, 1995). Significant induction of NO synthesis has been demonstrated in human peritoneal macrophages but not in human blood monocytes after identical treatment with cytokines (Weinberg *et al.*, 1995). Alternatively, human monocytes may require different signals to induce type II NOS expression. The production of NO by human blood monocytes has been reported to occur following treatment with granulocyte macrophage-colony stimulating factor, after co-culture with tumour cell lines and on infection with human immunodeficiency virus (HIV) type I (see Moncada, 1992; Bukrinsky *et al.*, 1995; James, 1995; Son & Kim, 1995). In the case of HIV, it has been suggested that increased production of NO within the central nervous system may contribute to

Table 6.2 *Agents that regulate the expression of type II NOS*

Activators/enhancers	Inhibitors
Lipopolysaccharide (LPS) (a,b,c)	Transforming growth factor β (TGF-β) (a,c)
Tumour necrosis factor-α (TNF-α) (a,b,c)	Glucocorticoids (a,c, Kunz *et al.*, 1996)
Interferon-γ (IFN-γ) (a,b,c)	IL-4 (a,c, James, 1995)
IL-1 (a,b,c, Geller *et al.*, 1995a)	IL-6 (Terenzi *et al.*, 1995)
IL-12 (Zhou *et al.*, 1995)	IL-8 (Sessa, 1994)
Granulocyte-macrophage colony stimulating factor (GMCSF) (a)	IL-10 (a,c, James, 1995)
Platelet activating factor (PAF) (Schmidt, Seifert & Bohme, 1989a)	Epidermal growth factor (EGF) (a, Terenzi *et al.*, 1995)
Tumour-derived recognition factor (TDRF) (Jiang *et al.*, 1995)	Platelet-derived growth factor (PDRF) (a, Sessa, 1994)
Prostaglandins (low concentration) (Milano *et al.*, 1995)	Prostaglandins (high concentration) (Milano *et al.*, 1995)
NO (low concentration) (Sheffler *et al.*, 1995)	NO (high concentration) (Sheffler *et al.*, 1995)
Leukotriene B_4 (Schmidt, Seifert & Bohme, 1989a)	Thrombin (a)
Cisplatin (a, Son & Kim, 1995)	Noradrenaline (via β) (c)
Taxol (Jun *et al.*, 1995)	ATP (via P2Y) (Denlinger *et al.*, 1996)
HIV proteins (a)	Nicotinamide (Akabane *et al.*, 1995)
Heat shock (a)	Glibenclamide (Wu, Thiemermann & Vane, 1995)
Hypoxia (Melillo *et al.*, 1995)	NF-$_{\kappa}$B activation inhibitors (Kwon *et al.*, 1995; Spink, Cohen & Evans, 1995)
	Serine/cysteine protease inhibitors (Griscavage, Wilk & Ignarro, 1995; Kim, Bergonia & Lancaster, 1995b)
	BH_4 synthesis inhibitor (Bogdan *et al.*, 1995)
	Tyrosine kinase inhibitor (a)

Agents that have been shown to induce or to inhibit the induction of type II NOS. For reviews see: (a) Kröncke *et al.*, 1995; (b) Nussler & Billiar, 1993; (c) Murphy *et al.*, 1993. Additional selected references have been given for each agent as required. Please note that the above agents have been investigated in studies of a variety of different cell types.

the pathogenesis of HIV-associated neurological disease (Bukrinsky *et al.*, 1995). In contrast, high levels of NO synthesis and type II NOS expression can readily be induced in human hepatocytes. It has been suggested that the ability of hepatocytes to express type II NOS may be of particular benefit in defence against blood-borne parasites, particularly the malaria parasite, that are transported to the liver (Nussler & Billiar, 1993; James, 1995).

6.3 Factors regulating the expression of type II NOS

Numerous agents have been investigated with regard to their ability to up regulate or downregulate the expression of type II NOS and many of these are listed in Table 6.2. It should be noted that agents that can inhibit NOS activity once the protein has been expressed are described separately in Chapter 10. The first agent to be used successfully to induce type II NOS expression was the bacterial endotoxin lipopolysaccharide (LPS). Experimentally, a combination of LPS and interferon-γ (IFN-γ) are most commonly used to induce type II NOS in murine cells, whereas in rat and human cells a cytokine mixture of IFN-γ, tumour necrosis factor-α (TNF-α) and interleukin-1β (IL-1β) is more effective (Kröncke *et al.*, 1995). It is clear that such agents act synergistically and also sequentially. IFN-γ is the most common primary signal for the activation of macrophages whereas TNF-α is a major secondary signal (see James, 1995 for review). Of the interleukins, IL-1β, IL-2 and IL-12 activate the induction of type II NOS whereas IL-4, IL-6, IL-8 and IL-10 have all been reported to inhibit induction. When one considers the physiological sources of these cytokines, then a pattern emerges with regard to the type of immune response that is characterized by the induction of type II NOS and NO-mediated cytotoxicity. IFN-γ is produced by T-lymphocytes including Th1 cells. TNF-α is also released by T-lymphocytes. Furthermore, macrophages, once they have been activated, are able to synthesize and release TNF-α where it can act in an autocrine manner. Similarly, IL-1β is produced by activated macrophages (Roitt, 1994). In addition to IFN-γ, Th1 cells secrete IL-2, while Th2 cells produce IL-4, IL-5 and IL-10 (Barnes & Liew, 1995). Thus activation of Th2 cells will result in the release of cytokines that inhibit the induction of type II NOS and favour the immune response of antibody production. In contrast, activation of Th1 cells results in the Th1-type of immune response characterized by the local activation of macrophages. A mixture of cytokines are produced under these circumstances which can actively stimulate the production of type II NOS and thus the synthesis of high levels of NO to exert its toxic effects (Barnes & Liew, 1995; James, 1995). The Th1-type response has generally been thought to protect against intracellular organisms while the Th2-type response acts against extracellular pathogens. Studies of parasitic infections have shown that such a distinction

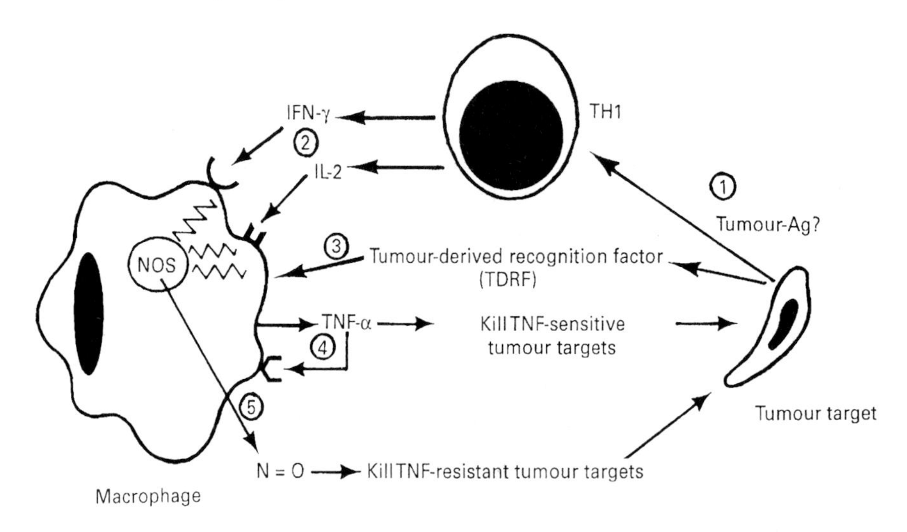

Fig. 6.1 Proposed mechanism for macrophage-mediated tumour cytotoxicity induced by interferon-γ (IFN-γ), interleukin-2 (IL-2) and tumour-derived recognition factor (TDRF). (1) Tumour target activation of T helper Th1 lymphocytes by presentation of tumour antigen; (2) activated Th1 lymphocyte secretion of lymphokines (IFN-γ and IL-2) to activate macrophages; (3) TDRF from tumour targets synergizes with IFN-γ and IL-2 to stimulate macrophages to produce tumour necrosis factor-α (TNF-α) which results in killing of TNF-sensitive tumour targets; (4) TNF-α autocrine feedback to macrophages through TNF receptors for NOS synthesis; (5) macrophages releasing NO to kill TNF-resistant tumour targets. (Modified from Jiang *et al.*, 1995; *Immunobiology*, **192**, 321–42, with permission.)

cannot always be made. Extracellular pathogens such as helminths (parasitic worms) can be as susceptible to attack by NO as can protozoa (James, 1995).

In addition to the inflammatory cytokines mentioned earlier, granulocyte macrophage-colony stimulating factor (released from T-lymphocytes, macrophages, fibroblasts, mast cells and the endothelium), leukotriene B_4 (released from mast cells on sensitization) and platelet activating factor can all upregulate type II NOS expression. Some soluble factors can also be derived from tumour cell lines. Macrophage deactivation factor produced by mastocytoma cells inhibits the production of NO. Tumour-derived recognition factor (TDRF) is produced by a variety of transformed cell lines and appears to act in synergy with IFN-γ, IL-2 and TNF-α, enhancing type II NOS mRNA synthesis and NO-mediated tumour cytotoxicity. In addition, tumour cell lines that are unable to produce TDRF are more resistant to killing by activated macrophages (Jiang, Stewart & Leu, 1995) (Fig. 6.1). In contrast, epidermal growth factor and platelet-derived growth factor inhibit induction of type II NOS. In addition, noradrenaline (acting via β-adrenoceptors) and ATP analogues (acting via P2Y-purinoceptors) both inhibit type II NOS induction (Table 6.2). Thus there is the potential for locally released agents, including neurotransmitters, to modify an immune response by regulating the expression of type II NOS. Similarly, glucocorticoids are powerful inhibitors of induction. This may not only play a physiological role but has also proved useful clinically in the treatment of inflammatory disorders (see Chapter 7). Glucocorticoids appear to affect type II NOS expression at both the transcriptional and translational levels (Kunz *et al.*, 1996). Immunosuppressants such as cyclosporin A and FK506, used in the prevention of transplant rejection, also inhibit the induction of type II NOS (Table 6.2). Cisplatin, an agent used in cancer therapy, upregulates type II NOS expression (Son & Kim, 1995) as does taxol, which appears to act by stimulating the release of TNF-α (Jun *et al.*, 1995). Interestingly, hypoxia has recently been reported to activate the synthesis of type II NOS. This is in marked contrast to type III NOS where chronic hypoxia downregulates its expression (see Chapter 3). Such induction of type II NOS may be of importance to the maintenance of NO synthesis under the hypoxic conditions that can exist in inflamed tissues and necrotic areas of neoplastic lesions (Melillo *et al.*, 1995). It should be noted that the list of agents that regulate the expression of type II NOS provided here is by no means complete, current research is revealing new agents all the time. Furthermore, the precise nature of the interrelationships between agents produced locally, that may be tissue specific, and the agents released from cells of the immune system have yet to be fully defined.

Clearly, it is important that once type II NOS has been induced there should be mechanisms that can stop the inflammatory process and prevent widespread tissue damage due to the toxicity of NO. In this context, transforming growth factor-β (TGF-β), prostaglandin E_2 (PGE_2) and NO itself are attractive candidates. All three are synthesized by the macrophages themselves once they have been activated and are thus readily available (Roitt, 1994). It is well established that TGF-β can reduce NO synthesis by destabilizing type II NOS mRNA (Nathan & Xie, 1994a). At low concentrations, PGE_2 appears to enhance NO synthesis but, at high concentrations, PGE_2 both inhibits TNF-α secretion and the expression of type II NOS (Milano *et al.*, 1995). Similarly, low levels of NO enhance type II NOS expression. High levels of NO inhibit both NOS activity and the induction of type II NOS expression (Sheffler *et al.*, 1995). Intriguingly, PGE_2 and TGF-β not only suppress the expression of type II NOS but also stimulate macrophages to express arginase (Boutard *et al.*, 1995; Modolell *et al.*, 1995). Arginase converts arginine to ornithine, which represents the first step in the pathway for the synthesis of polyamines, and polyamines can also inhibit the induction of type II NOS (Szabó *et al.*, 1994b). Since both arginase and NOS use arginine as a substrate, simultaneous expression of both enzymes would result in competition between the two pathways. It is becoming evident that type II NOS and arginase are regulated reciprocally in macrophages such that their expression can be separated in time (Modolell *et al.*, 1995). Polyamines promote proliferation and repair in various cell types and TGF-β is important for the stimulation of fibroblast division and the laying down of new extracellular matrix elements (Roitt, 1994; Boutard *et al.*, 1995). It can therefore be envisaged that the effects of PGE_2 and TGF-β could be to downregulate NO-mediated cytotoxicity and to upregulate mechanisms that can repair any tissue damaged by the inflammatory process.

6.4 Type II NOS and its gene

As has been described in detail in Chapter 2, NO is synthesized from arginine and molecular O_2 by type II NOS in a process involving NADPH as a co-substrate, FAD, FMN and BH_4 as co-factors and haem and calmodulin as prosthetic groups. Unlike type I and type III NOS, calmodulin binds tightly to type II NOS such that its binding does not require the presence of raised intracellular Ca^{2+} concentrations. However, Ca^{2+} may play a role in the regulation of type II NOS since it has recently been reported that Ca^{2+} ionophores can inhibit the induction of type II NOS (Jordan *et al.*, 1995) and reduce the stability of type II NOS mRNA (Geng & Lotz, 1995). BH_4 is involved in the dimerization of NOS and the amino acids essential for BH_4 binding and dimerization have been identified in type II NOS (Cho *et al.*, 1995). It has also been reported that inhibition of BH_4 synthesis suppresses the expression of type II NOS mRNA (Bogdan *et al.*, 1995). The findings of molecular biological studies investigating the type II NOS gene have been described in detail in several excellent reviews (Nathan & Xie, 1994a; Sessa, 1994; Kröncke *et al.*, 1995). Since type II NOS can be expressed in many different cell types and there is considerable species and tissue variation in the ways in which type II NOS expression can be induced, it has been a matter of controversy whether all the type II NOS proteins isolated are in fact the product of a single gene. When a single locus was identified on the human genome using both murine macrophage and human hepatocyte type II NOS cDNAs as probes, this question appeared to have been resolved (Chartrain *et al.*, 1994). However, additional type II NOS-like sequences have been reported although it is not known if these actually undergo transcription or represent pseudogenes (see Knowles & Moncada, 1994; Kröncke *et al.*, 1995). This question is further complicated by the fact that multiple forms of exon 1 have been demonstrated in human type II NOS mRNA transcripts due to a combination of multiple transcription initiation sites and alternative splicing. It has been suggested that these may play an important role in the regulation of translation of human type II NOS mRNA (Chu *et al.*, 1995). It has already been noted that, under certain conditions, the expression of

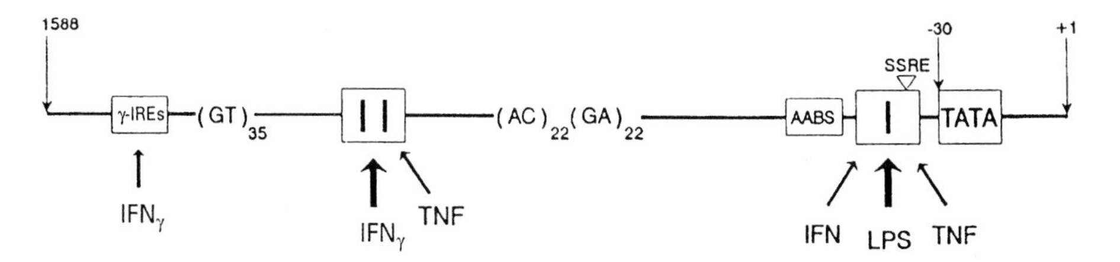

Fig. 6.2 Schematic structure of the 5′-flanking region of the type II NOS gene. The transcriptional start site is denoted as nucleotide position +1. A likely TATA box begins at −30. Several potential transcription factor binding sites (TNF, tumour necrosis factor; LPS, lipopolysaccharide; IFN-γ, interferon-γ) are indicated. Maximal expression of murine type II NOS depends on two discrete regulatory regions (I and II). On its own, region II cannot regulate type II NOS expression, indicating that it acts primarily as an enhancer region. The shear stress responsive element (SSRE) has only been described for the human type II NOS gene. (Reproduced from Kröncke *et al.*, 1995; *Biological Chemistry Hoppe-Seyler*, 327–43, with permission.)

type II NOS mRNA can be induced in human monocytes without expression of type II NOS protein, indicating that regulation at the translational level may be more tightly controlled than in rodents (Kröncke *et al.*, 1995). Two forms of type II NOS have also been reported in a murine macrophage cell line that could be distinguished by antigenic differences at the amino terminus and by the fact that one was cytosolic and the other membrane-bound (Ringheim & Pan, 1995). It has not been established whether these two forms were the products of two different mRNAs or if they had been subjected to post-translational modification.

Despite these problems, progress is being made with regard to the transcriptional control of type II NOS by studies of the promoter/enhancer region of the type II NOS gene (particularly of the mouse) and the intracellular mechanisms mediating the induction of transcription (see Nathan & Xie, 1994a; Kröncke *et al.*, 1995). The promoter/enhancer region of the type II NOS gene is complex, containing at least 22 potential binding sites for transcription factors. In addition to the putative TATA box, which is the main site for the initiation of transcription, two discrete regulatory regions in the murine gene have been described (Fig. 6.2). Region I contains specific binding sites for nuclear factor IL-6, transcription factor $NF_{\kappa}B$/rel and responsive elements for IFN-γ. This is the main site for the regulation of LPS-induced type II NOS expression. Region II contains an additional $NF_{\kappa}B$ site and four motifs for binding IFN-regulated transcription factors. Region II does not appear to be able to induce the expression of type II NOS independently, indicating that it is likely that it acts primarily as an enhancer rather than promoter. The proximal promoter region is well conserved in the human gene. However, there is more divergence in the distal elements. How this affects the induction of type II NOS in human cells compared with other species has yet to be clearly defined.

It is evident that the $NF_{\kappa}B$/rel transcription factor plays an important role in the induction of type II NOS (see Goldring *et al.*, 1995). Under resting conditions, $NF_{\kappa}B$ is bound to an inhibitory protein and is present in the cytoplasm in an inactive form. On stimulation of cells, $NF_{\kappa}B$ is liberated by the action of certain proteases and can translocate to the nucleus where it can regulate gene transcription (Chen *et al.*, 1995). $NF_{\kappa}B$ proteins are a heterogeneous group of proteins, and heterodimers containing mainly the rel family of proteins appear to be involved in the induction of type II NOS (Nathan & Xie, 1994a). Serine or cysteine protease inhibitors such as chloromethylketones and dithiocarbamate derivatives can prevent the activation of $NF_{\kappa}B$. Both of these classes of compounds have been shown to inhibit the induction of type II NOS in macrophages (Groscavage, Wilk & Ignarro, 1995; Kim *et al.*, 1995a; Kwon *et al.*, 1995). In the case of IL-1-stimulated induction of type II NOS, activation of tyrosine kinase appears to be required for the activation of $NF_{\kappa}B$ since inhibition of tyrosine kinase blocks IL-1-induced translocation of $NF_{\kappa}B$ (Kwon *et al.*, 1995). There is preliminary evidence that induction of type II NOS in non-immune cell types, such as vascular smooth muscle cells or renal epithelial cells, involves the activation of a different form of $NF_{\kappa}B$ protein that does not contain c-rel (Amoah-Apraku *et al.*, 1995; Spink, Cohen & Evans, 1995). The activation of $NF_{\kappa}B$ often relies on the production of reactive oxygen species (Kröncke *et al.*, 1995); however, exogenous NO has been reported to inhibit LPS/TNF-elicited $NF_{\kappa}B$ activation (Colasanti *et al.*, 1995). NO can also inhibit TNF synthesis by macrophages (Eigler, Moeller & Endres, 1995). Thus, there are negative feedback mechanisms whereby high levels of NO can inhibit the induction of type II NOS. IFN-γ induces the expression of type II NOS via the production of interferon regulatory factor 1 (IRF 1)which acts on interferon responsive elements in the promoter/enhancer region of the gene (Fig. 6.2). Nicotinamide has been shown to prevent the induction of interferon regulatory factor 1 (Akabane *et al.*, 1995). This has been proposed as the possible mechanism whereby nicotinamide can prevent NO-mediated killing of the pancreatic β cells and the development of diabetes mellitus (see Chapter 7). There is still some way to go before the intricacies of the transcriptional regulation of type II NOS are fully understood. Furthermore, the possibility that there are differences in the regulatory mechanisms between immune and non-immune cell types is an intriguing area that needs to be explored further. It remains to be determined whether the induction of type II NOS in non-immune cells serves a function other than host defence.

Table 6.3 *Pathogens and cellular targets of NO*

Viruses
Herpes simplex virus (a,d)
Coxsackie virus (a)
Vaccinia virus (d)
Ectromelia virus (d)
Bacteria
Francisella tularensis (a)
Mycobacterium tuberculosis (a,e)
Mycobacterium bovis (f)
Mycobacterium leprae (a,b)
Mycobacterium avium (e)
Listeria monocytogenes (a)
Chlamydia trachomatis (a)
Fungi
Cryptococcus neofomans (a,b)
Histoplasma capsulatum (g)
Parasites
Leishmania species (a,b,c)
Trypanosoma cruzi, musculi and *brucei* (a,c)
Plasmodium falciparum (a,c), *chabaudi* (h)
Toxoplasma gondii (b,c)
Schistosoma mansoni (b,c)
Entamoeba histolytica (a,i)
Mammalian cells
Tumour cells (a)

Pathogens that have been shown to induce the expression of type II NOS either *in vitro* or *in vivo* and are subject to the toxic effects of NO. Selected references are indicated in parentheses. (a) Lowenstein, Dinerman & Snyder, 1994; (b) Langrehr *et al.*,1993; (c) James, 1995; (d) Nathan, 1995; (e) Greenberg *et al.*, 1995a; (f) Yang *et al.*, 1995; (g) Zhou *et al.*, 1995; (h) Jacobs, Radzioch & Stevenson, 1995; (i) Lin *et al.*,1995.

6.5 Toxic effects of NO

Clearly, for NO to be a mediator in immunocytotoxicity it must not only be generated in the appropriate cells but must also exert a toxic effect in the target. That this is the case was demonstrated in early experiments when it was shown that inhibition of NO synthesis in activated macrophages also prevents bacterial cell death and killing or cytostasis of tumour cells in co-culture (see

Nussler & Billiar, 1993 for review). Since this time numerous pathogens have been shown to activate the induction of type II NOS and/or be susceptible to attack by NO either *in vitro* or *in vivo* (see Table 6.3). The most compelling evidence that NO plays a significant role in immunocytotoxicity has been the finding that knockout mice, in which the gene for type II NOS has been disrupted, are more susceptible to infection by the protozoan parasite *Leishmania major* (Wei *et al.*, 1995) and fail to restrain replication of the bacteria *Listeria* (MacMicking *et al.*, 1995). In addition, macrophages isolated from such mice are unable to prevent the proliferation of lymphoma cells.

Although NO is frequently described simply as a 'toxic molecule' this does not provide an adequate picture of the sequence of events that links the production of NO with its toxic effects. It is becoming evident that the actions of NO may be governed by the intracellular environment of both the generating cell and the target. The range of interactions that NO can undergo and the mechanisms by which they can occur *under biological conditions* have yet to be fully defined and are extremely difficult to study. However, this information is important because it is likely to provide some explanation for what determines the final outcome in the target. Not all microorganisms or parasites are susceptible to attack by NO. Furthermore, when toxic effects do occur they can be manifested in different ways. Cell death can occur by two fundamentally different processes: necrosis and apoptosis. Necrotic and apoptotic cell death can be distinguished by morphological and biochemical characteristics (see Brüne, Mohr & Messmer, 1996 for review). In necrosis, early changes are prominent in the cytoplasm rather than the nucleus, with swelling of the mitochondria. Membrane ion-pump activity disappears, which may be due to direct membrane damage or as a result of energy depletion. DNA is degraded following exposure to lysosomal nucleases which results in DNA fragments with a continuous spectrum of sizes. Ultimately, the cells lyse releasing cellular contents and debris which can provoke an inflammatory response in adjacent tissue. In contrast, DNA fragmentation occurs early in apoptosis before cell viability starts to decline. This occurs by the action of an endonuclease which cleaves DNA such that a characteristic stepwise 'ladder pattern' of fragments can be seen on electrophoresis. In many cases, the expression of p53, the product of a tumour suppressor gene, precedes apoptotic cell death. Both the nucleus and cytoplasm condense and eventually the cell forms multiple membrane-bound bodies. As a result, cellular contents are not released into the surrounding area to provoke an inflammatory response in adjacent tissue. Both necrotic and apoptotic cell death can occur as a consequence of an immune response and NO appears to have the capacity to mediate either event. In addition to causing necrosis, NO has been shown to induce apoptosis in a variety of cell types including macrophages, thymocytes, lymphocytes, chondrocytes and pan-

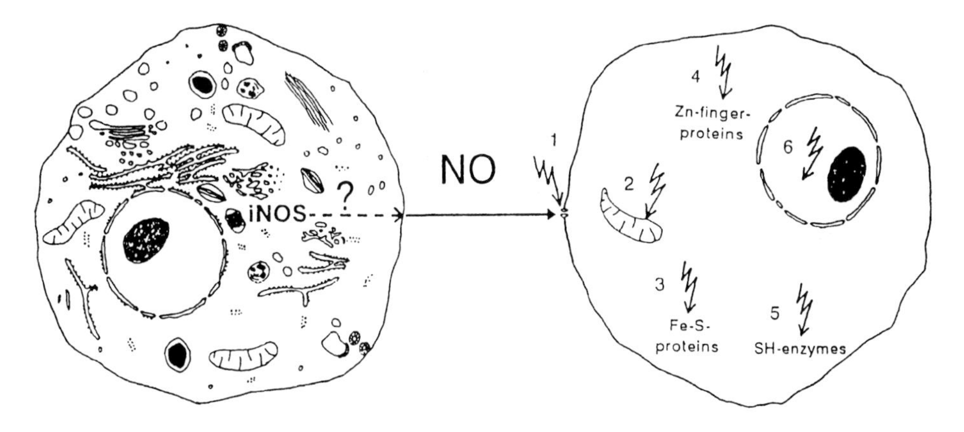

Fig. 6.3 Potential cytotoxic interactions of NO with cells. NO is synthesized in the cellular cytosol following induction of type II NOS expression. NO is released and acts in a paracrine manner; however, the precise form in which it is released and subsequently acts has yet to be fully characterized. NO can affect several different sites in the target cell: (1) transport proteins, ion channels and plasma membrane potential; (2) mitochondrial respiration and membrane potential; (3) Fe–sulphur clusters in proteins; (4) Zn-finger domains in proteins; (5) free thiol groups near the active sites of enzymes; and (6) DNA, damage to DNA resulting in activation of polyADP-ribose polymerase in the nucleus. NO can also interact with several different groups on ribonucleotidase to inhibit DNA synthesis. (Reproduced from Kröncke *et al.*, 1995; *Biological Chemistry Hoppe-Seyler*, **376**, 327–43, with permission.)

creatic β cells producing the ladder pattern of DNA fragments and in some cases the expression of p53 (Sarih, Souvannavong & Adam, 1993; Adler *et al.*, 1995; Fehsel *et al.*, 1995; Kaneto *et al.*, 1995; Kröncke *et al.*, 1995; Brüne *et al.*, 1996). Apoptotic cell death can be an important component of the immune response against intracellular organisms, particularly viruses, since it has the effect of preventing the spread of the pathogen. Indeed, the function of natural killer cells, which can express type II NOS, is to cause apoptosis. Not all cells die following exposure to the levels of NO synthesized by type II NOS. Particularly in the case of tumour cells NO only prevents proliferation resulting in cytostasis.

The potential targets for NO that contribute to cell death or cytostasis have been desribed in detail in Chapter 2 (for review see Henry *et al.*, 1993; Stamler, 1994; Kröncke *et al.*, 1995; Brüne *et al.*, 1996) and are summarized schematically in Fig. 6.3. Briefly, NO can inhibit the activity of ribonucleotidase, the rate-limiting enzyme in the DNA replication process. There is evidence that NO has this effect by interacting directly with a tyrosyl radical present on the R2 subunit of ribonucleotidase. This interaction causes the release of the tyrosyl radical from the enzyme. Loss of the tyrosyl radical after exposure to NO parallels inhibition of DNA synthesis and results in cytostasis (Henry *et al.*, 1993). NO can also influence the activity of ion channels and destroy the membrane potential of mitochondria (see Kröncke *et al.*, 1995). Energy depletion can be induced in a variety of different ways including: interaction with iron-sulphur groups on mitochondrial oxidoreductases and *cis*-aconitase; *S*-nitrosylation followed by ADP-ribosylation of thiol groups on the glycolytic enzyme glyceraldehyde-3-phosphate dehydrogenase; and activation of polyADP-ribose synthetase either directly or following damage to DNA. In most cases, the actions of NO in causing energy depletion are more likely to be mediated by peroxynitrite than free NO. Direct damage to DNA can result from deamination by NO if the thiol pool is depleted or by the mutagenic effects of peroxynitrite and the hydroxyl radical that can be generated by peroxynitrite. NO also causes the release of Zn^{2+} from a zinc finger repair enzyme (Kröncke *et al.*, 1995). Thus the form in which NO is present in the target cell is likely to determine the severity of the response. This has been illustrated well in a recent study investigating the effects on *Salmonella* of three NO donor compounds that supply NO in different redox forms (De Groote *et al.*, 1995). 3-Morpholinosydnonimine hydrochloride, which generates peroxynitrite, killed the bacteria whereas *S*-nitrosoglutathione, which can produce NO^+, only caused cytostasis. Remarkably, the free NO radical generated by a diethylenetriamine-NO adduct had no antibacterial activity. Similarly, exogenous administration of a spontaneous NO-releasing agent results in minimal toxicity to *Escherichia coli*. However, combined treatment with NO and hydrogen peroxide greatly potentiates toxicity (Pacelli *et al.*, 1995).

Peroxynitrite is formed from the interaction between NO and superoxide anions. Superoxide anions are themselves toxic, mediating iron-catalysed lipid peroxidation. Intriguingly, NO has been found to protect against superoxide anion-mediated injury under certain conditions. Some explanation for this paradox has been found very recently in a study which has shown that when there is excess production of either NO or superoxide anions over the other then the formation of peroxynitrite is inhibited (Miles *et al.*, 1996). However, when both NO and superoxide anions are present in equal amounts peroxynitrite can be generated. Moreover, if peroxynitrite is generated in the presence of free iron then this encourages the formation of the mutagenic hydroxyl radical from peroxynitrite (Miles *et al.*, 1996). Thus, the degree of toxicity that high levels of NO can induce may be related in part to whether high levels of superoxide anions are produced at the same time. In the case of activated macrophages, this appears to be the case since the production of NO and superoxide anions increases in parallel (Beckman *et al.*, 1994b).

Protective mechanisms and the intracellular environment of the target are also likely to affect the outcome of exposure to cytotoxic levels of NO. High levels of superoxide dismutase would prevent the formation of peroxynitrite. In the case of bacteria, activation of the redox-sensitive superoxide response regulon can confer resistance to the effects of NO (Nunoshiba *et al.*, 1995). Interestingly, pretreatment of hepatocytes to low doses of an NO donor induces resistance to killing by a subsequent higher dose of NO. Such resistance is associated with an upregulation of the heat shock protein, haem oxygenase, and increased levels of ferritin (Kim, Bergonia & Lancaster, 1995b). Ferritin sequesters iron, storing it in a non-redox form that does not potentiate the formation of the hydroxyl radical (Miles *et al.*, 1996). Experiments on fibroblast cell lines have demonstrated that resistance to NO-induced cytotoxicity is also related to the level of glutathione within the cell (Walker *et al.*, 1995). In addition to these defence systems, the energy stores of a cell may influence its ability to withstand the toxic effects of peroxynitrite on energy production. However, even in the absence of energy depletion peroxynitrite could still cause cell death if it damaged the DNA of the target. DNA damage is known to initiate apoptosis (Brüne *et al.*, 1996). In the case of parasitic infections, such as those caused by *Schistosoma*, susceptibility to the toxic effects of NO are likely to be governed by whether it is in an aerobic or anaerobic stage of its life cycle (James, 1995).

The interactions described here are those that contribute towards NO-mediated cytotoxicity in host defence. However, it should be recognized that NO can also be harmful to the host particularly when very high levels are produced or synthesis is sustained over a long period of time. Very high levels of NO are produced in septic shock and this is associated with severe hypotension which contributes significantly to mortality in this condition (see Chapter 7). In addition, NO can form nitrosamines and these are powerful mutagens and carcinogens (Tannenbaum *et al.*, 1994; Kerwin, Lancaster & Feldman, 1995). Thus, chronic infection or inflammatory conditions could result in accumulation of nitrosamines and carcinogenesis (Oshima & Bartsch, 1994; Tannenbaum *et al.*, 1994). It is known that some parasitic infections are a risk factor for the subsequent development of cancer. Increased urinary levels of nitrosoproline have been detected in certain infected individuals (Oshima & Bartsch, 1994). Finally, NO can also act as an immunoregulator and suppression of immune responses by NO may lead to increased susceptibility to infection in certain chronic conditions (see section 6.6).

6.6 NO and immunosuppression

While the majority of studies have considered the active role of NO in immunocytotoxicity, there is increasing evidence that NO can modulate immune function and in many cases produce immunosuppression (see Langrehr *et al.*,

1993; Kröncke *et al.*, 1995 for review). One of the first indications that this might be the case was in studies of the mixed lymphocyte reaction. Using splenocytes from mice, proliferation and cytolytic T-lymphocyte activity are observed in this reaction. However, the effect is greater in the presence of a NOS inhibitor. The inhibitory effect of NO on lymphocyte proliferation and activation appears to be greater in rats since little proliferation is observed unless a NOS inhibitor is present. Similar results have been obtained *in vivo* during allograft rejection where it has been shown that diminished cytolytic T-lymphocyte activity in the rat compared with the mouse was associated with enhanced NO synthesis (Langrehr *et al.*, 1993). Similarly, in a transfusion model of alloimmunization in mice, during the initial period before antibody production, there was induction of T-cell anergy and suppression of natural killer cell activity which was associated with the production of NO (Semple *et al.*, 1995). Bone marrow contains natural suppressor cells that can strongly inhibit proliferative responses in lymphocytes. Such suppressor activity can be inhibited by a NOS inhibitor and mimicked by an NO donor (Angulo *et al.*, 1995). NO produced by type II NOS in Kupffer cells has been reported to reduce IL-2 receptor expression and lymphokine-activated killer activity of splenocytes (Kurose *et al.*, 1995). In line with these findings, NO has been implicated in immunosuppression following bacterial or parasitic infection, burn injury and in cancer (see Abrahamsohn & Coffman, 1995; James, 1995; Kröncke *et al.*, 1995). NO has recently been reported to activate chloride currents in human T-lymphocytes, but the precise physiological significance of this in terms of immunoregulation has yet to be determined (Dong *et al.*, 1995). In a different system, NO synthesized by type III NOS in endothelial cells, in addition to regulating vascular tone, plays an anti-inflammatory role by preventing adhesion and infiltration of leucocytes (See Chapters 7 and 8).

Interestingly, NO can inhibit the proliferation of Th1 cells and their production of IL-2 and IFN-γ. In contrast, NO has no effect on Th2 cells (Barnes & Liew, 1995). Thus, NO appears to suppress the Th1-type response in which it is actively involved as a cytotoxic mediator. Type II NOS knockout mice react to infection with a significantly stronger Th1-type immune response and, despite this, are more susceptible to infection (Wei *et al.*, 1995). This indicates that loss of inducible NO synthesis prevents both resistance and immunosuppression. Whether immunosuppression is ultimately of benefit or detrimental to the host is not clear. In chronic conditions it may lead to increased susceptibility to infection. However, acutely, it may represent an important feedback mechanism that prevents an overactive inflammatory response.

6.7 Summary

The high levels of NO that are required for it to act as a cytotoxic mediator in an immune response are synthesized by the type II isoform of NOS which is not dependent on high intracellular Ca^{2+} for its activity. Type II NOS is not normally present constitutively but can be induced in a variety of immune and non-immune cell types. Induction of type II NOS expression can be triggered by many factors some of which are secreted as part of a Th1-type of immune response (IFN-γ, TNF-α, IL-1) that activates macrophages. The strict transcriptional control of type II NOS expression is reflected in the complexity of the promoter/ enhancer region of the type II NOS gene. Type II NOS expression can also be inhibited by several factors (e.g. TGF-β, NO, prostaglandins), which may represent important feedback control mechanisms. Once NO is synthesized it can cause cytostasis or cell death either by necrosis or apoptosis in the target. These toxic effects are achieved by the interaction of NO with several intracellular enzymes or with DNA itself. The nature of these interactions is dependent on the redox state of NO and is influenced by the chemical environment of both the generating cell and the target cell. In addition to its toxic effects, NO can act as an immunoregulator suppressing the Th1-type of immune response. This probably provides a feedback control mechanism whereby NO can prevent excessive inflammation since NO has the potential to be extremely toxic to the host. However, in cases of chronic infection or cancer, the immunosuppressive effect of NO could lead to increased susceptibility to infection.

6.8 Selected references

James, S. L. (1995). Role of nitric oxide in parasitic infections. *Microbiological Reviews*, **59**, 533–47.

Kröncke, K.-D., Fehsel, K. & Kolb-Bachofen, V. (1995). Inducible nitric oxide synthase and its product nitric oxide, a small molecule with complex biological activities. *Biological Chemistry Hoppe-Seyler*, **376**, 327–43.

Langrehr, J. M., Hoffman, R. A., Lancaster, J. R. & Simmons, R. L. (1993). Nitric oxide: a new endogenous immunomodulator. *Transplantation,* **55**, 1205–12.

Section 2

Pathological implications of nitric oxide

Introduction

As fast as research has provided evidence that NO may act as a mediator in a wide range of physiological processes, so NO has been implicated in an ever increasing number of pathological disorders. Given the properties of NO, this is not surprising. It can easily be envisaged that loss or inactivation of NO that is synthesized constitutively will result in the dysfunction of NO-dependent mechanisms that form part of normal physiological processes. This has been most clearly demonstrated with respect to the endothelium, where a deficit in responses mediated by EDRF had been observed in a variety of cardiovascular disorders even before EDRF had been identified as NO. Equally, the high levels of NO produced following induction of type II NOS are toxic and have the potential to damage healthy tissue. Thus, NO may contribute to a number of apparently disparate disorders that all have a common element of abnormal immune or inflammatory responses. However, increased levels of NO cannot be universally regarded as harmful. Apart from its role in host defence, NO has also been shown to promote cell survival under certain circumstances, rather than contribute to cell death. As a consequence, phrases such as 'the double-edged sword' are frequently applied to NO in health and disease. This emphasizes the importance of establishing the precise contribution that NO makes to a pathological process before considering how to manipulate NO levels as part of an attempt at therapy. In this section the pathological implications of NO synthesized by the different isoforms of NOS will be considered separately. Thus, the role of NO following induction of type II NOS in immune and inflammatory disorders will be described in Chapter 7. Cardiovascular disorders and the role of type III NOS within the endothelium and EDNO will be discussed in Chapter 8. Finally, changes that can occur in type I NOS both in the central and the peripheral nervous systems and their contribution to neuropathic disorders and nerve cell death will be described in Chapter 9.

7

Type II NOS and immune and inflammatory disorders

7.1 Septic shock

Sepsis is a systemic response to infection that occurs when toxic bacterial products are present in the circulation. This initiates a sequence of events involved in host defence including the activation of a number of different cell types and the induction of a variety of cytokines. Faced with a continued challenge, the immune response can escalate leading to profound hypotension (septic shock) and, ultimately, multiple organ failure and death (Natanson *et al.*, 1994). Septic shock can be induced in animals by the administration of bacterial lipopolysaccharide (LPS). Apart from some differences with regard to time scale, the features of endotoxaemia in animal models largely resemble those reported to occur in patients with septic shock. These include: a decrease in mean arterial pressure; an increase in pulmonary arterial pressure and vascular resistance; a decrease in p_aO_2 and an increase in p_aCO_2 values; an increase in heart rate; and a decreased pressor response to noradrenaline (Natanson *et al.*, 1994; Szabó, Southan & Thiemermann, 1994a; Weitzberg *et al.*, 1995).

Since type II NOS is known to be induced as part of the immune response (see Chapter 6) and NO is a potent vasodilator (see Chapter 3), it has been proposed as a prime candidate for causing the hypotension associated with sepsis. LPS treatment *in vivo* increases Ca^{2+}-independent NOS activity and type II NOS mRNA expression in a wide range of peripheral tissues from rats (Salter, Knowles & Moncada, 1991; Liu *et al.*, 1993). Increased plasma levels of nitrite/nitrate (used as an indicator of NO synthesis) have been found in both animal models and patients with septic shock (Nussler & Billiar, 1993; Klemm *et al.*, 1995). In culture, cytokine/LPS treatment induces type II NOS mRNA expression in a variety of cell types including human vascular smooth muscle cells, which may account for both the reduction in peripheral vascular resistance and also the reduced pressor responses to noradrenaline in septic shock (MacNaul & Hutchinson, 1993). Glucocorticoids, which have been used in the treatment of endotoxaemia, prevent the induction of type II NOS by LPS (Moncada, Palmer & Higgs, 1991). Knockout mice, lacking expression of the type II NOS gene, do not display the same fall in blood pressure as the wild type when subjected to LPS under anaesthesia. However, mutants are as susceptible to liver damage as the wild type in endotoxaemia (Macmicking *et al.*, 1995). Endotoxaemia clearly involves an array of mediators and cytokines, and NO is unlikely to be responsible for all its harmful effects. However, the degree of hypotension that occurs is closely associated with the prognosis. In view of the mortality rate (up to 70%) from septic shock (Petros *et al.*, 1994), the effect of NOS inhibitors on septic shock has already been examined clinically as well as in experimental models.

In a randomized double-blind placebo-controlled study, low doses of the NOS inhibitor L-NMMA were shown to increase blood pressure. However, cardiac output was decreased indicating that an adverse effect of L-NMMA might be to worsen tissue perfusion (Petros *et al.*, 1994). Similar results have been observed in animal studies (Klemm *et al.*, 1995). Furthermore, L-NAME and high doses of L-NMMA can eventually enhance the fall in blood pressure in rats with septic shock (Nava, Palmer & Moncada, 1991; Teale & Atkinson, 1994). In addition, NOS inhibitors can actually increase tissue damage in sepsis indicating that some NO synthesized may have a protective effect in maintaining the blood supply to peripheral organs (Moncada *et al.*, 1991). Further clinical studies are required to assess fully the adverse effects of NOS inhibitors and the dosage required for effective treatment. Methylene blue has also been used clinically to treat hypotension in septic shock since it can inhibit cGMP production caused by activation of guanylate cyclase by NO. Methylene blue does increase mean arterial pressure. Although cardiac function was also improved, O_2 delivery to tissues was not significantly enhanced in patients with septic shock treated with methylene blue (Preiser *et al.*, 1995). Due to the small numbers of patients studied it is not possible to assess any effect on mortality by these treatments at this stage.

Treatment with L-NAME, L-NMMA or methylene blue will inhibit the production or the effects of NO synthesized by all the isoforms of NOS. Some experimental studies have investigated the effects of NOS inhibitors that are relatively specific for the type II isoform (see Chapter 10) induced in septic shock. Canavanine, aminoguanidine and substituted isothioureas have all been shown to increase arterial pressure and increase survival in rodents with septic shock (Szabó *et al.*, 1994a; Teale & Atkinson, 1994; De Kimpe *et al.*, 1995; Thiemermann *et al.*, 1995). In the case of isothioureas, increased pressor responses to noradrenaline and decreased signs of liver failure were also observed with treatment. These inhibitors have yet to be investigated clinically.

Combined treatment with a non-selective NOS inhibitor administered systemically and NO administered as a gas by inhalation has been studied in pigs with endotoxic shock (Klemm *et al.*, 1995; Weitzberg *et al.*, 1995). The rationale behind what at first appears a surprising approach is that systemic NOS inhibitors can raise mean arterial pressure but exacerbate the pulmonary hypertension associated with septic shock. In contrast, inhalation of NO gas can decrease pulmonary vascular resistance without systemic effects on blood pressure (See Chapter 8). Combined treatment not only increases mean arterial pressure but also attenuates the rise in pulmonary vascular resistance and protects against the fall in p_aO_2 and the rise in p_aCO_2 that occur in untreated animals or in animals treated by NOS inhibition alone (Klemm *et al.*, 1995; Weitzberg *et al.*, 1995). Therefore, this approach may have the additional benefit of improving O_2 supply to peripheral organs. However, the studies cited here are in conflict with regard to whether treatment has an adverse effect on cardiac output. Finally, decreased hypotension and renal dysfunction with increased survival rates have been reported to occur in rats with endotoxic shock treated with an imidazolineoxyl *N*-oxide derivative (Yoshida *et al.*, 1994). This is a novel approach in which the agent acts by scavenging the high levels of NO produced by type II NOS and promotes its conversion to nitrite. The same agent does not appear to inactivate NO produced by the endothelium since it has no effect on mean arterial pressure in control animals.

7.2 Localized inflammatory disorders

Septic shock is an acute and extreme form of a *systemic* inflammatory response. However, the fact that NO synthesis by type II NOS can be induced during host defence (see Chapter 6) indicates that NO may also play a role in disorders associated with *localized* chronic inflammation and contribute to the cell damage that can occur in such conditions (Nussler & Billiar, 1993).

7.2.1 Asthma

NOS is present constitutively in the lung and low levels of NO may be synthesized by several different cell types under normal conditions. NOS is present in endothelial cells within the pulmonary circulation and EDNO is important in maintaining blood flow in the lungs. Inhibitory NANC nerves supplying the airways also express NOS. NO is a bronchodilator and it has been proposed that NO synthesized by such nerves can modulate reflex bronchoconstriction (Barnes & Belvisi, 1993; Zoritch, 1995). The epithelial lining of the airways inhibits bronchoconstriction, with damage or removal of the epithelium resulting in hyperresponsiveness to bronchoconstrictors. It has been suggested that epithelial cells produce an epithelial-derived relaxing factor analagous to EDRF in endothelial cells. However, this factor has yet to be identified as NO (see Zoritch, 1995 for review). Type II NOS can be induced in murine and human epithelial cells by a

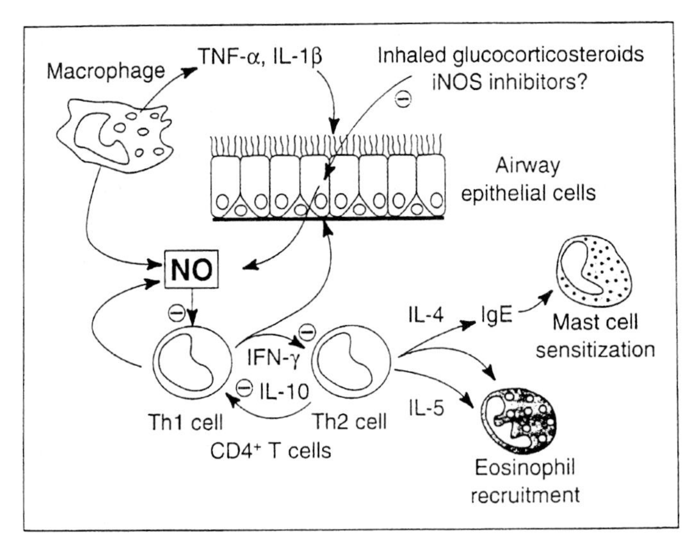

Fig. 7.1 A schematic representation of the possible role of NO in asthmatic inflammation. NO produced by epithelial cells and macrophages in asthma has an inhibitory effect on T helper (Th1) cells, allowing the activation of Th2 cells, which are normally suppressed by interferon-γ (IFN-γ). This results in increased secretion of interleukin 4 (IL-4) and IL-5, which leads to immunoglobulin E (IgE) formation and eosinophil recruitment into the airways. The production of NO is catalysed by NOS, which can be induced (iNOS or type II NOS) in epithelial cells by macrophage-derived tumour necrosis factor α (TNF-α) and IL-1β, as well as by IFN-γ produced by Th1 cells. Inhaled glucocorticosteroids may provide a mechanism for inhibiting cytokine-induced expression of type II NOS by epithelial cells and, thereby, reduce airway inflammation. (Reproduced from Barnes & Liew, 1995; *Immunology Today*, **16**, 128–30, with permission.)

mixture of cytokines *in vitro* (Kharitonov *et al.*, 1995). Asthmatic patients exhale significantly higher levels of NO and type II NOS expression is increased in biopsy specimens from asthmatics (Hamid *et al.*, 1993; Kharitonov *et al.*, 1995) indicating that type II NOS can also be induced in humans *in vivo*.

High levels of NO synthesized by type II NOS have been implicated in the hyperaemia and resultant plasma exudation that occurs in the airways in asthma (Barnes & Belvisi, 1993; Barnes & Liew, 1995; Zoritch, 1995). NO has differing effects on different cell types involved in host defence and this has been used to propose a mechanism for the inflammation that is associated with asthma, which can account for the pathological changes characteristic of this condition (Barnes & Liew, 1995) (Fig.7.1). The basis for this mechanism is that NO has an inhibitory effect on T helper 1 (Th1) cells which allows the activation of Th2 cells. Th2 cells are normally suppressed by INF-γ which is secreted by Th1 cells. Th2 cells secrete

IL-4 and IL-5 which lead to mast cell sensitization by IgE formation and eosinophil recruitment within the airways, respectively. Ultimately, high levels of NO may be cytotoxic to epithelial cells resulting in shedding of the epithelium (Barnes & Liew, 1995). In view of the recent increased incidence of asthma, it is interesting to note that epithelial damage due to ozone is thought to be mediated by NO (Zoritch, 1995).

As a consequence, NOS inhibitors could represent a novel therapeutic approach for the inflammation that is associated with asthma, although this needs to await the development of inhibitors specific for the type II isoform since both type I and type III NOS are present constitutively in the lung (Barnes & Liew, 1995). The sequence of events illustrated in Fig. 7.1 could also explain the beneficial effects of inhaled glucocorticosteroids since these are known to inhibit the induction of type II NOS (Moncada *et al.*, 1991; Barnes & Liew, 1995). Exhaled NO and type II NOS expression in biopsied specimens are not increased in asthmatics receiving glucocorticosteroids (Hamid *et al.*, 1993; Kharitonov, Yates & Barnes, 1996). Although it is a bronchodilator, it is not thought likely that NO inhalation will be of greater benefit than the currently used nebulized β-agonists in the treatment of asthma attacks (Barnes & Liew, 1995).

7.2.2 Arthritis and inflammatory disorders of the gastrointestinal tract and kidney

In addition to asthma, NO has been investigated in connection with a variety of other inflammatory disorders. In humans, there is an eightfold increase in type II NOS activity in intestinal mucosa from patients with ulcerative colitis but not in patients with Crohn's disease (Boughton-Smith *et al.*, 1993). Increased type II NOS expression has been demonstrated immunohistochemically in damaged epithelia in both ulcerative colitis and Crohn's disease (Singer *et al.*, 1995). Plasma nitrate levels are increased in irritable bowel syndrome (Dykhuizen *et al.*, 1995). The nitrite/nitrate levels are also significantly increased in the synovial fluid from patients with arthritis (Nussler & Billiar, 1993; Stefanovic-Racic, Stadler & Evans, 1993). Similarly, studies of experimental models of arthritis (Connor *et al.*, 1995), colitis (Grisham, Specian & Zimmerman, 1994; Hogaboam *et al.*, 1995; Ribbons *et al.*, 1995), ileitis (Miller *et al.*, 1993) and glomerulonephritis (Bachmann & Mundel, 1994) have all demonstrated increased levels of NO synthesis by various techniques. In the case of arthritis, both articular chondrocytes and synovial fibroblasts can express type II NOS after cytokine treatment *in vitro*. NO has been implicated in the destruction of the pericellular and extracellular matrix of cartilage in arthritis (Stefanovic-Racic *et al.*, 1993). Aminoguanidine and *N*-iminoethyl lysine (relatively selective inhibitors of type II NOS) suppress joint inflammation in adjuvant-induced arthritis in rats without affecting blood pressure (Connor *et al.*, 1995).

NOS inhibitors can also reduce some, if not all, the pathological changes that occur in experimental colitis (Grisham *et al.*, 1994; Hogaboam *et al.*, 1995) and ileitis (Miller *et al.*, 1993). Interestingly, in the last study, it was noted that a non-selective NOS inhibitor actually caused a local inflammatory response in the gut in control animals. This indicates that, under normal conditions, NO can be anti-inflammatory. It is likely that part of this effect was due to the inhibition of type III NOS present constitutively in the vasculature. Continuous release of NO reduces the adhesive properties of the endothelium (see Chapter 3) and prevents the adhesion of leucocytes that forms an early stage of the inflammatory process. In addition, L-NAME increases endothelial permeability in the microvasculature of the intestine (Kubes & Granger, 1992). In acute inflammation of the gut induced by castor oil, L-NAME can inhibit the diarrhoea but does not prevent the intestinal damage produced by this treatment (Mascolo *et al.*, 1993; Capasso *et al.*, 1994). The gastrointestinal damage that can occur during treatment with non-steroidal anti-inflammatory drugs such as flurbiprofen can actually be prevented by an NO donor (Wallace, Reuter & Cirino, 1994). Whether or not NO synthesized by constitutive NOS is pro- or anti-inflammatory under acute conditions may depend on the vascular bed. NO donors and NOS inhibitors have different effects on endothelial permeability in different vascular beds (see Chapter 3). In addition, NO contributes to the process of neurogenic inflammation in certain regions (see Chapter 9). Thus, the effectiveness of non-selective NOS inhibitors in inflammation is likely to vary in different organs.

7.3 Localized autoimmune disorders

NO has also been implicated in the development of some autoimmune disorders, including diabetes mellitus, and the demyelinating neuropathic disorders multiple sclerosis and Guillain-Barré syndrome (see Vladutiu, 1995 for review). Insulin-dependent diabetes mellitus (IDDM) is considered to be an autoimmune disease. The development of IDDM is characterized by initial infiltration of the pancreas by lymphocytes. This is followed by a decrease in insulin secretion and, eventually, the pancreatic islet β cells are selectively destroyed (Corbett & McDaniel, 1992). There is increasing experimental evidence that NO is the mediator of β cell destruction and that increased NO synthesis is induced by IL-1β secreted by lymphocytes (Kolb & Kolb-Bachofen, 1992). In pancreatic islet cell cultures, IL-1β treatment has been shown to result in increased production of NO, as detected by EPR spectroscopy, and the accumulation of cGMP (Corbett *et al.*, 1991, 1992b). Both of these events can be blocked by L-NMMA and by the protein synthesis inhibitor cyclohexamide, indicating that induction of NOS expression is taking place (Corbett *et al.*, 1992c). Studies of insuloma cell lines and purified pancreatic β cells indicate that NOS expression is induced in the β cell itself and that α cells do not express NOS when treated with IL-1β (Corbett & McDaniel, 1992; Corbett *et al.*, 1992c). Paradoxically, IL-1 treatment *in vivo* is antidiabetogenic in non-obese spontaneously diabetic mice. It has been suggested that this effect occurs by the induction of NO synthesis in macrophages, which suppresses lymphocyte proliferation thus preventing the infiltration of the pancreas. If lymphocyte infiltration is able to take place, then the local effect of IL-1β is to cause rather than prevent the development of IDDM (Corbett & McDaniel, 1992). In spontaneously diabetic rats, urinary excretion of nitrate is enhanced prior to the onset of IDDM and administration of aminoguanidine, L-NMMA or L-NAME either delays the onset or reduces the incidence of IDDM (Lindsay *et al.*, 1995; Wu, 1995). Interestingly, streptozotocin, which has long been used to induce diabetes in animals, is now known to contain a nitroso moiety that can liberate NO (Turk *et al.*, 1993).

NO appears to exert its toxic effects by several of the mechanisms described in Chapter 2. Endogenous NO produced after the induction of type II NOS inhibits mitochondrial oxidoreductase activity probably by interacting with the iron-sulphur centres on these enzymes (Corbett *et al.*, 1992b). Streptozotocin activates the polyADP-ribose system, causes DNA cleavage and can trigger apoptotic cell death of pancreatic β cells (Heller *et al.*, 1994; Kaneto *et al.*, 1995). At the same time intracellular levels of NAD^+ are depleted. It has been known for some time that the concomitant administration of nicotinamide can prevent the diabetogenic effect of streptozotocin and this may occur by restoring NAD^+ levels. Furthermore, nicotinamide has been reported to partially prevent or delay the onset of IDDM in children predisposed to diabetes mellitus (Corbett & McDaniel, 1992; Heller *et al.*, 1994). It is a characteristic of IDDM that the insulin-secreting β cells are *selectively* destroyed. The particular susceptibility of β cells to the toxic effects of NO may be related to the fact that these cells only contain low levels of mitochondrial Mn superoxide dismutase and glutathione peroxidase. These enzymes form part of a defence system against reactive oxygen species. Increased levels of superoxide anions would tend to favour the formation from NO of peroxynitrite and the hydroxyl radical, which are more toxic than NO itself (Corbett & McDaniel, 1992; Heller *et al.*, 1994). The evidence for a role for NO in the development of IDDM has been obtained almost exclusively from culture studies or animal models of diabetes mellitus. It should be noted that a similar role in human patients has yet to be established (Welsh, Eizirik & Sandler, 1994).

In multiple sclerosis and the equivalent animal model, experimental autoimmune encephalomyelitis (EAE), peripheral blood leucocytes invade the brain and spinal cord resulting in the expression of a variety of cytokines within the lesion that are known to induce type II NOS (Brosnan *et al.*, 1994). It has been hypothesized (Sherman, Griscavage & Ignarro, 1992) that induction of type II NOS in cells within the central nervous system would lead to high levels of NO synthesized locally, which would be toxic to the myelin-producing oligodendrocytes. In addition, peroxynitrite and the hydroxyl radical, which can be produced from NO (see Chapter 2), could damage the lipid myelin directly by causing peroxidation. Indirect evidence for this hypothesis includes the facts that corticosteroids are used clinically during relapses in multiple sclerosis and that glucocorticoids can inhibit the induction of type II NOS. Similarly, transforming growth factor-β_1 can reduce the effects of EAE *in vivo* and the induced cytotoxicity of microglia towards oligodendrocytes in culture (see Sherman *et al.*, 1992 for review). Transforming growth factor-β_1 inhibits the induction of type II NOS and decreases the stability of type II NOS mRNA (see Chapter 2). Oligodendrocytes appear to be particularly susceptible to the cytotoxic effects of NO compared with microglia (Mitrovic *et al.*, 1994). In culture, type II NOS can be induced in both astrocytes and microglia derived from rodents (Murphy *et al.*, 1993). However, with human cells, only astrocytes appear to be able to express type II NOS after treatment with cytokines *in vitro*. In specimens obtained at autopsy, astrocytes have been shown to be heavily stained for NADPH-diaphorase activity both within and around the edge of lesions in multiple sclerosis (Brosnan *et al.*, 1994).

Aminoguanidine has been reported to ameliorate EAE induced in mice (Cross *et al.*, 1994). However, in rats with EAE, L-NMMA was without effect and aminoguanidine actually enhanced the neurological deficit (Zielasek *et al.*, 1995). Whether this represents a species difference, a difference in the method of EAE induction or the involvement of NO in an, as yet, undetermined mechanism in EAE has yet to be established. Depending on the method of induction, both L-NMMA and aminoguanidine have been reported to suppress symptoms in experimental autoimmune neuritis in rats, an animal model of Guillain-Barré syndrome (Zielasek *et al.*, 1995).

It is not surprising that NO has also been implicated in the immune responses involved in rejection following transplantation and in graft-versus-host disease (Langrehr *et al.*, 1993; Nussler & Billiar, 1993; Stefanovic-Racic *et al.*, 1993). Elevated levels of nitrosylhaemoglobin have been detected by EPR spectroscopy in instances of both heart allograft rejection (Lancaster *et al.*, 1992) and graft-versus-host disease (Langrehr *et al.*, 1992). Elevated serum levels of nitrite/nitrate precede graft rejection and such increases can be prevented by an immunosuppressive drug (Stefanovic-Racic *et al.*, 1993). The precise role that NO plays in the complex sequence of events that occur during rejection and graft-versus-host disease is not yet known. However, there is preliminary evidence that aminoguanidine can promote survival in lethal graft-versus-host disease in mice (Hoffman *et al.*, 1995).

7.4 Selected references

Barnes, P. J. & Liew, F. Y. (1995). Nitric oxide and asthmatic inflammation. *Immunology Today*, **16**, 126–30.

Macmicking, J. D., Nathan, C., Hom, G., Chartrain, N., Fletcher, D. S., Trumbauer, M., Stevens, K., Xie, Q.-W., Sokoi, K., Hutchinson, N., Chen, H. & Mudgett, J. S. (1995). Altered responses to bacterial infection and endotoxic shock in mice lacking inducible nitric oxide synthase. *Cell*, **81**, 641–50.

Nussler, A. K. & Billiar, T. R. (1993). Inflammation, immunoregulation and inducible nitric oxide synthase. *Journal of Leukocyte Biology*, **54**, 171–8.

Vladutiu, A. O. (1995). Role of nitric oxide in autoimmunity. *Clinical Immunology and Immunopathology*, **76**, 1–11.

8

Type III NOS and cardiovascular disorders

8.1 Introduction

It has long been recognized that the endothelium is of critical importance to the maintenance of normal cardiovascular function and endothelial dysfunction has been demonstrated to occur in a wide range of cardiovascular disorders. The use of nitroglycerine in the treatment of angina represents the earliest form of NO therapy although its effectiveness was recognized long before its mechanism of action was understood. It is now clear that changes in endothelium-dependent mechanisms mediated by EDNO synthesized by type III NOS occur in many cardiovascular disorders. It should be emphasized that NO is only one of a number of mediators and modulators that contribute to endothelial cell function. The changes that occur in EDNO and many other agents present in the endothelium and their contribution to vascular dysfunction have been reviewed extensively (Harrison *et al.*, 1991; Lincoln, Ralevic & Burnstock, 1991b; Lüscher *et al.*, 1991; Moncada, Palmer & Higgs, 1991; Rubanyi, 1993; Lüscher & Noll, 1994; Pearson, 1994). This chapter will focus on atherosclerosis, hypertension, reperfusion injury and diabetic angiopathy.

8.2 Atherosclerosis

Endothelial cell injury is regarded as the critical initiating event for the atherogenic process. Branch points in the vasculature are particularly prone to continuing minor endothelial damage since they are areas of turbulent flow. Circulating monocytes become activated and adhere to the injured vessel surface and then migrate towards the intima. Such cells can take up extracellular lipids developing into foam cells. Of particular significance to this process is the accumulation of a form of low-density lipoprotein (LDL), oxidized LDL, that is generated in atherosclerosis. After infiltration, vascular smooth muscle cells proliferate and migrate resulting in intimal thickening and, over a period of time, the atherosclerotic plaque develops (Rubanyi, 1993; Lüscher & Noll, 1994). Such diseased vessels become the site for thrombus formation and subsequent vasospasm (Pearson, 1994). When one considers the various functions of EDNO, described in detail in Chapter 3, it can be seen that under normal conditions EDNO is anti-atherogenic. EDNO reduces the adhesive properties of the endothelium, preventing the adhesion of monocytes to its surface. In the coronary circulation, EDNO helps maintain endothelial barrier function and EDNO is known to inhibit vascular smooth muscle cell proliferation. Under acute conditions, NO donors have been reported to attenuate the toxic effects of oxidized LDL (Struck *et al.*, 1995). EDNO inhibits platelet aggregation and adhesion, preventing thrombus formation. Agents released from aggregating platelets, which normally cause vasodilatation via the production of EDNO, can cause vasoconstriction in disease. Epidemiological data have indicated that oestrogen may protect against atherosclerotic heart disease. The incidence of atherosclerosis is significantly less in premenopausal women than in men of the same age. The protective effect of oestrogen may occur by its ability to upregulate type III NOS expression and EDNO synthesis (see Chapter 3).

It has been known for some time that endothelium-dependent relaxations mediated by EDNO are impaired in atherosclerosis. This occurs both in humans and in animal models of hypercholesterolaemia. Reduced endothelium-dependent responses can be observed before the development of overt morphological damage and responses to vasodilators acting directly on the smooth muscle are not impaired (Moncada *et al.*, 1991; Rubanyi, 1993). Dietary arginine has been shown to improve NO-dependent vasodilatation, to attenuate enhanced endothelial adhesiveness and to prevent the development of intimal lesions in hypercholesterolaemic rabbits (Tsao *et al.*, 1994; Wang *et al.*, 1994b). It has been speculated that the high arginine content of nuts may contribute to the reduced incidence of coronary artery disease in heavy consumers of nuts (Cooke *et al.*, 1993). Acute administration of arginine i.v. has also been reported to improve endothelium-dependent responses in humans with hypercholesterolaemia (Cooke *et al.*, 1992). However, another study failed to confirm this effect although arginine augmented responses in normal subjects (Casino *et al.*, 1994).

It is generally regarded that the beneficial effects of arginine are due to it increasing the synthesis of NO. While this may be the case, it is by no means certain that NO synthesis is decreased in hypercholesterolaemia despite the fact that endothelium-dependent responses are diminished. This was originally called into question in a study (Minor *et al.*, 1990) in which basal and stimulated release of EDNO were assessed both by a bioassay and a chemiluminescence assay following reduction of the superfusate under acidic conditions (see Chapter 13). As expected, the bioassay revealed a significant decrease in the release of EDNO from aortae of hypercholesterolaemic rabbits, as assessed by its ability to relax the detector vessel. In marked contrast, the chemiluminescence assay, which meas-

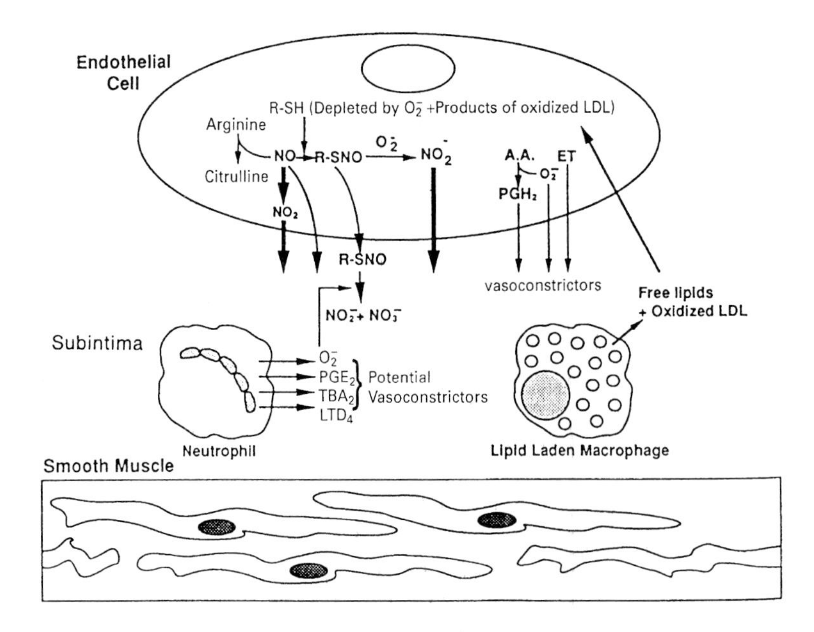

Fig. 8.1 Schematic representation of the possible interactions between endothelium-derived relaxing and constricting factors and factors released from infiltrating cell types in atherosclerosis. AA, Arachidonic acid; ET, endothelin; LTD_4, leukotriene D_4; PGE_2, prostaglandin E_2; PGH_2, prostaglandin H_2; TBA_2, thromboxane A_2. (Reproduced from Harrison *et al.*, 1991; In *Cardiovascular Significance of Endothelium-derived Vasoactive Factors*, pp. 263–80. Futura, with permission.)

ured nitrogen oxides (both NO and its reaction products), demonstrated that basal and stimulated release were significantly higher in hypercholesterolaemia. Thus NO synthesis may actually be increased but the NO that is formed may be inactivated in terms of its vasodilator properties. A scheme has been proposed for how this may occur (Harrison *et al.*, 1991) and this is illustrated in Fig. 8.1. It is known that superoxide production is increased in atherosclerosis (Ohara, Peterson & Harrison, 1993) and superoxide dismutase enhances endothelium-dependent relaxations to a greater extent in hypercholesterolaemic rabbits (White *et al.*, 1994). This indicates that in atherosclerosis the endothelial cell may be under oxidant stress. As has been described in detail in Chapter 2, under these conditions NO is more likely to form nitrite, nitrate and peroxynitrite than be released as free NO or a nitrosothiol. In addition, incubation of endothelial cells in culture with oxidized LDL results in a decrease in the bioactivity of the EDNO released upon agonist stimulation without altering the levels of nitrogen oxides released (Myers *et al.*, 1994). This scheme could also account for the inactivation of other mechanisms mediated by NO in the endothelium. At this stage this sequence of events has not been established unequivocally. Oxidized LDL has also been reported to decrease the expression of type III NOS mRNA in endothelial cells in culture (Liao *et al.*, 1995). Induction of type II, Ca^{2+}-independent NOS activity has been demonstrated to occur in hypercholesterolaemia (Lang, Smith & Lewis, 1993). Arterial smooth muscle cells express type II NOS mRNA following endothelial injury (Hansson *et al.*, 1994). This isoform has also been localized in vascular smooth muscle and macrophages in atherosclerotic plaques in humans (Buttery *et al.*, 1995). The significance of these findings has yet to be fully evaluated but, under conditions of oxidant stress, this might be expected to contribute to vascular damage.

Coronary artery disease is commonly treated either by balloon angioplasty or arterial/vein bypass grafts. However, intimal hyperplasia and restenosis can occur rapidly after either of these treatments. Since balloon angioplasty causes endothelial injury, it can be seen that it could initiate the same sequence of events as occurs in the atherogenic process (Rubanyi, 1993). Although the endothelium rapidly regenerates after angioplasty, it is known that endothelium-dependent relaxations can still be impaired after regeneration has taken place (Shimokawa, Aarhus & Vanhoutte, 1987). Repeated endothelial removal exacerbates this effect (Niimi, Azuma & Hirakawa, 1994). Since renewal of the endothelium occurs normally throughout life, such functional alterations during regeneration could contribute to the increased incidence of coronary artery disease with age. In view of this, it has been investigated whether replacement of NOS during balloon angioplasty may be of use in the prevention of restenosis. The transfer of the type III NOS gene into the vascular wall has been achieved *in vivo* in rats. Such treatment has been shown to increase NO synthesis and to inhibit intimal hyperplasia following balloon angioplasty (von der Leyen *et al.*, 1995). Relaxation responses to the calcium ionophore A23187 were also improved after gene therapy. However, whether relaxation responses after agonist stimulation of endothelial receptors were similarly improved was not investigated.

Administration of arginine before and after surgery for a vein bypass graft in rabbits inhibits intimal hyperplasia and enhances endothelium-dependent relaxation responses to ACh and 5-HT (Davies *et al.*, 1994). The heterogeneity of endothelium-dependent responses mediated by NO in different vessels also has implications with regard to the type of vessel that is used for autologous grafts (see Yang & Lüscher, 1993 for review). The saphenous vein, internal mammary artery and right gastroepiploic artery have all been used in coronary bypass operations. The venous circulation of humans *in vivo* appears to have a lower basal release of EDNO compared with that of the arterial circulation (Moncada *et al.*, 1991). In isolated vessels, the basal release of EDNO is greater in the internal mammary artery than in the saphenous vein, as is the release of EDNO stimulated by ACh or histamine. While the gastroepiploic artery is similar to the internal mammary artery with respect to these properties, both the gastroepiploic artery and saphenous vein contract in the presence of aggregating platelets whereas the internal mammary artery relaxes. Thus, the ability of the graft to maintain the anti-atherogenic effects of EDNO and to inhibit vasoconstriction upon platelet aggregation may well be determined by the intrinsic properties of the endothelium of the graft (Yang & Lüscher, 1993).

8.3 Hypertension

Although the integrity of the endothelium is well preserved in hypertension, morphological changes do occur. Endothelial cells have a higher replication rate and display a greater variability in size and shape, occasionally bulging into the lumen. Increased adhesion of circulating monocytes also occurs (Lüscher *et al.*, 1991). The basal release of EDNO is clearly important in maintaining vasodilator tone and administration of L-NAME results in an increase in mean arterial pressure (Ribeiro *et al.*, 1992). Impaired endothelium-dependent responses have been demonstrated to occur in large conduit arteries and small resistance arteries in several different vascular beds in a variety of genetic and animal models of hypertension (Lüscher *et al.*, 1991; Moncada *et al.*, 1991). In the heart, increased Ca^{2+}-dependent NOS activity and enhanced basal release of EDNO have been observed in spontaneously hypertensive rats which may represent a compensatory mechanism to protect this organ against the effects of hypertension (Kelm *et al.*, 1995).

In the forearm arterial bed in humans, patients with essential hypertension have been reported to have blunted responses to L-NMMA (Calver *et al.*, 1992a). However, another study found no significant change in vasodilator responses (Cockcroft *et al.*, 1994). When responses to L-NMMA in normotensive, treated hypertensive and untreated hypertensive patients were compared, it was found that there was no significant difference between those who were normotensive and those with treated hypertension or between those with treated and those with untreated hypertension. However, there was a significant difference between normotensive and untreated hypertensive patients. When data from all of the patients were combined there was a significant correlation between individual blood pressure measurements and the response to L-NMMA, irrespective of the group (Calver, Collier & Vallance, 1994). One of the inferences that has been drawn from such studies is that changes in NO-mediated responses may occur as a consequence of hypertension rather than contributing to its cause. In a genetic study, no evidence could be found for a link between the gene for type III NOS and essential hypertension (Bonnardeaux *et al.*, 1995). How hypertension could cause such changes has yet to be established. Shear stress both activates type III NOS and upregulates its expression (see Chapter 3). Thus, it might be expected that increased shear stress due to increased peripheral resistance would enhance EDNO-dependent mechanisms. Whether *chronic* shear stress eventually impairs the production of EDNO requires further investigation. Long-term changes in EDNO-dependent mechanisms induced by hypertension could be of greater significance at a later stage in the disease process. In humans, hypertension is a major risk factor in the development of atherosclerosis.

NO has been suggested to play a number of roles in renal function including renal blood flow, renin secretion and pressure-induced natriuresis and diuresis (Bachmann & Mundel, 1994; Cowley *et al.*, 1995, Salazar & Llinás, 1996). Pressure-induced natriuresis and diuresis form a mechanism whereby increases in arterial pressure can be counteracted by a reduction in blood volume. The pressure-induced natriuretic response is blunted in animal models of hypertension and, in the spontaneously hypertensive rat, this effect can be observed *before* the development of hypertension. NOS inhibitors are known to prevent pressure-induced natriuresis (Cowley *et al.*, 1995). In view of this, NO-dependent mechanisms in the kidney may be of particular significance in the development of hypertension. However, such mechanisms have received little direct investigation in humans. One study has demonstrated that L-arginine infusion can reduce blood pressure to a similar extent in both control and hypertensive patients. In contrast, L-arginine only increased renal blood flow in normotensive subjects. Since the hypertensive group consisted of patients with only very mild hypertension, it was speculated that changes in renal endothelial function mediated by EDNO contribute to, rather than result from, hypertension (Higashi *et al.*, 1995). In chronic renal failure, plasma levels of the asymmetrical form of dimethyl-arginine (see Chapter 10) are markedly increased. This is an endogenous inhibitor of NOS and the concentrations reached are sufficient to inhibit NO synthesis *in vivo*. This has been speculated to contribute to the hypertension associated with renal failure (Vallance *et al.*, 1992).

8.4 NO inhalation and pulmonary hypertension

Pulmonary hypertension can occur in a variety of disorders including congenital heart disease, persistent pulmonary hypertension of the newborn and adult respiratory distress syndrome, which can result in severe hypoxaemia (Lunn, 1995). Much of the current research on NO in this field is concerned with the use of NO as therapy rather than its role in pathogenesis. Under normal conditions, blood vessels supplying regions of the lung that are underventilated will constrict, decreasing perfusion by a process known as hypoxic pulmonary vasoconstriction. This process is important because it preserves ventilation–perfusion matching and

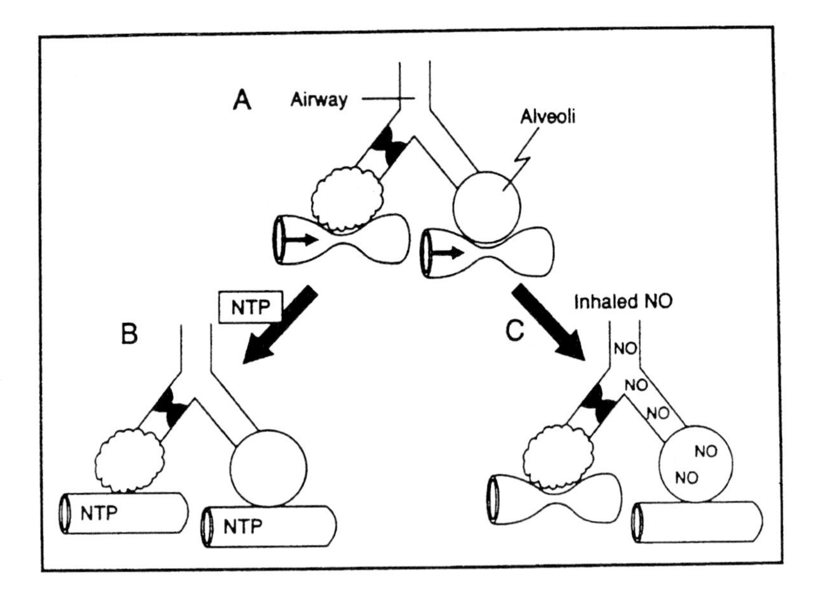

Fig. 8.2 A: Schematic representation of vasoconstriction of the pulmonary bed with normal and collapsed alveoli. B: Nitroprusside (NTP) causes non-selective vasodilatation of all pulmonary arteries, which may worsen ventilation–perfusion ($\dot{V}/\dot{Q}$) matching. C: Inhaled NO dilates only ventilated alveoli, an outcome that improves $\dot{V}/\dot{Q}$ matching. (Reproduced from Lunn, 1995; *Mayo Clinical Proceedings*, **70**, 247–55, with permission.)

can limit the amount of hypoxaemia that can occur during lung collapse. EDNO is important in maintaining blood flow in the lungs and, since hypoxia reduces EDNO release, it has the effect of matching perfusion with ventilation (Zoritch, 1995). The use of NO donors administered systemically in the treatment of pulmonary hypertension is not recommended because they would increase perfusion in all regions of the lung whether or not they are ventilated. This can have the effect of decreasing p_aO_2 despite decreasing pulmonary vascular resistance. In contrast, NO administered by inhalation will only reach those parts of the lung that are ventilated (Fig. 8.2). NO can diffuse readily and reach the abluminal surface of the pulmonary vasculature to cause vasodilatation (Loskove & Frishman, 1995). Since NO is rapidly inactivated by haemoglobin in the circulation, the effect of NO is confined to the vascular smooth muscle adjacent to the alveolar unit. Thus, inhaled NO can decrease pulmonary vascular resistance and decrease hypoxaemia without having any effect on mean arterial pressure or cardiac output (Loskove & Frishman, 1995; Lunn, 1995).

In view of the seriousness of severe pulmonary hypertension, inhalation NO therapy has already been used clinically in congenital heart disease, adult respiratory distress syndrome and persistent pulmonary hypertension of the newborn (see Loskove & Frishman, 1995; Lunn, 1995 for review). In general, inhaled NO has been shown to reduce pulmonary arterial pressure and improve oxygenation in some patients. However, at this stage, there are insufficient data to conclude whether this form of therapy decreases mortality or morbidity. Higher nitrogen oxides (e.g. NO_2) formed from reactions of NO gas with oxygen are extremely toxic and so stringent safety precautions are required. In infants, the possibility that inhalation of NO could lead to increased methaemoglobin in the blood needs to be investigated. Finally, a rebound severe pulmonary spasm can occur after interruption of NO inhalation. This may be due to the inhibition by NO of constitutively expressed NOS activity by its interaction with haem (see Chapter 2).

8.5 Diabetic angiopathy

One of the most common complications of diabetes mellitus is vascular disease. A reduction of endothelium-dependent responses mediated by NO has been reported to occur in animal models and in humans with diabetes mellitus (Calver, Collier & Vallance, 1992b; Lawrence & Brain, 1992; McVeigh *et al.*, 1992; Cameron & Cotter, 1994). As observed in atherosclerosis (Section 8.2), impaired endothelium-dependent relaxation in diabetes mellitus does not appear to be due to decreased EDNO synthesis but rather occurs as a consequence of inactivation of EDNO by reactive oxygen species such as superoxide anions (Langenstroer & Pieper, 1992). Clearly, the development of vascular disease in diabetes mellitus is likely to involve similar mechanisms to those described in Section 8.2 for non-diabetic patients. However, diabetics are known to have an increased incidence of atherosclerosis and hypertension and there is increasing evidence that hyperglycaemia, *per se*, can initiate a variety of metabolic changes within the endothelium that will predispose the vessel to abnormal EDNO function. Such changes are also likely to explain why the microvasculature is also affected in diabetes (microangiopathy), a feature not normally observed in non-diabetic patients.

The types of metabolic changes that can occur in hyperglycaemia tend to produce conditions of oxidant stress within the endothelium (Giugliano, Ceriello & Paolisso, 1995) and these have been summarized in Fig. 8.3. Hyperglycaemia activates the polyol pathway. Increased oxidation of sorbitol by sorbitol dehydrogenase results in an increase in the intracellular $NADH/NAD^+$ ratio, a redox imbalance that is referred to as hyperglycaemic pseudohypoxia (Ruderman, Williamson & Brownlee, 1992; Williamson *et al.*, 1993). The cyclooxygenase pathway is also activated by hyperglycaemia and increased levels of NADH favour the synthesis of prostaglandin H_2 and the generation of superoxide anions (Tes-

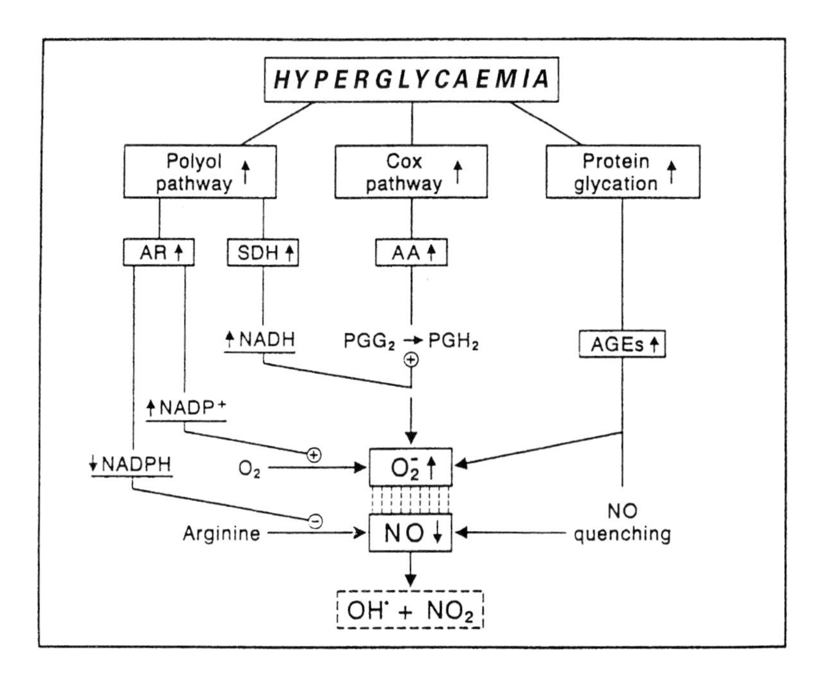

Fig. 8.3 Possible mechanisms by which hyperactivity of some metabolic pathways strictly related to hyperglycaemia may lead to upset of the balance between superoxide anions and NO. AR, aldose reductase; SDH, sorbitol dehydrogenase; Cox, cyclo oxygenase; AA, arachidonic acid; PGG_2, prostaglandin G_2; PGH_2, prostaglandin H_2; $OH^{\bullet}$, hydroxyl radical; NO_2, nitrogen dioxide. Formation of $OH^{\bullet}$ and NO_2 from decomposition of peroxynitrite has yet to be demonstrated *in vivo.* (Reproduced from Giugliano *et al.*, 1995; *Metabolism*, **44**, 363–8, with permission.)

famariam, 1994). With the activation of the polyol pathway, aldose reductase activity is increased. Since aldose reductase is an NADPH-dependent enzyme, it has been suggested that its activation could result in a decrease in NO synthesis since NOS is also dependent on NADPH. However, there is little evidence, at least in the short term, that hyperglycaemia inhibits EDNO synthesis. Indeed, the reverse is more likely to be the case. Hyperglycaemia increases Ca^{2+} mobilization in endothelial cells and activates both NOS and the arginine transporter *in vitro* (Wascher *et al.*, 1994; Sobrevia, Yudilevich, & Mann, 1995). Therefore, the harmful effect of aldose reductase activation may be due rather to the increased synthesis of $NADP^+$, which again favours the production of superoxide anions. Treatment of diabetic rabbits with an aldose reductase inhibitor restores endothelial function (Tesfamariam *et al.*, 1993). Finally, prolonged hyperglycaemia can result in the non-enzymatic glycosylation of proteins and the formation of advanced glycation end-products (AGEs). AGEs have been shown to quench NO and inhibit endothelium-dependent vasodilatation (Bucala, Tracey & Cerami, 1991). Treatment with aminoguanidine can prevent vascular dysfunction in diabetes mellitus (Tilton *et al.*, 1993). Since aminoguanidine can prevent the formation of AGEs and inhibit aldose reductase activity (Kumari *et al.*, 1991; Tilton *et al.*, 1993), it is not clear whether aminoguanidine exerts its beneficial effects by these mechanisms or by its ability to inhibit NOS.

8.6 Ischaemia/reperfusion injury

Myocardial cell death can occur after coronary artery occlusion. Some of the damage occurs as a consequence of reperfusion, which results in the infiltration of neutrophils into the site of injury (Loskove & Frishman, 1995). Reperfusion injury has been attributed to endothelial dysfunction in the coronary arteries. In a study of the time course of events, endothelium-dependent relaxations of coronary arteries, mediated by EDNO, were not reduced during the period of ischaemia. However, on reperfusion such responses rapidly became impaired and this impairment preceded neurophil infiltration into the myocardium and myocardial cell necrosis (Tsao *et al.*, 1990). Administration of an NO donor at the time of reperfusion after a 1-h period of ischaemia has been shown to prevent the accumulation of neutrophils and myocardial necrosis (Lefer *et al.*, 1993). NO is therefore considered to be cardioprotective and is thought to act by inhibiting neutrophil aggregation and adhesion to the endothelium. It should be noted that other factors produced by the endothelium, such as prostacyclin and adenosine, can also be protective during ischaemia and reperfusion. Adenosine attenuates neutrophil adhesion and platelet aggregation and inhibits superoxide generation by neutrophils. Thus, adenosine administration has been shown to exert beneficial effects in ischaemia/reperfusion (Lefer & Lefer, 1993).

Whether or not the synthesis of EDNO is reduced or EDNO is inactivated during reperfusion has yet to be established unequivocally. A 10-fold increase in levels of NO has been detected, by EPR spectroscopy, in the rat heart after 30 min of ischaemia, although the site of synthesis was not determined (Zweier, Wang & Kuppusamy, 1995). However, infusion of superoxide dismutase before reperfusion has been shown to prevent endothelial dysfunction in both myocardial and mesenteric ischaemia/reperfusion injury (Lefer & Lefer, 1993; Loskove & Frishman, 1995). Superoxides are known to inactivate EDNO (Rubanyi & Vanhoutte, 1986) and neutrophils synthesize superoxides (Lefer *et al.*, 1993). Such mechanisms are not only of significance in myocardial infarction but may also play a role during preservation and reperfusion of organs during transplantation, particularly with regard to their early failure after surgery (Langrehr *et al.*, 1993). Supplementation of the preservation medium with the NO donor nitroglycerine has been reported to improve survival and decrease neutrophil accumulation in both the heart and the lungs after transplantation in animal models (Pinsky *et al.*, 1993; Naka *et al.*, 1995).

8.7 Selected references

Giugliano, D., Ceriello, A. & Paolisso, G. (1995). Diabetes mellitus, hypertension, and cardiovascular disease: which role for oxidative stress? *Metabolism*, **44**, 363–8.

Loskove, J. A. & Frishman, W. H. (1995). Nitric oxide donors in the treatment of cardiovascular and pulmonary diseases. *American Heart Journal*, **129**, 604–13.

Moncada, S., Palmer, R. M. J. & Higgs, E. A. (1991). Nitric oxide: physiology, pathophysiology and pharmacology. *Pharmacological Reviews*, **43**, 109–42.

Rubanyi, G. M. (1993). The role of the endothelium in cardiovascular homeostasis and diseases. *Journal of Cardiovascular Pharmacology*, **22** (Suppl 4), S1–14.

9

Type I NOS and disorders of the central and peripheral nervous systems

9.1 The brain

The role of NO within the brain in neurotoxicity and neuroprotection has been described in detail in Chapter 4. One of the reasons why it was originally suggested that NO synthesized within neurones is neuroprotective was because NADPH-diaphorase-positive neurones appear to be spared in neurodegenerative disease and following NMDA-mediated nerve cell death (see Vincent & Hope, 1992; Zhang & Snyder, 1995 for review). However, since this time, it has been demonstrated that other neurones which do not contain NOS are also spared (Vincent, 1994). It is now recognized that NO mediates many of its toxic effects in the brain by its interaction with superoxide anions to form peroxynitrite (Lipton *et al.*, 1993). Certain populations of neurones are now thought to be resistant to degeneration in NMDA-induced neurotoxicity because they contain manganese superoxide dismutase, which removes superoxide anions, rather than because they can synthesize NO with type I NOS (Vincent & Hope, 1992).

One of the major causes of neuronal cell death in the brain is ischaemia due to stroke and a scheme illustrating the part that type I NOS has been suggested to play in this process has been given in Fig. 9.1. In this scheme, ischaemia leads to an increase in glutamate release. Glutamate stimulates NMDA-receptors which leads to an increase in intracellular Ca^{2+} in the target neurone. This allows the Ca^{2+}-dependent activation of type I NOS resulting in the synthesis of increased levels of NO. In addition, the expression of type II NOS may be induced in astrocytes and contribute to increased production of NO within the brain (Choi, 1993). NO subsequently causes energy depletion and neuronal death by a variety of interactions described in detail in Chapter 2. These include reactions with iron-sulphur centres on certain mitochondrial enzymes and *cis*-aconitase leading to reduced energy production, and interaction with nucleophilic groups leading to DNA damage. Certain actions of NO can act as a brake on this sequence of events. NO can downregulate NMDA-receptors, reducing the effect of glutamate. In addition, NO can cause vasodilatation via the production of cGMP and thus increase cerebral blood flow (Choi, 1993) (Fig. 9.1). This latter interaction probably ac-

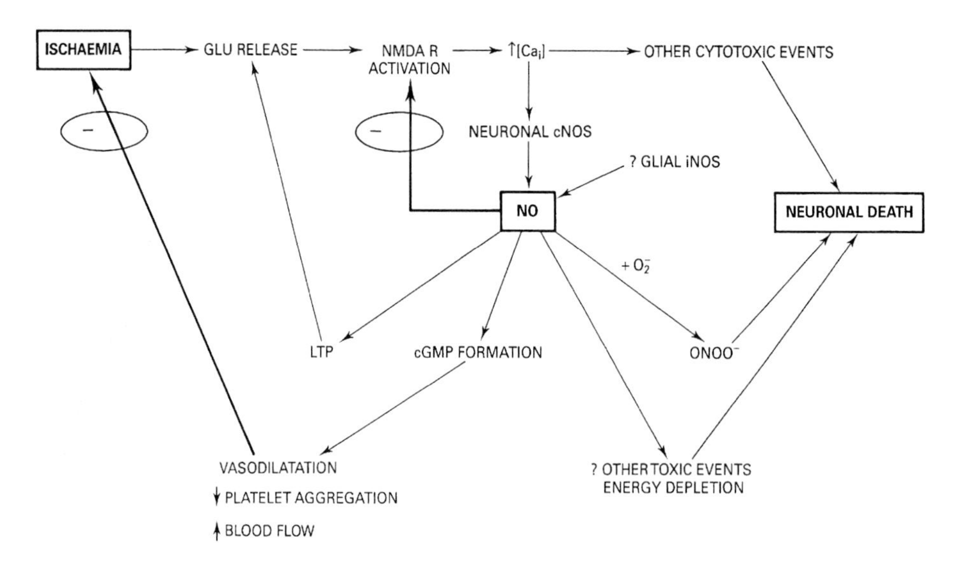

Fig. 9.1 Schematic representation of possible relationships between NMDA-receptor activation, NO formation and cell injury in the brain following ischaemia due to stroke. NO synthesis by type I NOS (neuronal cNOS) in the neurones may be stimulated by the activation of NMDA-receptors by glutamate released following ischaemia. NO is postulated to be cytotoxic primarily through combination with superoxide to form the destructive free radical, peroxynitrite. In addition NO itself may directly cause injury by impairing mitochondrial energy production. Two cytoprotective 'brakes' on the injury are also shown: (1) NO-mediated NMDA-receptor downregulation and (2) NO-mediated vasodilatation and reduced platelet aggregation leading to increased blood flow. The possibility that the induction of type II NOS in glia (iNOS) also contributes to the production of NO is indicated. (Adapted from Choi, 1993; *Proceedings of the National Academy of Science USA*, **90**, 9741–3, with permission.)

counts for the apparently contradictory results that have been obtained in studies of the effects of non-specific NOS inhibitors on damage caused by ischaemia. Low doses have been reported to block neuronal damage following middle cerebral artery occlusion in animals. In contrast, high doses exacerbate neuronal damage in experimentally induced stroke. This is probably due to the fact that such inhibitors are also able to inhibit the production of EDNO by type III NOS in the endothelium and thus have the effect of decreasing cerebral blood flow enhancing ischaemic damage (see Zhang & Snyder, 1995 for review). The ability of non-selective NOS inhibitors to act at more than one site for NO synthesis should also be taken into account when comparing the results from *in vivo* and *in vitro* studies. As examples, inhibition of NOS activity in central neurones in culture results in protection against neuronal injury due to hypoxia or NMDA-induced toxicity from which it may be deduced that NO is neurotoxic (Wallis, Panizzon & Wasterlain, 1992). In contrast, NOS inhibition *in vivo* can potentiate NMDA-induced neuronal damage, from which it has been concluded that NO does not play a large role in neurotoxicity (Haberny, Pou & Eccles, 1992). However, such results could equally be due to the additional effect of NOS inhibitors on cerebral blood flow. More recently, experiments have been performed that are able to distinguish between NO produced by type I or type III NOS in the brain and this has provided compelling evidence that NO synthesized by type I NOS in neurones during a stroke is neurotoxic. In knockout mice lacking the expression of type I but not type III NOS, ischaemic damage following induction of stroke is significantly reduced (Huang *et al.*, 1994). In addition the increase in NO levels that can be measured in the brain *in vivo* by spin-trapping during stroke does not occur in mice lacking type I NOS (Mullins *et al.*, 1995). 7-Nitroindazole (7-NI), a NOS inhibitor that inhibits type I NOS *in vivo* without affecting type III NOS as assessed by its lack of pressor activity, reduces infarct size following ischaemia (Urbanics *et al.*, 1995).

9.2 The sensory nervous system

In the skin, NO synthesized by constitutive NOS has been shown to contribute to the inflammatory response under acute conditions (Ialenti *et al.*, 1992). In addition, there is increasing evidence that NO is involved in neurogenic inflammation. During neurogenic inflammation, substance P (SP) and calcitonin gene-related peptide (CGRP) are released from sensory nerves mediating oedema formation and vasodilatation, respectively. 7-NI inhibits neurogenic inflammation induced by saphenous nerve stimulation. Since 7-NI has no effect on the actions of SP or CGRP injected directly into the skin, it has been concluded that NO can act prejunctionally by promoting the release of these sensory transmitters. L-NAME inhibits CGRP release induced by capsaicin (Kajekar, Moore & Brain, 1995). *In vivo*, 7-NI inhibits type I NOS but does not appear to inhibit type III NOS, having no effect on blood pressure (Moore *et al.*, 1993a). Thus, the source for the NO that modulates neurogenic inflammation is from the activity of type I NOS.

Although the aetiology of migraine has yet to be established unequivocally, there is considerable experimental evidence that migraine pain is due to neurogenic inflammation of the sensory trigeminovascular system supplying the dural and meningeal arteries (Moskowitz *et al.*, 1989). Neither SP nor CGRP causes significant pain when injected locally. Thus, the increased release of CGRP that occurs in migraine is unlikely to be the cause of the pain. Nitroglycerine, an NO donor, causes headache and migraineurs are more sensitive to its effect. Histamine also causes headache in migraineurs (see Olesen, Thomsen & Iversen, 1994 for review). The effect of nitroglycerine is unlikely to be due to activation of histamine release from mast cells since H_1 receptor blockade does not prevent nitroglycerine-induced headache (Lassen *et al.*, 1995). However, histamine can stimulate the synthesis of NO in cranial arteries (Olesen *et al.*, 1994). Other agents can be released during reactive hyperaemia in migraine, including ATP, which has been suggested to act on nociceptive endings of local sensory nerve fibres initiating pain (Burnstock, 1996).

The role that NO plays in pain has yet to be established and is likely to be complex. Type I NOS and NADPH-diaphorase activity have been localized in subpopulations of primary sensory neurones (see Chapter 5). In addition, type I NOS has been localized in some spinothalamic neurones and in neuronal processes in close apposition to the central projections of primary sensory neurones. Thus, in addition to its peripheral effects, NO may also contribute to the central processing of afferent information (Lee *et al.*, 1993). However, NOS inhibitors do not appear to alter baseline reflexes. In contrast, NOS inhibitors *do* abolish facilitation of nociceptive reflexes (see Meller & Gebhart, 1993 for review). Nociceptive reflexes can be enhanced following peripheral nerve injury or the induction of neurogenic and non-neurogenic inflammation in the skin, leading to hypersensitivity and hyperalgesia. The changes that occur during the development of hyperalgesia have been likened to synaptic plasticity during long-term potentiation in the brain (Meller & Gebhart, 1993). It is known that the induction and maintenance of central sensitization requires the activation of NMDA-receptors in the spinal cord (Woolf & Thompson, 1991). By analogy, it has been suggested that the effects of NMDA-receptor activation are mediated through the production of NO. Systemic administration of 7-NI and L-NAME has been shown to be antinociceptive, inhibiting the late, but not the early, phase hindpaw licking time following peripheral injection of formalin (Morgan *et al.*, 1992; Moore *et al.*, 1993a). Similarly, L-NAME (i.v.) reduces the prolonged second peak of firing, but not the short-duration first peak of firing, of neurones in the dorsal horn of the spinal cord induced by formalin injection (Haley, Dickenson & Schachter, 1992a). Thus, NO contributes to the increased excitability of sensory neurones and the pain associated with the later stages of peripheral inflammation in the skin. NOS inhibitors, especially 7-NI which has no pressor effect *in vivo*, are now being considered in research into the treatment of pain, particularly that associated with chronic inflammation (Morgan *et al.*, 1992; Moore *et al.*, 1993a).

L-NAME is also antinociceptive when it is administered intracerebroventricularly (Babbedge, Hart & Moore, 1993b). In addition, mechanical reflexes can be inhibited by L-NAME in intact rats with inflammation but not in rats with no inflammation or in spinalized rats with inflammation (Grisham, Specian & Zimmerman, 1994). Taken together these studies indicate a supraspinal role for NO in enhanced nociceptive reflexes. However, direct administration of L-NAME to the spinal cord via intrathecal catheters also produces antinociception following injection of formalin and blocks NMDA-induced hyperalgesia (Malmberg & Yaksh, 1993). Thus, NO may act at more than one site and further research is required to elucidate the precise involvement of NO in the development of hyperalgesia. The production of increased levels of NO in hyperalgesia may not simply be due to the activation of type I NOS by NMDA-receptor stimulation. Peripheral axotomy in the rat results in the increased expression of type I NOS and type I NOS mRNA in dorsal root ganglia, indicating that the induction of type I NOS can also occur following nerve injury (Zhang *et al.*, 1993; Hökfelt *et al.*, 1994). Surprisingly, unilateral hindpaw inflammation produces a bilateral increase in NADPH-diaphorase staining of neurones in the dorsal horn of the spinal cord (Solodkin, Traub & Gebhart, 1992). The significance of this finding in terms of hyperalgesia, which is only observed in the ipsilateral hindpaw, has yet to be determined.

9.3 Somatic-motor nervous system

Unlike primary sensory neurones and autonomic neurones, motoneurones rarely express type I NOS under normal conditions. However, type I NOS expression can be induced in motoneurones following nerve injury although this process depends on the type of injury or the stage of development at which it occurs. During the early stages of postnatal development, immature motoneurones are highly dependent on their skeletal muscle target for survival (Greensmith, Hasan & Vrbová, 1994). If the sciatic nerve is crushed in neonatal rats a large proportion of the motoneurones die whereas little motoneurone death occurs if the same injury is inflicted at 5 days of age. However, if an NMDA-receptor agonist is administered 7 days after a nerve crush at 5 days of age, then 60% of the motoneurones die indicating that nerve injury has increased the susceptibility of immature neurones to NMDA-induced neurotoxicity. In contrast, in the adult rat, motoneurones do not die following nerve crush even after treatment with an NMDA-receptor agonist (Greensmith *et al.*, 1994). Peripheral axotomy of the sciatic nerve in the neonate has been shown to induce the expression of NADPH-diaphorase in sciatic motoneurones, which is maintained throughout the period during which motoneurone death occurs (Clowry, 1993). Thus the death of target-deprived immature motoneurones following nerve injury may involve the induction of NOS within the motoneurone and could be mediated, at least in part, by the toxic effects of NO.

In adult rats, neither peripheral axotomy nor ventral root transection results in significant motoneurone death or induces the expression of NADPH-diaphorase in the motoneurone (Clowry, 1993; Wu, 1993). In contrast, spinal root avulsion does cause motoneurone death in the adult. Furthermore, NADPH-diaphorase activity can be observed in motoneurones following spinal root avulsion (Wu, 1993). NADPH-diaphorase staining of injured motoneurones coincides with immunoreactivity using an antibody raised against type I NOS (Wu, Li & Schinco, 1994b). Administration of a NOS inhibitor significantly reduces the motoneurone death induced by spinal root avulsion (Wu & Li, 1993). Thus the induction of type I NOS and the synthesis of NO have been suggested to contribute to the degeneration that occurs after spinal root avulsion (Wu *et al.*, 1994b). The mechanisms by which the expression of type I NOS is induced in injured motoneurones has yet to be determined.

Although type I NOS is rarely expressed in adult motoneurones under normal conditions, it is highly expressed in one of their peripheral targets. Type I NOS has been localized in fast-twitch skeletal muscle fibres but not in slow-twitch fibres (Kobzik *et al.*, 1994). Unlike type I NOS in neurones, type I NOS is membrane bound in skeletal muscle, being localized in the sarcolemma. It has recently been reported that the N-terminal region of type I NOS, which is longer than the other two isoforms, contains a sequence which enables NOS to complex with dystrophin and that this complex formation accounts for its binding to the membrane in skeletal muscle (Brenman *et al.*, 1995). Dystrophin is the protein mutated in Duchenne muscular dystrophy. There is a loss of both type I NOS expression and NOS activity in skeletal muscle from mice lacking the dystrophin gene. In addition, type I NOS is cytoplasmic in skeletal muscle from humans with Duchenne muscular dystrophy and NOS activity is markedly reduced (Brenman *et al.*, 1995). The functional implications of this loss of NO synthesis in fast-twitch skeletal muscle fibres in muscular dystrophy have yet to be evaluated.

9.4 The autonomic nervous system

Type I NOS is widely represented in the autonomic nervous system under normal conditions (see Chapter 5) and there are preliminary indications that the loss of type I NOS in disease may contribute to abnormal visceral function, particularly with respect to NOS-containing nerve fibres supplying sphincters. The major abnormalities observed to occur in mice with targeted disruption of the type I NOS gene are grossly enlarged stomachs together with hypertrophy of the pyloric sphincter. It has been noted that this resembles infantile pyloric stenosis, a disorder in humans in which there is a lack of NADPH-diaphorase-positive neurones in the pylorus (Huang *et al.*, 1993). Similarly, in achalasia, a disorder of the gastro-oesophageal junction, a complete lack of NOS activity has been observed together with a loss of NOS-immunoreactivity in the myenteric plexus (Mearin *et al.*, 1993). Under normal conditions, relaxation responses to electrical field stimulation of the internal anal sphincter are thought to be mediated by NO. Such responses are reduced in children with severe idiopathic constipation (Patel *et al.*, 1995).

NO contributes to penile erection in two ways. NO synthesized by type III NOS in the endothelial lining of the corpus cavernosal erectile tissue mediates endothelium-dependent relaxation. In addition, autonomic nerves supplying erectile tissue contain type I NOS. Stimulation of these nerves results in the production of cGMP and neurogenic vasodilatation (Lincoln, Crowe & Burnstock, 1991a; Rajfer *et al.*, 1992; Lincoln, Hoyle & Burnstock, 1995). Studies of erectile tissue obtained from patients with impotence have shown that NO formation and the relaxation response to nerve stimulation are reduced in impotence (Pickard *et al.*, 1994; Pickard, Powell & Zar, 1995). Diabetes mellitus is the most common cause of autonomic neuropathy and the incidence of impotence is increased in male diabetics. Both nerve-mediated and endothelium-dependent relaxations are impaired in impotence due to diabetes mellitus (Lincoln *et al.*, 1991a). It seems probable that the same metabolic changes that have been implicated in diabetes-induced changes in endothelial function (see Chapter 8) also contribute to altered function of NO synthesized by type I NOS in autonomic neurones. However, it should be noted that autonomic nerves contain more than one neurotransmitter and it is unlikely that erectile dysfunction is due exclusively to changes in NO. Vasoactive intestinal polypeptide, which is localized in the same nerve population that contains type I NOS, is also known to be reduced in diabetic impotence (Lincoln *et al.*, 1987).

Impaired relaxant responses to stimulation of NO-synthesizing nerves have been reported to occur in the anococcygeus muscle from diabetic rats. However, the decreased response may be due to an alteration in the ability of cGMP to relax the smooth muscle rather than a deficit in NO, since the postjunctional responses to an NO donor and 8-bromo-cGMP are similarly reduced (Way & Reid, 1994). A preliminary study has demonstrated that soluble Ca^{2+}-dependent NOS activity measured *in vitro* is increased in a myenteric plexus/smooth muscle preparation from the ileum of rats 8 weeks after the induction of diabetes mellitus (Lincoln *et al.*, 1993). This study indicates that type I NOS expression is increased in diabetes mellitus but does not necessarily demonstrate that NO synthesis is actually enhanced *in situ*. BH_4 synthesis is known to be decreased in experimental diabetes mellitus (Hamon, Cutler & Blair, 1989). NOS activity is not increased in the myenteric plexus from diabetic rats if BH_4 is omitted from the incubation medium in the NOS assay (unpublished observations), indicating that endogenous BH_4 may have been rate limiting *in vivo*. Why type I NOS expression should be upregulated in diabetes mellitus is not clear. It is possible that this occurs as a consequence of nerve degeneration in diabetes mellitus rather than contributing to its cause. Degeneration of enteric nerves is evident in the myenteric plexus at this stage of diabetes mellitus (Loesch *et al.*, 1986) and type I NOS is upregulated in both sensory and motor neurones following injury (see Sections 9.2 and 9.3).

9.5 Type I NOS in development and programmed cell death

In development of the nervous system ordered patterns of neuronal connectivity are established and 50% of mammalian neurones die during during the natural course of development. In view of the suggested role for NO in synaptic plasticity and in neuronal cell death in the adult, it has been speculated that transient expression of type I NOS during development may contribute to these changes in the developing nervous system (Lepoivre *et al.*, 1990; Zhang & Snyder, 1995). There is preliminary evidence that this is the case; however, a considerable amount of research is still required to establish the precise mechanisms involved. Transient expression of type I NOS in populations of neurones has been demonstrated in embryonic and early postnatal development of the brain (Brüning, 1993; González-Hernández *et al.*, 1994; Zhang & Snyder, 1995) and dorsal root ganglia (Wetts & Vaughn, 1993; Ward, Shuttleworth & Kenyon, 1994a). In the rat, nearly all embryonic sensory ganglion cells contain type I NOS whereas only 1% contain type I NOS in the adult (Zhang & Snyder, 1995). In an *in vitro* model of early postnatal development of the cerebellar cortex, inhibition of NOS and inactivation of NO with haemoglobin both prevent granule cell migration and the differentiation of Bergmann glia (Tanaka *et al.*, 1994). NO has been reported to amplify Ca^{2+}-induced gene transcription in neuronal cells, indicating that NO does have the capacity to regulate transcription (Peunova & Enikolopov, 1993). Type I NOS is transiently expressed in early postnatal life in a subpopulation of ventral horn cells in the vicinity of immature motoneurones. Inhibition of NOS activity blocks the molecular maturation of the motoneurones, indicating that NO within the ventral horn may contribute to a late phase of motoneurone differentiation (Kalb & Agostini, 1993).

In vitro, NO can modify GAP-43 and SNAP-25, proteins involved in neurite outgrowth and synaptogenesis, respectively (Hess *et al.*, 1993). During development of the visual system in the chick, inappropriate ipsilateral retinotectal connections are eliminated such that in the adult only connections to the contralateral side are present. NOS is transiently expressed in the optic tectum at the same stage of development at which such ipsilateral connections are eliminated. Furthermore, inhibition of NOS at this stage results in the preservation of ipsilateral connections beyond the developmental stage at which they would normally disappear (Williams, Nordquist & McLoon, 1994; Wu, Williams & McLoon, 1994a). It has therefore been suggested that NO has a role in the development of the proper pattern of neuronal connections in the visual system. At this stage a large number of retinal ganglion cells die. However, it has not been determined whether NO contributes directly to neuronal cell death or whether it occurs indirectly as a consequence of NO preventing the establishment of synaptic contact with the target cell. The fact that the development of mice lacking expression of the type I NOS gene is largely normal would appear to argue against a major role for type I NOS in the development of the nervous system. However, it has been pointed out that the transcription of the type I NOS gene could be regulated by exons in early development that are different from the one that was disrupted in this knockout mouse (Nathan & Xie, 1994a).

9.6 Selected references

Choi, D. W. (1993). Nitric oxide: foe or friend to the injured brain? *Proceedings of the National Academy of Science USA*, **90**, 9741–3.

Meller, S. T. & Gebhart, G. F. (1993). Nitric oxide (NO) and nociceptive processing in the spinal cord. *Pain*, **52**, 127–36.

Vincent, S. R. (1994). Nitric oxide: a radical neurotransmitter in the central nervous system. *Progress in Neurobiology*, **42**, 129–60.

Zhang, J. & Snyder, S. H. (1995). Nitric oxide in the nervous system. *Annual Reviews in Pharmacology and Toxicology*, **35**, 213–33.

Summary and future directions

It can be seen that changes in the function of all three isoforms of NOS may play a part in a wide range of pathological disorders. The induction of type II NOS occurs normally as part of host defence mechanisms. However, type II NOS is also induced in septic shock and in localized immune and inflammatory disorders such as diabetes mellitus, asthma and arthritis and the NO produced is likely to contribute to the tissue damage that occurs in these conditions. NO synthesized by type III NOS in the endothelium normally has a protective role, being anti-atherogenic, anti-inflammatory and reducing reperfusion injury following ischaemia. However, under conditions of oxidant stress, such as occur in atherosclerosis and diabetic angiopathy, changes in EDNO function, but not necessarily a reduction in the synthesis of EDNO, contribute to the development of vascular disease. NO synthesized during transient expression of type I NOS in embryonic and early postnatal development has been suggested to contribute to the establishment of ordered patterns of neuronal connectivity and to the death of a large number of neurones that occurs normally as part of development. However, in the adult, NMDA-mediated activation of type I NOS has been implicated in neuronal damage in the brain following stroke and in hyperalgesia following sensory nerve injury. Type I NOS can also be upregulated in motoneurones after nerve injury where it has been suggested to mediate motoneurone death. In the autonomic nervous system, a reduction in the expression or activity of type I NOS has been observed in several conditions with impaired relaxation of sphincters.

Research in the future is likely to focus on the manipulation of NO levels *in vivo* and whether this can be used therapeutically. Such research will need to take into account not only the nature of the changes in NO function that occur in a particular pathological process but also the fine balance that exists between the protective actions of NO and its toxic effects. In addition, it should not be assumed that NO is the only factor contributing to any pathological disorder. NO donors have long been used in the treatment of angina. Recent research also indicates that the effects of increasing dietary administration of arginine in preventing or treating vascular disease deserves further investigation particularly in conjunction with antioxidants. Inhalation of NO gas has already been used clinically in the treatment of pulmonary hypertension although the beneficial or adverse effects have yet to be assessed sufficiently. This method of administration can target NO to the pulmonary vasculature in the ventilated sites of the lung. Clinically, non-specific NOS inhibitors have already been used for the treatment of septic shock. For more localized disorders a major problem for the future lies in how to target NOS inhibitors both to the specific isoform of NOS involved and to the particular site that is affected in the pathological process. Such studies will be of importance for the selective reduction of the toxic effects of NO and the concurrent preservation of the protective actions of NO in other regions.

Section 3

Experimental approaches

10

Molecular biology and biochemistry

10.1 Introduction

Since the first discovery that NO can act as a physiological messenger, the mechanisms involved in NO synthesis, the different isoforms of the enzyme responsible for synthesis, NOS, and many of the intracellular targets of NO have been characterized. The genes encoding for the three isoforms of NOS have been identified and localized in the human genome. Such a wealth of information (summarized in Chapter 2) has been obtained in a remarkably short period of time and is derived largely from the concerted application of a variety of biochemical and molecular biological techniques by a large number of different laboratories. The number of publications on this subject in recent years prevents a systematic survey of all the literature in the field. Therefore, different technical approaches will be described individually and selected studies will be cited to demonstrate their application. The nature of the information that can be obtained from each technique will be discussed and areas for future research highlighted. It should be emphasized that, as with many other areas of research, the study of NO, its synthesis and its physiological actions crosses many disciplines. Thus, the molecular biological and biochemical approaches described here should be regarded as complementary to the microscopical and pharmacological approaches discussed in later chapters (Chapters 11 and 12).

10.2 Studies of NOS genes and their mRNAs

This approach is the most rapidly expanding area in biological research and is unquestionably providing a great deal of information that was previously inaccessible. The experimental protocols involved are complex and highly specialized and it is beyond the scope of this book to provide such practical detail. Any researcher wishing to enter this field would be best served by seeking the advice of experts working in laboratories specifically set up for this purpose. However, in order to gain the maximum benefit from the literature it is necessary to appreciate the types of experiments that can now be done and the nature of the information that can be derived from their application to the investigation of the biology of NO.

10.2.1 Targeted disruption of NOS genes

With advances in molecular genetics, it is now possible to target specific genes in embryonic stem cells and subsequently generate mice which fail to express the targeted gene. Such knockout mice represent important *in vivo* models with which to investigate the essential functions of the gene product. Knockout mice in which the gene for type I (Huang *et al.*, 1993), type II (MacMicking *et al.*, 1995; Wei *et al.*, 1995) and type III (Huang *et al.*, 1995) NOS are disrupted have all now been generated and in all cases the mutants are viable and fertile. Mice lacking type I NOS were the first to be produced and this was achieved by targeting the first exon of the type I NOS gene. Although type I NOS was shown to be absent from neurones and nerve fibres that normally contain the enzyme in both the CNS and the periphery, most of the tissues appeared to be normal on microscopical examination. One exception to this was the grossly enlarged stomach with hypertrophy of the pyloric sphincter and circular muscle observed in mutant mice (Huang *et al.*, 1993). This finding has emphasized the role for NO as a NANC inhibitory transmitter in the stomach (see Chapter 5). More recently, behavioural abnormalities consisting of an increase in aggression and inappropriate sexual behaviour in male but not female mutant mice have been reported (Nelson *et al.*, 1995). In the CNS, long-term potentiation in hippocampal neurones has been shown to be similar in wild-type and knockout mice (O'Dell *et al.*, 1994). The lack of an effect of disruption of the type I NOS gene on long-term potentiation in the hippocampus has now been shown to be due to the fact that these neurones, in both wild-type and mutant mice, express type III NOS rather than type I NOS (O'Dell *et al.*, 1994).

In wild-type mice, acute inhibition of NOS alters the threshold for isofluorane anaesthesia and the cerebrovascular response to CO_2 inhalation, indicating that NO is involved in these processes under normal conditions. In mice lacking type I NOS gene expression, inhibition of NOS no longer has any effect despite the fact that responses to these stimuli appear normal (Ichinose, Huang & Zapol, 1995; Irikura *et al.*, 1995). This demonstrated that NO produced by either type I or type III NOS was not responsible for the normal responses in the knockout mice. These findings illustrate an important point that applies to all studies of gene knockouts. It is not uncommon for disruption of functionally important genes to result in no or minimal phenotypic change in the resultant mutant (Shastry, 1994). Both of the above-cited studies indicate that mechanisms independent of NO can compensate for the chronic lack of NO induced by disruption of the type I NOS gene, thus demonstrating true physiological redundancy. Clearly, a great deal of work is required before various functions in type I NOS knockout mice are fully characterized. In addition, the interpretation of the results from these studies will need to take into account the possibility that, in certain physiological processes, NO-dependent mechanisms may not be the only mechanisms that can mediate the appropriate response. This is of particular relevance to the nervous system where it is known that an individual population of nerves can contain more than one transmitter and co-transmission is an established phenomenon.

Knockout mice generated following disruption of the type II and type III genes have only recently been available for study. Type II NOS knockout mice have been reported to be susceptible to infection by the protozoan parasite *Leishmania major* whereas the wild-type is highly resistant (Wei *et al.*, 1995). In addition, the fall in blood pressure and early death induced by bacterial endotoxin lipopolysaccharide (LPS) is markedly attenuated in mutant mice compared with the wild-type (MacMicking *et al.*, 1995). Thus a clear role for type II NOS in immunocytotoxicity (see Chapter 6) and endotoxic shock has been demonstrated. However, certain morphological damage following LPS treatment still occurred in mutant mice, indicating that other mechanisms also contribute to the overall process. A preliminary study has also reported a predictable response to targeted disruption of the type III NOS gene. The importance of endothelium-derived NO in basal vascular tone (see Chapter 3) has been confirmed by the elevated blood pressure present in mice lacking type III NOS. In addition, aortic rings isolated from type III knockout mice fail to demonstrate ACh-induced endothelium-dependent relaxation (Huang *et al.*, 1995).

10.2.2 Recombinant DNA techniques

In addition to the targeted disruption of genes, it is also possible to clone DNA encoding for NOS and subsequently express the DNA in cell lines by transfection. The recombinant enzyme produced by such a process can then be subjected to detailed analysis. Although the application of such techniques in the study of NOS is still relatively novel, recombinant type I (Charles *et al.*, 1993; Roman *et al.*, 1995), type II (Lyons, Orloff & Cunningham, 1992; Moss *et al.*, 1995) and type III (Busconi & Michel, 1995) isoforms of NOS have all been successfully produced and have been shown to possess properties similar to those of the native enzymes. This approach will undoubtedly be of key importance to the complete elucidation of the mechanisms by which NO is synthesized and how its synthesis is regulated.

The study of recombinant NOS has several advantages. Firstly, cell lines can be made to overexpress individual isoforms of NOS providing a source for the large amounts of enzyme that are required for detailed analysis. This is of particular importance for type III NOS, the investigation of which has lagged behind that of type I and type II NOS largely due to the difficulties involved in isolating sufficient quantities from native or cultured endothelial cells (Busconi & Michel, 1995). Secondly, each isoform can now be expressed in the same cell line allowing their properties to be compared in the same cellular environment (Wolfe *et al.*, 1995). This is likely to assist in the detection of subtle differences in the enzymatic mechanisms of each isoform. In addition, it is possible to express recombinant NOS in a specific cell line in which the intracellular environment has been altered. An example of this is the expression of recombinant type II NOS in a cell line that is deficient of BH_4. This has provided evidence that BH_4 is required for the assembly of NOS into a dimer and that dimerization is necessary for NO synthesis to occur (Tzeng *et al.*, 1995). Thirdly, specific domains of the NOS molecule can be expressed individually and mutations can be made at specific sites of the molecule in order to examine the contribution that such sites make to the overall process of NO synthesis. In this way, the importance of the myristoylation site at the N-terminal end of type III NOS in determining the membrane localization of the enzyme in endothelial cells has been established (Sessa, Barber & Lynch, 1993). The application of these techniques is not confined to increasing our understanding of basic mechanisms. The development of gene transfer techniques also opens up the possibility of gene therapy. This approach to the treatment of disease has a long way to go before it is established as feasible or effective. However, there is one exciting report which provides indications of its potential in vascular disease associated with a loss of type III NOS in the endothelium (von der Leyen *et al.*, 1995) (see Chapter 8).

10.2.3 Regulation of NOS gene transcription

Many agents are now known to induce the expression of type II NOS in a variety of cell types (see Chapter 6), often acting in synergy with an interferon or with bacterial LPS. Such findings have been made by studies examining the effects of agents either *in vivo* or *in vitro* on the expression of NOS mRNA or on NO synthesis. Recently, these studies have been extended to examine the promoter/enhancer region of the type II NOS gene which is responsible for the control of transcription. It is likely that such experiments will enable the mechanisms of induction of type II NOS to be investigated on a more systematic basis. The fact that type II NOS can be induced by many diverse stimuli is reflected by the complexity of the promoter region of its gene which has yet to be fully characterized. The presence of consensus sequences for the binding of a variety of transcription factors have been reported. A unique NF_kB sequence responsible for LPS-inducibility of type II NOS has been identified. IRF 1 and IRF 2 are transcription factors important in gene regulation by INF-γ and binding sites for both factors have been demonstrated to be present in the promoter region of the type II NOS gene (Nathan & Xie, 1994a). A preliminary study has shown that IRF 2 is a suppressive transcription factor that competes for the IRF 1 binding sites and thus may be important in keeping promoter activity low under basal conditions (Evans, Spink & Cohen, 1995). Importantly, there is now preliminary evidence for significant differences between the promoter regions for type II NOS in human and murine genes (Geller *et al.*, 1995b; Spitsin, Koprowski, & Michaels, 1995). These are likely to provide some insight into the difficulties there have been to date in inducing type II NOS in human macrophages.

Although type I and type III NOS are both expressed constitutively (unlike type II NOS), it is important to recognize that the expression of these isoforms can also be regulated at the level of transcription. In the case of type III NOS, evidence has already accumulated that NOS expression in endothelial cells can be regulated by a variety of stimuli including shear stress, hypoxia and sex steroids (see Chapter 3). Analysis of the promoter/enhancer region of the type III NOS gene in human umbilical vein (Marsden *et al.*, 1992) and bovine aortic endothelial cells (Venema *et al.*, 1994) has revealed that it is also extremely complex, containing numerous putative transcription factor binding sites. Although there is high sequence homology between the type III NOS gene in these two studies, some differences have been noted to occur in the region regulating the initiation of transcription. Whether this represents a species difference or a potential difference in the transcriptional regulation of type III NOS in two different vascular beds is an interesting question that remains to be answered (Venema *et al.*, 1994). Characterization of the regulatory regions of the type I NOS gene has yet to be published. It is likely that such analysis will be of importance in determining the regulation of changes in NO synthesis by type I NOS already observed in a variety of conditions including pregnancy (Weiner *et al.*, 1994), nerve injury and nerve development (see Chapter 9).

10.2.4 NOS mRNA expression

Clearly, the study of the NOS family of genes cannot be carried out in isolation from the study of their products. The development of probes specific to the different isoforms of NOS has enabled the study of their respective mRNAs by hybridization techniques. Such techniques are becoming more accessible to researchers and their application provides a variety of information of importance to research into the biology of NO:

1. At its simplest, detection of a NOS mRNA confirms that transcription of the gene has occurred in a given situation or cell type. This is now used routinely in the investigation of the induction of NOS both *in vitro* and *in vivo*. It is also useful in the screening of tissues for both the constitutive and the induced expression of NOS. A wide range of tissues can readily be examined at specific stages of development and in a variety of experimental conditions *in vivo* including animal models of disease. In this way, it was first discovered, surprisingly, that skeletal muscle highly expresses the type I NOS mRNA that was originally thought to be specific for neurones (Nakane *et al.*, 1993).
2. Levels of NOS mRNA can be quantified. This enables the relative amounts of NOS mRNAs and the upregulation or downregulation of constitutive NOS mRNAs to be assessed quantitatively. When changes in NO synthesis have been detected, it is necessary to determine whether this occurs via transcriptional regulation of the synthesis of the NOS protein or via post-translational modification of NOS enzyme activity. In the case of increased NO synthesis during pregnancy or following oestrogen treatment, it has been shown that this is due to increased production of type I and type III NOS mRNA rather than by simple post-translational activation of the constitutive NOS enzymes (Weiner *et al.*, 1994).
3. *In situ* hybridization enables NOS mRNAs to be localized at the light microscope level in individual cell types within a given tissue (Bredt *et al.*, 1991a). This represents a useful technique that can be carried out in parallel with immunohistochemical and histochemical procedures for the localization of NOS itself (see Chapters 12 and 16). It can also be important in confirming the specific isoform that is being expressed rather than relying on there being no cross-reactivity between the antibodies used for immunohistochemistry.

It should be emphasized that the demonstration of NOS mRNA cannot be assumed to indicate that NO synthesis is taking place or indeed that NOS itself is present. Type II NOS mRNA can be expressed in human cells following activation by cytokines whereas the translated product and NO synthesis are not detectable (Attur *et al.*, 1995). Similarly, in the absence of BH_4, both NOS mRNA and NOS may be present within a cell but NO will not be synthesized (Tzeng *et al.*, 1995). Thus, wherever possible, it is important to investigate NOS activity in addition to the NOS family of genes and their mRNAs (see next section).

10.3 Biochemical studies of NO synthesis

In contrast to the study of the transcriptional and translational control of NOS expression and, indeed, to the study of NO itself, the investigation of NOS activity and the manipulation of NO synthesis is technically easier to perform. Measurement of NOS activity also provides a sensitive and accurate method for assessing changes in NO synthesis in any given experimental or pathological situation. A detailed description of the experimental protocol for measuring NOS activity is given in Chapter 14. The large number of studies using a biochemical approach to the study of NOS is also a reflection of the nature of NO as a physiological messenger. Unlike classical messengers, NO is not stored or transported by specialized mechanisms. Thus, stored levels of NO cannot be measured and activation of an NO-dependent mechanism is achieved by stimulation of NO synthesis rather than by direct stimulation of NO release. Similarly, NO does not act on specific receptors to exert its effects, rendering the standard pharmacological approach using specific receptor agonists and antagonists ineffective. As a result, biochemical research into the manipulation of both NOS activity and of the intracellular targets for NO is continuing to play a key role in investigating basic mechanisms, in providing the appropriate pharmacological tools with which to investigate NO function and in the development of specific inhibitors for use in therapy.

10.3.1 Manipulation of the synthesis of NO and of its effects

NO is synthesized during the conversion of arginine to citrulline by NOS and, in the majority of cases, exerts its physiological effects by the activation of soluble guanylate cyclase. Numerous biochemical studies have investigated the effects of a wide variety of different classes of compounds on different parts of the arginine–NO–cGMP pathway and on different sites on the NOS molecule. Table 10.1 provides a summary of the results of these studies together with selected references. The information that can be derived from such research is proving useful in a variety of different ways.

1. The demonstration of the inhibitory effects of agents that act on specific sites on the NOS molecule has provided the evidence that the binding of NADPH, FMN, FAD, BH_4, calmodulin and haem are all of critical importance to the mechanism by which NO is synthesized (see Chapter 2). Ca^{2+} chelators can also be used to demonstrate whether NO synthesis is Ca^{2+}-dependent and thus distinguish the type II isoform, which is Ca^{2+}-independent, from the type I and type III isoforms (see Chapter 14). The ability of compounds such as nitroblue tetrazolium to accept electrons produced by the CPR domain of NOS enables the NADPH-diaphorase activity of NOS to be localized histochemically (see Chapters 12 and 16). It should be noted that, by their very nature, many of these agents will not be specific for NOS. Thus, diphenyleneiodonium inhibits all

Table 10.1 *Summary of compounds that act at the various sites indicated in the arginine–NO–cGMP pathway together with their effects*

Site	Compound	Effects	References
I. **Arginine uptake**	(a) Lysine, ornithine, homoarginine, SDMA	Inhibit arginine uptake, can reduce NO synthesis, no direct effect on NOS	Bogle *et al.*, 1992, 1996; Baydoun *et al.*, 1995; Closs *et al.*, 1995
	(b) ADMA*, NMMA*, NIO, NAA	NOS inhibitors that also inhibit arginine uptake	Bogle *et al.*, 1992; Lopes *et al.*, 1994; Baydoun *et al.*, 1995; Closs *et al.*, 1995
II. NOS molecule			
i. CPR domain – NADPH-, FAD-, FMN-binding sites	(a) Diphenylene iodonium	Inhibits all NADH- or NADPH-utilizing enzymes including NOS	Nathan, 1992; Knowles & Moncada, 1994
	(b) Nitroblue tetrazolium	Removes electrons produced by NADPH-diaphorase activity of NOS	Hope *et al.*, 1991
ii. Calmodulin-binding site	(a) Trifluoperazine, W-7, W-13, chlorpromazine, calmidazolium, antifungal imidazoles	Calmodulin antagonists; inhibit NOS activity by preventing calmodulin binding to NOS	Bredt & Snyder, 1990; Förstermann *et al.*, 1991a; Wolff, Datto & Samatovicz, 1993a
	(b) EDTA, EGTA	Ca^{2+} chelators: prevent Ca^{2+}-dependent calmodulin binding	Bredt & Snyder, 1990; Pollock *et al.*, 1991
iii. BH_4-binding site	(a) DAHP, methotrexate, *N*-acetyl-5-HT	Reduce availability of co-factor by inhibiting BH_4 synthesis or recycling. Can prevent NOS dimerization	Gross & Levi, 1992; Schmidt *et al.*, 1992c; Knowles & Moncada, 1994
	(b) 7-Nitroindazole	Inhibits BH_4 binding to NOS	Mayer *et al.*, 1994

iv. Thiol group	Ebselen or carboxyebselen	Bind to thiol group on NOS and inhibit activity	Zembowicz *et al.*, 1993; Hatchett *et al.*, 1994, Hattori *et al.*, 1994
v. Haem domain	(a) CO, NO, methylene blue troleandomycin, dihydroergotamine	Bind directly to haem – inhibit NO synthesis but not NADPH-diaphorase activity of NOS	Assreuy *et al.*, 1993; Buga *et al.*, 1993; Marletta, 1993; Mayer, Brunner & Schmidt, 1993; Rengasamy & Johns, 1993; Dudek *et al.*, 1995
	(b) Thiocitrulline	Bind to arginine-binding site but also affect environment of haem	Abu-Soud *et al.*, 1994a; Frey *et al.*, 1994
	indazoles		Mayer *et al.*, 1994; Wolff, Lubeskie & Umansky, 1994
	imidazoles		
	ethylisothiourea		Garvey *et al.*, 1994
	aminoguanidine		Wolff & Lubeskie, 1995
	NAME, NMMA*		Abu-Soud *et al.*, 1994a; Olken, Osawa & Marletta, 1994
vi. Arginine-binding site	(a) NOARG, NAME, NMMA*, ADMA*, NAA, NAPNA	Arginine analogues: L-enantiomers inhibit NOS activity, D-enantiomers do not	Moncada, Palmer & Higgs, 1991; Babbedge *et al.*, 1993c; Knowles & Moncada, 1994; Komori, Wallace & Fukuto, 1994; MacAllister *et al.*, 1994a; Yokoi *et al.*, 1994; Traystman *et al.*, 1995
	(b) Canavanine	All display some structural similarity to arginine, bind to arginine-binding site and inhibit NOS activity	Knowles & Moncada, 1994
	Indazoles, e.g. 7-nitroindazole		Babbedge *et al.*, 1993a; Moore *et al.*, 1993a; Mayer *et al.*, 1994; Wolff *et al.*, 1994

Table 10.1 *(cont.)*

Site	Compound	Effects	References
vi. Arginine-binding site *(cont.)*	(c) Imidazoles	All display some structural similarity to arginine, bind to arginine-binding site and inhibit NOS activity	Wolff *et al.*, 1993a,b, 1994; Mayer *et al.*, 1994
	(d) Guanidino compounds, e.g. aminoguanidine		Griffiths *et al.*, 1993; Misko *et al.*, 1993a; MacAllister, Whitley & Vallance, 1994b; Laszio, Evans & Whittle, 1995; Wolff & Lubeskie, 1995
	(e) Imino compounds, e.g. NIO, NIL, iminobiotin		McCall *et al.*, 1991; Moore *et al.*, 1994; Sup, Gordon Green & Grant, 1994
	(f) Thiocitrullines		Abu-Soud *et al.*, 1994a; Frey *et al.*, 1994; Furfine *et al.*, 1994; Narayanan *et al.*, 1995
	(g) Isothioureas		Garvey *et al.*, 1994; Szabó, Southan & Thiemermann, 1994a; Nakane *et al.*, 1995; Southan, Szabó & Thiemermann, 1995
III. NO	(a) Superoxides and superoxide generators, free radical scavengers, (e.g. hydroquinone, ascorbate), PTIOs, haemoglobin	React with NO and therefore prevent some mechanisms mediated by NO	Hobbs, Tucker & Gibson, 1991; Moncada *et al.*, 1991; Yoshida *et al.*, 1994

Superoxide dismutase	(b)	Removes O_2^- which can inactivate NO; therefore can enhance NO-dependent mechanisms	Moncada *et al.*, 1991
IV. Effects of NO mediated by cGMP	(a) Methylene blue, ODQ	Prevent NO-induced production of cGMP by inhibiting guanylate cyclase	Rajfer *et al.*, 1992; Garthwaite *et al.*, 1995
	(b) IBMX, M&B22948	Inhibit degradation of cGMP, can enhance effects of NO	Rajfer *et al.*, 1992; Boulton *et al.*, 1994
	(c) 8-bromo-cGMP	Membrane-permeant analogue of cGMP, can mimic effects of NO	Boulton *et al.*, 1994

*Indicates that these arginine analogues are naturally occurring. Note that the L-isomers of the arginine analogues listed in the table are active whereas the D-isomers are not. Abbreviations: ADMA, asymmetrical N^G,N^G-dimethyl-L-arginine; CO, carbon monoxide; DAHP, 2,4-diamino-6-hydroxypyrimidine; IBMX, 3-isobutyl-1-methylxanthine; M&B22948, 2-*o*-propoxyphenyl-8-azapurine-6-one; *N*-acetyl-5-HT, *N*-acetyl-5-hydroxytryptamine; NAA, N^G-amino-L-arginine; NAME, N^G-nitro-L-arginine methyl ester; NAPNA, N^G-nitro-L-arginine-*p*-nitroanilide; NIL, L-N^6-(1-iminoethyl)-lysine; NIO, L-N^5-(1-iminoethyl)-ornithine; NMMA, N^G-monomethyl-L-arginine; NOARG, N^G-nitro-L-arginine; ODQ, 1*H*-[1,2,4]oxadiazolo-[4,3-*a*]quinoxalin-1-one; PTIOs, 2-phenyl-[4,4,5,5]-tetramethylimidazoline-1-oxyl-3-oxide and its derivatives; SDMA, symmetrical N^G, N^G-dimethyl-L-arginine.

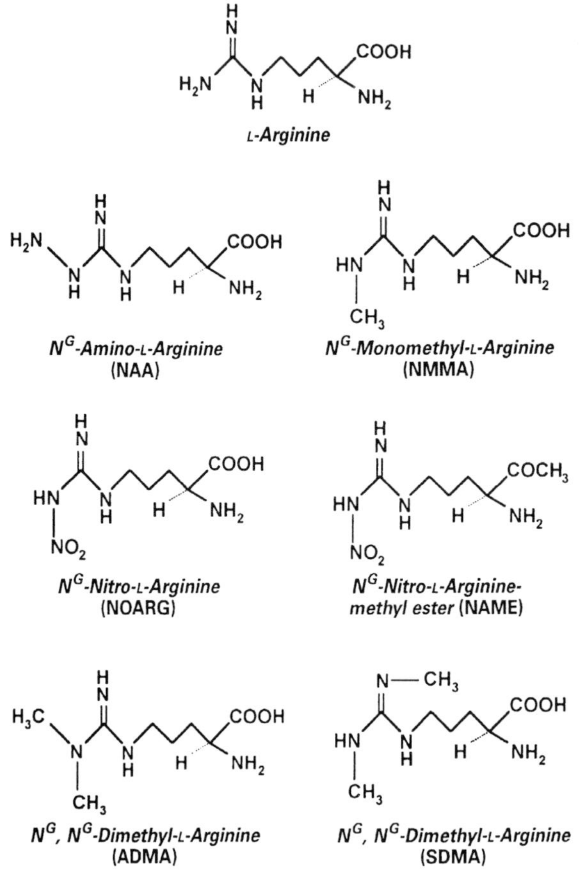

L-Arginine

N^G-Amino-L-Arginine (NAA)

N^G-Monomethyl-L-Arginine (NMMA)

N^G-Nitro-L-Arginine (NOARG)

N^G-Nitro-L-Arginine-methyl ester (NAME)

N^G, N^G-Dimethyl-L-Arginine (ADMA)

N^G, N^{G}-Dimethyl-L-Arginine (SDMA)

Fig. 10.1 Structure of L-arginine and its analogues. The L-isomers of all the analogues shown inhibit NOS activity, whatever the isoform, with the exception of SDMA which does, however, inhibit arginine uptake. The D-isomers are inactive, indicating that binding is stereo specific. NMMA and ADMA are naturally occurring inhibitors.

NADPH- or NADH-utilizing flavoproteins, calmodulin antagonists inhibit all calmodulin-dependent enzymes and agents binding directly to haem can affect the activity of many haem-containing enzymes including guanylate cyclase. Conversely, certain drugs used therapeutically, such as imidazoles (antimycotic), dihydroergotamine (antimigraine) and troleandomycin (antibiotic) have now been shown to inhibit NOS activity (see Table 10.1). It is likely that other drugs will need to be investigated in this respect.

2. It was soon discovered that a variety of arginine analogues can inhibit NO synthesis. The structures of the most commonly used arginine analogues are given in Fig. 10.1. These analogues inhibit all the isoforms of NOS and are

generally regarded to be specific for NOS. However, there has been little systematic investigation of their effects on other enzymes (Nathan, 1992; Peterson *et al.*, 1992). Inhibition of NOS is stereospecific, and D-enantiomers of these arginine analogues fail to have any inhibitory effect. Initially, when added simultaneously with arginine, all these arginine analogues compete with arginine for binding on the NOS molecule. However, following initial competitive binding, the inhibition by some analogues is only slowly reversible or irreversible (Nathan, 1992; Knowles & Moncada, 1994). Thus addition of arginine *after* an arginine analogue does not always reverse the inhibitory effect of the analogue. L-NMMA is unusual in that it also acts as a partial substrate for NOS (Olken, Osawa & Marletta, 1994). Both L-NMMA and asymmetrical dimethylarginine (ADMA) occur naturally and can be metabolized (MacAllister *et al.*, 1994a). More recently, other compounds which have some structural similarity to arginine, generally containing a guanidino group, have been investigated for their ability to inhibit NOS (Table 10.1, Fig. 10.2). Much of this research is being undertaken to identify inhibitors that are more specific for individual isoforms of NOS (see Section 10.3.2). For an excellent review of the actions of arginine analogues and arginine-related compounds on NOS activity see Kerwin *et al.* (1995).

3. Several of the compounds listed in Table 10.1 now provide the pharmacological tools with which to investigate NO function (see Chapters 11 and 15). Thus, arginine analogues are now routinely used to determine whether a particular response requires the synthesis of NO for it to occur. Similarly, agents that can inactivate NO, such as haemoglobin or free radical scavengers, can be used as partial evidence that NO is mediating a response. Lastly, a variety of agents that can manipulate cGMP levels are now available for demonstrating that, once it is synthesized, NO produces a physiological response by the activation of soluble guanylate cyclase (Table 10.1). Research is continuing in this area to refine these tools such that their effects can be made more specific. Methylene blue, commonly used to inhibit soluble guanylate cyclase, also inhibits NOS. ODQ, a compound that has recently been shown to inhibit soluble guanylate cyclase without having any effect on NOS (Garthwaite *et al.*, 1995), is likely to supersede the use of methylene blue. Similarly, PTIOs may become important in pharmacological studies since they have been reported to act as free radical scavengers that are specific to free NO (Yoshida *et al.*, 1994).
4. Biochemical studies have also investigated the possibility of regulating NO production by agents acting at sites other than the NOS molecule (Table 10.1). These include restricting the availability of arginine (by inhibiting uptake) and of the co-factor BH_4 (by inhibiting synthesis or recycling) (see Table 10.1). A number of compounds, which have no direct effect on NOS activity, have been

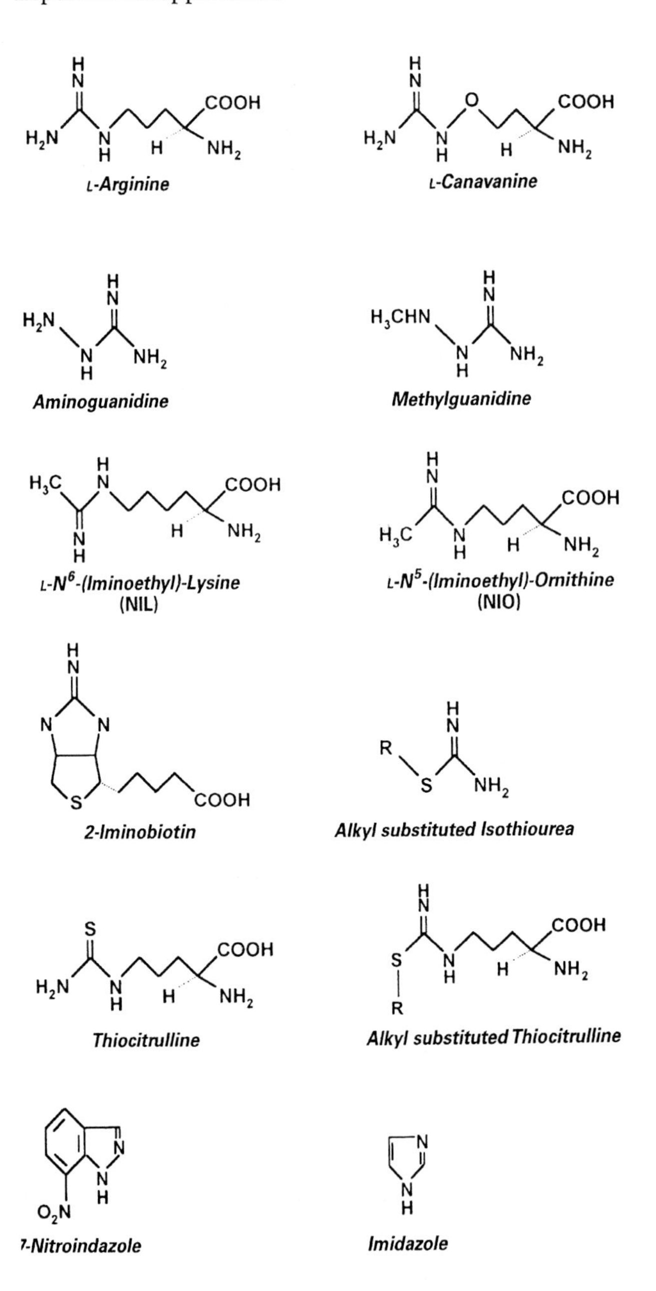

Fig. 10.2 L-Arginine and compounds with certain similar structural elements that have all been shown to inhibit NOS activity by binding to the arginine-binding site of the enzyme. See text for information concerning which inhibitors are relatively specific for individual isoforms of NOS.

shown to inhibit arginine uptake. In certain cases, inhibition of arginine uptake has been shown to inhibit NO production by intact cells as a consequence (Lopes *et al.*, 1994). This raises the possibility of manipulating NO synthesis by agents that do not even need to enter the cell. Some arginine analogues that inhibit NOS can also inhibit arginine uptake competitively. It is likely that these analogues enter the cell via the arginine carrier. Exceptions to this are L-nitroarginine (L-NOARG) and its methyl ester (L-NAME) which, unlike arginine, do not carry a positive charge at physiological pH (Bogle *et al.*, 1992; Knowles & Moncada, 1994). These two analogues have been shown to enter the cell by a separate mechanism and are thus still effective as inhibitors in intact cells or *in vivo*. Reducing BH_4 levels can also inhibit NO synthesis (Gross & Levi, 1992). Initially, it was assumed that arginine and BH_4 would only be rate limiting during the synthesis of large amounts of NO by type II NOS. It has been shown that BH_4 synthesis is an absolute requirement for cytokine-induced NO production (Gross & Levi, 1992). However, inhibition of BH_4 synthesis can also inhibit the production of NO by type III NOS in intact endothelial cells (K. Schmidt *et al.*, 1992).

10.3.2 Specific inhibitors for individual isoforms of NOS

10.3.2.1 Type II NOS

Much of the research into specific inhibitors is being focused on inhibitors that are specific for type II NOS since these would be potential therapeutic agents for use in the treatment of septic shock. The majority that have been investigated to date bear some structural similarity to arginine (Fig. 10.2) and have been shown to bind at the arginine site on the NOS molecule (Table 10.1). L-Canavanine, aminoguanidine, L-*N*-(iminoethyl)-ornithine (L-NIO), L-*N*-(iminoethyl)-lysine (L-NIL), 2-iminobiotin and phenylimidazole have all been reported to be more potent as inhibitors of type II NOS compared with type I and type III NOS *in vitro* (Knowles & Moncada, 1994; Moore *et al.*, 1994; Sup, Gordon Green & Grant, 1994; Wolff, Lubeskie & Umansky, 1994; Wolff & Lubeskie, 1995). The degree of specificity varies between these compounds and, in the majority of cases, effective inhibition of type II NOS can be achieved with concentrations in the low micromolar range. L-Canavanine is a broad spectrum inhibitor that also inhibits other arginine-utilizing enzymes such as arginase (Hrabák, Bajor & Temesi, 1994). Despite being equally effective as L-NAME at inhibiting type II NOS, aminoguanidine is 100-fold less potent at producing a pressor response *in vivo* (Corbett *et al.*, 1992a). Thus, aminoguanidine may be useful as a specific inhibitor of type II NOS *in vivo*. However, the effect of aminoguanidine on type I NOS has not been investigated either in intact cells or *in vivo*. Similarly, it is not known whether imino-containing compounds are still selective for type II NOS when their effects are compared with type I and type III NOS in intact cells. More recently, the inhibitory effects of substituted isothioureas (Fig. 10.2) have been investigated. These are extremely potent NOS inhibitors. Ethylisothiourea has an IC_{50} value of 13 nM for type II NOS *in vitro*. In addition it is 20–30 times more potent at inhibiting type II NOS than either type I or type III NOS (Nakane *et al.*, 1995). Many isothioureas also produce a pressor response on *in vivo* administration. However, aminoethylthiourea only produces a very weak pressor response relative to its potent inhibition of type II NOS in intact cells (Southan, Szabó & Thiemermann, 1995).

10.3.2.2 Type I and type III NOS

Fewer studies have investigated whether there are inhibitors specific for either type I or type III NOS. Discovery of such inhibitors could prove very useful, for example, in distinguishing whether a response detected in a highly vascularized tissue is due to the synthesis of NO by endothelial cells or by neurones. This would be of particular significance to studies of the CNS. Ebselen has been reported to inhibit type III NOS preferentially relative to both type I and type II NOS *in vitro* (Zembowicz *et al.*, 1993). This drug is unusual in that it acts on a thiol group on the NOS molecule (Table 10.1). However, another study has reported that, within a certain concentration range, ebselen preferentially inhibits type II NOS (Hattori *et al.*, 1994). *In vitro*, ethylthiocitrulline, unlike thiocitrulline or methylthiocitrulline, is much more potent at inhibiting type I NOS compared with both type II and type III NOS (Furfine *et al.*, 1994). Thus, it has been suggested that ethylthiocitrulline could be useful as a specific inhibitor of NO-dependent neuronal mechanisms *in vivo*. Unfortunately, ethylthiocitrulline has also been shown to have a strong pressor effect *in vivo* (Narayanan *et al.*, 1995) which may limit its usefulness. Interestingly, two agents, nitroarginine *p*-nitroanilide (Babbedge *et al.*, 1993c) and 7-nitroindazole (Moore *et al.*, 1993a), have been shown to inhibit responses mediated by NO synthesized in sensory nerves *in vivo* without increasing blood pressure. *In vitro*, neither compound inhibits type I NOS to a greater extent than type III NOS, nor are they more potent than L-NAME. However, *in vivo* it is clear that they are much weaker as pressor agents than L-NAME, while still maintaining their effect on NOS in sensory neurones. The reason for such selectivity for type I NOS *in vivo* but not *in vitro* has not been established. It has been suggested that the inhibitory effect of 7-nitroindazole may be reduced by xanthine oxidase which is present in higher concentrations in intact endothelial cells than in neurones (P.A. Bland-Ward & P.K. Moore, personal communication).

The search for inhibitors specific for individual isoforms of NOS still has some way to go. Much of this research is in its infancy and, where relative specificity has been demonstrated *in vitro*, it has not always been confirmed that such specificity is maintained *in vivo*. In addition, the reasons why certain agents appear to be more effective at inhibiting a particular isoform of NOS are poorly defined. Until the arginine-binding sites of the different isoforms and their relationship to the haem domain of NOS have been characterized completely, the search for specific inhibitors cannot be systematic.

10.4 Studies of molecular targets for NO

Characterization of the reactions that NO can undergo and identification of its targets under physiological and pathophysiological conditions is perhaps the greatest challenge in the field of NO research. Based on its chemical and physical properties, many of the potential reactions for NO can be predicted on a theoretical basis. However, actual identification of the reaction products can be extremely difficult and often requires specialized and sophisticated techniques. Understanding the literature in this area can also be daunting particularly for workers who are more at home with biological terms and concepts.

The properties of NO are such that is has the capacity to react with the following: other free radicals; molecular O_2 and superoxides; nucleophilic groups such as amines and thiols; and transition metals such as manganese, copper or iron. Its ability to bind with haem on haemoglobin or guanylate cyclase forms the basis of two methods used for measuring NO levels (see Chapter 13). Most of these types of reactions have been shown to occur within the cell. The results of such reactions also account for some of the effects that NO is known to have both as a physiological messenger and as a toxic agent (see Chapter 2). However, if one considers the number of biological molecules that contain transition metals or thiol groups, then it is possible that we have only scratched the surface with regard to the targets for NO and its effects both as an intracellular and an intercellular messenger.

10.4.1 Transition metals

With regard to the investigation of the interaction of NO with transition metals, electron paramagnetic resonance (EPR) spectroscopy will probably play an important role. NO is paramagnetic and reacts with transition metals to form paramagnetic species that are readily detectable by EPR spectroscopy. Indeed, NO was used as a probe for detecting a variety of metalloproteins long before it was known to exist as a biological molecule. Henry *et al.* (1993) provide an excellent review of EPR spectroscopy and its use in investigating both the established and potential targets for NO by its interaction with transition metals. They also highlight several areas worthy of future investigation. NO is known to interact with prostaglandin H synthases, involved in the synthesis of prostaglandins, and lipoxygenases, involved in the inflammatory process, and ceruloplasmin, an oxidase present in plasma. However, the metabolic effects of such interactions are unknown. It has been recognized that NO may be involved in the regulation of iron metabolism and that during immunocytotoxicity induced by activated macrophages there is a massive loss of intracellular iron. The mechanisms by which this could occur have yet to be elucidated. Ferritin, transferrin and iron-responsive binding protein are all involved in controlling iron metabolism and are all potential targets for NO (Henry *et al.*, 1993; Mülsch, 1994).

Superoxide dismutases (SOD), which remove superoxide anions, occur in various isoforms containing copper, zinc, manganese or iron. It has not been demonstrated whether NO interacts with SOD *in vivo* and, if so, how this influences SOD activity. The toxicity of NO is, in part, governed by its ability to react with superoxides to form peroxynitrite (see Chapter 2). If NO does influence SOD activity, then this could represent a mechanism whereby NO could regulate its own toxicity. Finally, there is the intriguing question of haem oxygenase. Haem oxygenase binds haem and, with the concerted activity of CPR, catalyses the oxidoreduction of haem producing CO in the process. The haem oxygenase–CPR system bears a strong resemblance to the combined haem and CPR domains of NOS. In the brain, haem oxygenase is co-localized with soluble guanylate cyclase. CO is known to activate soluble guanylate cyclase as does NO and has also been proposed as a messenger in the brain (Verma *et al.*, 1993). CO also inactivates NOS (Olken *et al.*, 1994). If NO modulates haem oxygenase activity then this raises the possibility of cross-regulation of the two enzyme systems.

10.4.2 Thiols

NO has the potential to react with low molecular weight thiols and with thiol groups on proteins to produce nitrosothiols. Nitrosothiols are not paramagnetic and are therefore not detectable by EPR spectroscopy. Methods for detecting nitrosothiols include UV, IR and NMR spectroscopy and, most recently, electrospray ionization-mass spectrometry (Mirza, Chait & Landers, 1995). Using such techniques, it has been demonstrated, under controlled chemical conditions, that a variety of peptides and proteins can react with NO solutions to produce nitrosothiols (Stamler *et al.*, 1992; Mirza *et al.*, 1995). Not all the thiol groups on proteins may be able to react in this way. A small GTP-binding protein, p21ras, contains five free cysteine residues and yet only forms one nitrosothiol group on chemical reaction (Mirza *et al.*, 1995). Clearly, it is necessary to demonstrate that such reactions can also occur under physiological conditions and this has been shown to be the case with serum albumin (Stamler *et al.*, 1992).

Increasing levels of reducing agents have been reported to prevent *S*-nitrosylation of somatostatin *in vitro*. This may suggest that, within the cell, endogenous reducing agents may be able to regulate the *S*-nitrosylation of proteins (Mirza *et al.*, 1995). In addition, nitrosothiols are not stable. They can release NO and also NO can effectively be transferred from one molecule to another by transnitrosation reactions (Mülsch, 1994; Stamler, 1994). Thus, all nitroso compounds should be regarded as being in dynamic equilibrium. The direction of reactions will be determined by the *relative* stability of the products. Attempts are now being made to assign a scale of transnitrosation potentials such that the direction of transfer might be predicted (Wang *et al.*, 1995).

It should be noted that even if *S*-nitrosylation of a particular molecule is shown to occur *in vivo*, the physiological significance of the reaction still remains to be determined. The reaction of NO with low molecular weight thiols may simply represent a protective mechanism preventing the accumulation of high concentrations of free NO within the cell. Other reactions may form part of a process whereby NO can be stored and/or transported. Modification of the function of a particular protein by *S*-nitrosylation will only occur if the thiol group that reacts is at a critical site.

10.5 Summary

Studies of knockout mice in which there is targeted disruption of the genes for type I, type II or type III NOS can provide valuable information on the essential functions of NO *in vivo*. Investigation of the genes and mRNAs for the different isoforms are of importance to the elucidation of the regulation of NOS expression and the factors involved in such regulation. Measurement of NOS activity represents the simplest method to assess changes in NO synthesis in a varitety of physiological, experimental and pathological conditions. Biochemical studies of NOS inhibitors continue to play a role in: (1) research on the mechanisms by which NO is synthesized; (2) manipulation of NO synthesis *in vitro* and *in vivo*; and (3) providing the pharmacological tools with which to investigate NO function. Finally, biochemical investigation of the interactions that NO can undergo is important for understanding the mechanisms by which NO can exert its physiological effects. Much of this work is still in its infancy. Therefore, the molecular biological and biochemical studies will continue to play a key role in research into the biology of NO.

10.6 Selected references

Henry, Y., Lepoivre, M., Drapier, J. C., Ducrocq, C., Boucher, J. L. & Guissani, A. (1993). EPR characterization of molecular targets for NO in mammalian cells and organelles. *FASEB Journal*, **7**, 1124–34.

Kerwin, J. F., Lancaster, J. R. & Feldman, P. L. (1995). Nitric oxide: a new paradigm for second messengers. *Journal of Medicinal Chemistry*, **38**, 4343–62.

Marletta, M. A. (1993). Nitric oxide synthase structure and mechanism. *Journal of Biological Chemistry*, **268**, 12231–4.

Nathan, C. & Xie, Q.-W. (1994). Regulation of biosynthesis of nitric oxide. *Journal of Biological Chemistry*, **269**, 13 725–8.

11

Pharmacology

11.1 Introduction

The experimental approaches to determine whether NO is a vasoactive substance released from endothelial cells, i.e. EDRF, or whether it is a neurotransmitter are broadly similar. In both cases NO has to be shown to be synthesized and released from a specific source, then the released NO has to be shown to act on its target in a predictable way. The criteria that form the basis for defining neurotransmitters, including NO, are given in Chapter 5. Agents used in the determination of NO as a neurotransmitter or as an EDRF are summarized in Table 11.1.

11.2 Synthesis and release of NO

For NO, the enzyme responsible for its synthesis, NOS, can be shown to be present in nerves and endothelial cells by using immunohistochemical methods or, under defined conditions, NADPH-diaphorase activity (see Chapters 12 and 16). Thus nerves and endothelial cells with the potential to release NO can be identified.

In vitro, the simplest way to show that neurotransmitter release is responsible for an observed response is to use tetrodotoxin (TTX). This compound, isolated from the liver of the fugu fish (Japanese puffer fish), blocks voltage-dependent sodium ion channels, which in neurones are responsible for the conduction of action potentials. If TTX abolishes responses to EFS it can be said that the responses are indeed neurogenic and that they do involve a transmission process. Nicotine, or the nicotinic agonist 1,1-dimethyl-4-phenylpiperazinium (DMPP) can also be used to stimulate intramural neurones (provided of course that they possess nicotinic receptors, which they usually do because they are most often autonomic postganglionic nerves), and, by being blocked by TTX, responses to nicotinic stimulation can be shown to be neurogenic. However, whatever the stimulus, TTX will not reveal anything about the identity of the transmitter substance.

To show that endothelial cells are responsible for a particular response to a particular stimulus the best approach is to remove the endothelium. In isolated

Table 11.1 *Agents used in the evaluation of NO as a neurotransmitter or as an endothelium-derived relaxing factor*

Action potential blocker
Tetrodotoxin (0.1–1.0 μM)

Authentic NO
NO gas dissolved in water (≈5 μM)
Sodium nitrite, acidified solution (1–100 μM)

NO donors
Glyceryltrinitrate (nitroglycerine, GTN)
Isoamylnitrite (0.1–10 μM)
3-Morpholinylsydnoneimine (SIN-1, 0.1 μM)
S-Nitroso-L-cysteine (3 μM)
S-Nitrosoglutathione (1 μM)
S-Nitroso-*N*-acetyl-L,L-penicillamine (SNAP, 10 μM)
Sodium nitroprusside (SNP, 100 μM)
Diethylamine/NO (DEA/NO, 1 μM)

NO synthesis inhibitors (NOS antagonists)*
Ca^{2+}-free conditions
Diphenyleneiodonium (DPI, 0.1 μM)
N^G-Monomethyl-L-arginine (L-NMMA, 100 μM)
N^G-Nitro-L-arginine (L-NNA, = L-NOARG, = NOLA, 10–100 μM)
N^G-Nitro-L-arginine methyl ester (L-NAME, 10–100 μM)
7-Nitroindazole (100 μM)

NO trappers
2-(4-Carboxyphenyl)-4,4,5,5-tetraethylimidazoline-1-oxyl-3-oxide (carboxy-PTIO, 100 μM to 1 mM)
Hydroxocobalamin (30 μM)
Oxyhaemoglobin (10 μM)

Superoxide anion inhibitor (NO inactivation inhibitor)
6,7-Dimethoxy-8-3′,4′,5-trihydroxyflavone (3 μM)
Superoxide dismutase (100–500 U ml^{-1})

Superoxide anion generator (NO inactivation promoter)
Hydroquinone (50 μM)
Pyrogallol (100 μM)

Superoxide dismutase inhibitor (NO inactivation promoter)
Diethyldithiocarbamate (3 mM)

Guanylate cyclase inhibitors
LY 83583 (6-anilino-5,8-quinolenedione, 10 μM)
1*H*-[1,2,4]Oxadiazolo[4,3-a]quinoxalin-1-one (ODQ, 1 μM)
Methylene blue (30 μM)

Phosphodiesterase V inhibitors
SK&F96231 (1–10 μM)
Zaprinast (M&B22948, 1–10 μM)

Guanosine 3′,5′-cyclic monophosphate analogue
8-Bromo-cyclic 3′,5′-guanosine monophosphate (100 μM)

Concentrations given are rough guides to working concentrations and ought to be determined experimentally for each type of preparation.
*Other types of NOS inhibitors are available. The ones listed here are recommended for pharmacological determination of a putative role of NOS in neurotransmission or as an endothelium-derived relaxing factor – see text and Chapters 2, 10 and 12.

vessels this is usually accomplished by mechanical means, by gently rubbing the endothelial surface with a saline-soaked cotton wool swab, or by drawing cotton or silk thread through the lumen, or even by using wooden sticks to rub off the endothelium. If isolated vessels are being perfused, or if a whole vascular bed is being studied, either *in situ* or *in vitro*, the endothelium can be removed or sufficiently damaged by perfusing the preparation with a detergent such as sodium deoxycholate, CHAPS, Triton X-100 or saponin. It may be possible to de-endothelialize a vascular bed by passing air or CO_2 through it for a short period: both these means have the effect of drying, and killing, the endothelial cells. Ethanol (96%) can be used to the same effect. Details of these methods are given in Chapter 15. It is necessary to check that following the chosen method the endothelium has indeed been removed. This can be done by histochemical staining for endothelial cells and microscopic examination, but more commonly the lack of response to ACh (assuming that prior to endothelial removal there was a response to ACh) is taken as an index of endothelial absence. In addition to ACh, bradykinin, the calcium ionophore A23187 and shear stress are often used to stimulate the release of EDRF from endothelial cells.

NO is not stored within neurones or endothelial cells and, because it diffuses freely, it does not have a Ca^{2+}-dependent release mechanism. Nevertheless, the activity of both neuronal type I NOS and endothelial type III NOS is Ca^{2+}-dependent (see Chapters 2 and 3). In isolated organ experiments the omission of Ca^{2+} will have adverse effects on the tissue; for example, it will inhibit contractile activity of smooth muscle and prevent secretion from exocrine and endocrine tissue. It may be possible to carry out experiments under Ca^{2+}-free conditions, especially direct NO assays or superfusion cascade assays (see Chapters 13 and 15), and show that release of NO is abolished. For example, Chakder and Rattan (1993a) have used a chemiluminescence method to measure release of NO from isolated preparations of opossum internal anal sphincter in which the muscle bath perfusate is taken very rapidly into a de-gassing chamber, from which the NO driven out of solution is carried into the analyser. It is worth pointing out that these authors routinely included superoxide dismutase in their physiological saline, to protect released NO from conversion to nitrite by superoxide radicals. Assays for NO using specialized microelectrodes or microsensors have been developed (Shibuki, 1990; Malinski & Taha, 1992; Malinski *et al.*, 1993a,b; Ichimori *et al.*, 1994; Kanai *et al.*, 1995) and are proving to be useful for determining local NO release from neurones or endothelial cells.

NOS itself is the object of many investigations aimed at determining the release of NO. The natural substrate is L-arginine: when applied to an organ bath responses to neural or endothelial stimulation are often potentiated simply because of the greater substrate availability. The binding of arginine is stereo-specific and its enantiomer D-arginine is without this effect. More useful are competitive antagonists of the L-arginine-binding site, especially L-NMMA, L-NNA and L-NAME, and they can severely inhibit, or even abolish, neurogenic responses and endothelium-dependent responses. These antagonists are also stereo specific: their D-enantiomers have no effect. Because they are competitive they can be displaced by L-arginine, and application of L-arginine, but not D-arginine, should reverse the inhibition. Other types of NOS antagonist are available, including diphenylene iodonium (DPI), 7-nitroindazole and aminoguanidine. DPI blocks the NADPH-binding site on NOS, thereby inhibiting its flavoprotein function, while the latter two are nitrogen-containing compounds that interact with the arginine-binding site (see Fukuto & Chaudhuri, 1995). Inhibition of constitutive NOS (types I and III) can also be accomplished by inhibiting calmodulin activity with agents such as calmidazolium, W-7 or fendiline. The organoselenium compound ebselen is a thiol-modifying agent that inhibits NOS activity, especially type III NOS. Finally, agents that bind haem groups can also inhibit NOS activity, and include such compounds as NO itself, CO, miconozole and cyanide (Fukuto & Chaudhuri, 1995). Tannin appears to be a selective inhibitor of type III NOS, but its mode of action has not been fully characterized (Chiesi & Schwaller, 1995). Of all the types of NOS inhibitor the most useful for determining whether or not NOS is involved in an observed pharmacological or physiological response are the arginine analogues.

Figs. 11.1–11.3 illustrate a series of experiments carried out using isolated preparations of rat gastric fundus (Currò, Volpe & Preziosi, 1996). The production of L-citrulline was taken as an index of NOS activity, and citrulline production evoked by EFS is shown to be blocked by TTX, Ca^{2+}-free conditions and L-NAME. Also, L-arginine but not D-arginine reverses the inhibition due to L-NAME.

11.3 Pharmacology of NO action

For a free radical, NO is remarkably stable (Archer, 1993) but in biological systems it tends to have a short half-life due to the presence of molecular oxygen, superoxide anions and haem proteins. Thus, when it is released from a single nerve terminal it can only diffuse to nearby neighbouring cells before losing activity. In aqueous solution the dismutation of NO follows this scheme:

$$4NO + O_2 + 2H_2O \rightarrow 4NO_2^- + 4H^+ \qquad 11.1$$

and

$$-d[NO]/dt = 4k[NO]^2[O_2] \qquad 11.2$$

where $k = 2.4 \times 10^6\ M^{-2} \cdot s^{-1}$ at 37 °C (Lewis & Deen, 1994). Thus the reaction is second order with respect to NO, first order with respect to O_2, and the rate

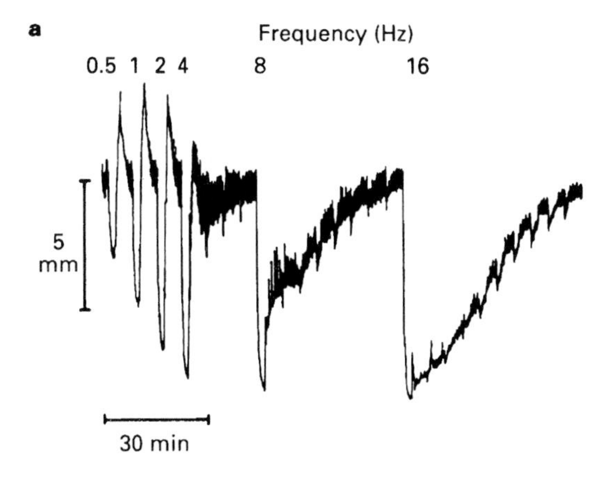

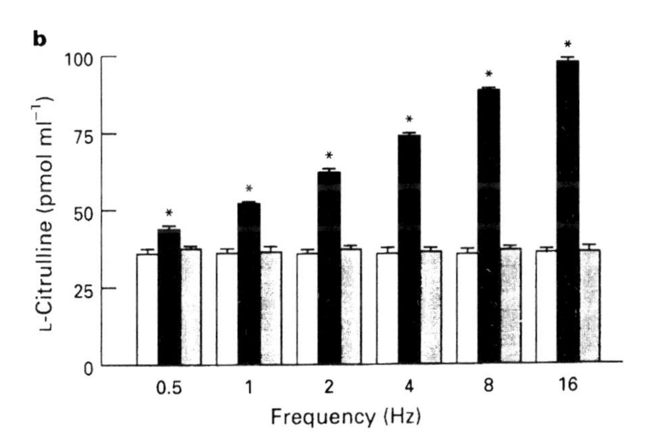

Fig. 11.1 Isolated rat gastric fundus. Prepared strips of gastric fundus were mounted in 5-ml organ baths, and the mechanical activity was recorded under auxotonic conditions in Krebs' solution containing atropine (1 μM) and guanethidine (5 μM). A standard level of tone was induced by 5-hydroxytryptamine (5-HT, 3 μM): L-citrulline was measured as an index of NOS activity and NO production. (a) Frequency–relaxation relationship: electrical field stimulation (EFS) was applied for 2 min at each of the frequencies indicated. (b) Frequency-L-citrulline production relationship: the concentration of L-citrulline in the incubation medium was measured before (open columns), during (black columns) and after (grey columns) application of EFS. $^{*}P < 0.001$, relative to prestimulation levels. (Reproduced with permission from Currò, Volpe & Preziosi, 1996.)

constant is third order. The significance of this is that at low concentrations of NO, such as those produced by neurones or endothelial cells, the half-life of NO could be over 100 s, allowing time for reaction with haem proteins (e.g. guanylate cyclase) but at high concentrations of NO, as produced by macrophages, there will be a rapid reaction with O_2 (Ford, Wink & Stanbury, 1993; Vanderkooi, Wright &

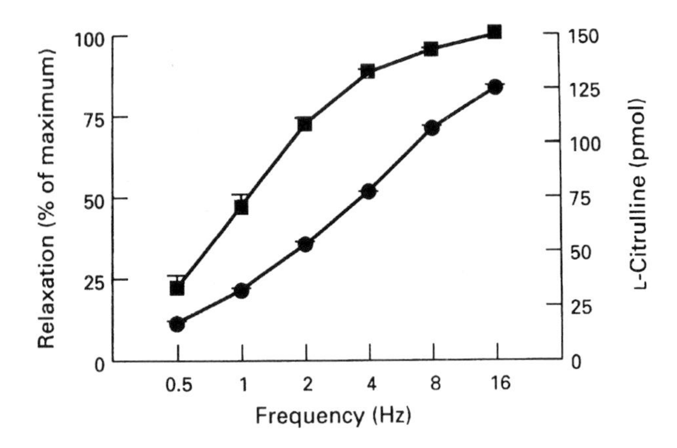

Fig. 11.2 Isolated rat gastric fundus (as in Fig. 11.1). Mean frequency–response relationships for relaxation (■) and L-citrulline production (●). Points shown mean ± SEM, $n = 6$–8. Relaxant responses were expressed relative to responses at 16 Hz. (Reproduced with permission from Currò, Volpe & Preziosi, 1996.)

Erecinska, 1994). In terms of diffusion path length, with a diffusion coefficient of 10^{-7} cm^2·s^{-1} the distance would be of the order of 100 μm for low concentrations of NO, and of the order of 10 μm at high concentrations (Vanderkooi *et al.*, 1994). Using a model of endothelial cell NO production, it has been estimated that if the half-life of NO is anything greater than 25 ms then the NO released from a given cell will diffuse distances of the order of 100 μm (Lancaster, 1994), and that if the NO that is released in the lumen is rapidly scavenged by haemoglobin, then the diffusion of NO into the blood vessel wall becomes reduced. However, none of these models takes into account interactions of NO with superoxide or haem proteins, including myoglobin, or vitamins that might interact with NO and effectively remove it from the system. Using porphyrinic microsensors in the rabbit aorta it has been shown that there is a significant loss of NO released from the endothelium, due to chemical reactions (Malinski *et al.*, 1993a), and in superfusion experiments the activity of NO (0.22 nmol) when passed over tissue declines with a half-life of approximately 4 s, whereas in physiological saline without exposure to tissue the half-life is approximately 30 s (Palmer, Ferrige & Moncada, 1987). Thus, the degradation of NO will restrict its active domain to the locality of its site of release. Although in the CNS NO could realistically diffuse into a volume equivalent to a sphere with a radius of 10 μm (Garthwaite, 1993), and this volume would encompass hundreds or thousands of synapses in the cerebral cortex, it is still small enough for NO to be regarded as a transmitter rather than a hormone. In autonomically innervated tissues, a distance of 10 μm from a release site is likely to be no more than a couple of parenchymal cells away at most.

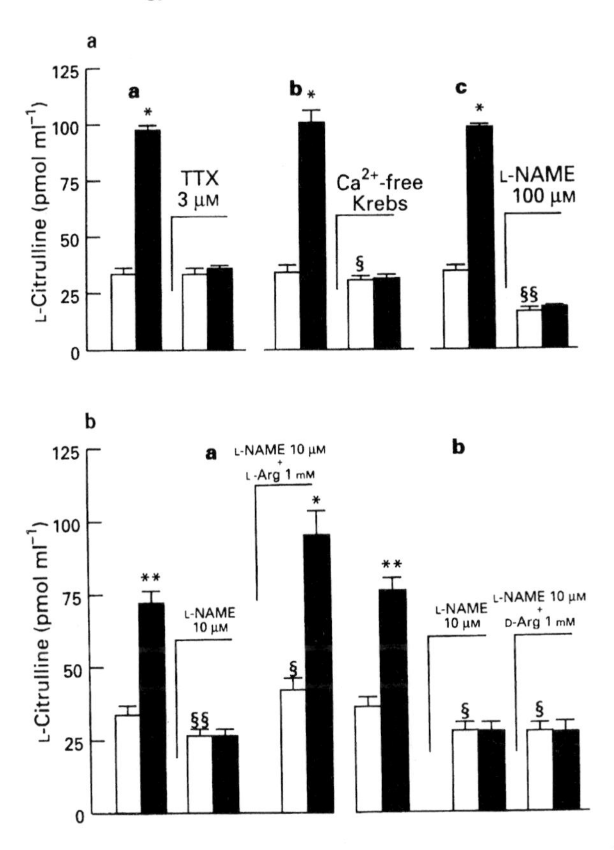

Fig. 11.3 Isolated rat gastric fundus (as in Fig. 11.1). a. Effects of tetrodotoxin (TTX, 3 μM), Ca^{2+}-free conditions and L-NAME (100 μM) on production of L-citrulline evoked by EFS at 16 Hz. Significant differences between corresponding prestimulation and stimulation values: $^{*}P < 0.001$; significant differences with respect to control prestimulation values: §$P < 0.05$, §§$P < 0.001$. b. Effects of L- and D-arginine, on inhibition of production of L-citrulline by L-NAME (10 μM). Open columns show prestimulation levels of L-citrulline in the incubation medium, filled columns show L-citrulline levels during EFS. Significant differences between corresponding prestimulation and stimulation values: $^{*}P < 0.01$, $^{**}P < 0.001$; significant differences with respect to control prestimulation values: §$P < 0.01$, §§$P < 0.001$. All panels, n=4–6. (Reproduced with permission from Currò, Volpe & Preziosi, 1996.)

Because NO is so freely diffusible there is usually little problem with restricted access to its target, but because of its relative insolubility and of the potential danger in handling NO, particularly its reaction with atmospheric oxygen yielding the extremely noxious nitrogen dioxide, it is not a popular substance to try and apply to a tissue in an organ bath. Two methods have been adopted for the application of authentic NO. The safest is to dissolve sodium nitrite in concen-

Table 11.2 *NO donors, classes and examples*

Organic nitrates
Glyceryltrinitrate (nitroglycerine, GTN)
Organic nitrites
Isoamylnitrite (amylnitrite, isopentylnitrite), isobutylnitrite
Ferrous nitro complexes
Sodium nitroprusside (SNP)
Sydnonimines*
Molsidomine
3-Morpholinylsydnoneimine (SIN-1)
***S*-Nitrosothiols**
S-Nitroso-*N*-acetyl-D,L-penicillamine (SNAP)
S-Nitroso-L-cysteine (*S*-nitrosocysteine)
S-Nitrosoglutathione
Nucleophile adducts
NONOates, dithyleamine/NO, spermine/NO

*Sydnonimines need to be degraded in the liver, and their metabolites undergo chemical hydrolysis to release NO. SIN-1 (linsidomine) is the metabolite of molsidomine.

trated HCl (pH < 2.0), which generates NO in solution. The better method, but potentially more dangerous, is to make a saturated solution of NO by bubbling pure NO gas through de-ionized, de-oxygenated water. Protocols for both these methods are given in Chapters 13 and 15.

NO donors may also be employed: the pros and cons of using donors and solutions of NO have been reviewed by Feelisch (1991). In general terms NO donors are compounds whose pharmacological activity is determined by the release NO from their intramolecular structure. There are six main classes of NO donor compound, based on their chemical structure and mode of action (Bauer, Booth & Fung, 1995) (Table 11.2). The most recent class is that of the nucleophile/NO adducts (Maragos *et al.*, 1991), which consists of complexes of NO with nucleophilic compounds; these compounds may also be referred to as NONOates. In solution they decompose non-enzymatically, and in so doing they release NO. In rabbit aorta the NONOates cause relaxation, with a close correlation between their potency and their rate of NO production (Maragos *et al.*, 1991; Morley & Keefer, 1993). Nucleophile adducts can be synthesized that have relatively rapid or slow rates of NO generation; for example, diethylamine/NO has a half-life of approximately 2 min while spermine/NO has one of nearly 40 min. The former compound produces responses that peak quickly and are not sustained, while the latter compound produces responses that take several minutes to reach a maximum level, and which are subsequently maintained (Morley *et al.*, 1993), thus opening up a therapeutic potential of these compounds (Hanson *et al.*, 1995).

Organic nitrates have to react with specific thiol compounds, such as cysteine or *N*-acetylcysteine, in order to generate NO (see Feelisch 1991,1993). In the presence of cysteine the rate of formation of NO from a series of organic nitrates is closely correlated with the level of activation of soluble guanylate cyclase. Organic nitrites also need to react with thiol compounds to generate NO, but the requirement is not as specific as that for the nitrates. *S*-Nitrosothiols are formed, and these unstable compounds degenerate to release NO (Feelisch, 1991,1993; Bauer *et al.*, 1995). *S*-Nitrosothiols *per se* can be used as NO donors. The ability of these compounds to release NO in solution does not correlate well with their ability to cause vasorelaxation (Kowaluk & Fung, 1990). In solution, NO released from SNAP is only detectable from concentrations of SNAP that are well in excess of those that cause maximal or near-maximal relaxation of isolated aorta (Kowaluk & Fung, 1990; Salas *et al.*, 1994). Membrane fractions prepared from blood vessels catalyse the generation of NO from SNAP, so the site of production of NO, which goes on to stimulate guanylate cyclase, is likely to be intracellular rather than extracellular (Kowaluk & Fung, 1990). Similarly, the NO generated from the ferrous nitro complex SNP is insufficient to account for its pharmacological activity, and it undergoes NADPH-catalysed conversion, involving membrane-bound enzymes (Feelisch, 1993; Bauer *et al.*, 1995) The localization of degeneration of SNP is so widespread that it can appear that the conversion to NO is spontaneous (Bauer *et al.*, 1995). SIN-1 is a metabolite of the sydnonimine, molsidomine. *In vivo* molsidomine is enzymatically degraded in the liver to yield linsidomine, the ring structure of which is unprotected and which, in the presence of molecular oxygen, undergoes non-enzymatic cleavage to yield NO (Feelisch, 1991,1993; Bauer *et al.*, 1995).

In pharmacological experiments any of the NO donors can be used if the aim is to mimic the effect of EFS or EDRF. However, each compound is associated with practical difficulties that may affect the choice of which to use (Feelisch, 1993). Nitrovasodilators all generate high levels of metabolites, especially NO_2^-. They also are highly lipophilic, with poor aqueous solubility, and can migrate into plastics; this is particularly so for nitroglycerine. Care should be taken, especially in perfusion systems using plastic tubing, because there may be a loss in the concentration of the compound as well as the risk of subsequent contamination (Feelisch, 1993). Aqueous solutions of SNP are sensitive to light and temperature, and decompose to release cyanide. *S*-Nitrosothiols need to be kept cold and protected from light, both as stock solutions and when applied to an organ bath or in perfusates.

Sydnonimine solids should be stored cool, dry, light-protected, and preferably in an atmosphere of N_2 (Feelisch, 1993).

11.4 Pharmacology of NO inactivation

As already indicated, there does not appear to be a specific mechanism for the inactivation of NO, nor does it need one. In aqueous medium NO reacts readily with O_2 to form NO_2^-, with superoxide radicals to form peroxynitrite ($ONOO_3^-$), with hydrogen peroxide (H_2O_2) to form nitrite (NO_2^-) or nitrate (NO_3^-), and with proteinaceous haem groups to form nitrosylated proteins.

The enzyme SOD removes superoxide anions (O_2^-) so that they are not available to react with NO (see Chapter 2). Thus, incubation of tissues with SOD increases the available NO (Gryglewski, Palmer & Moncada, 1986; Rubanyi & Vanhoutte, 1986). The superoxide anion appears to be a major determinant of NO activity in several tissues, including the canine ileocolonic junction (Bult *et al.*, 1990), rat gastric fundus (Boeckxstaens *et al.*, 1991a) and rat colon (Hata *et al.*, 1990). In some tissues SOD has little or no effect against responses to NANC nerve stimulation, e.g. mouse anococcygeus muscle or bovine retractor penis (Hobbs, Tucker & Gibson, 1991; Martin, McAllister & Paisley, 1994), which argues against an inactivating role of superoxide anions in this organ. However, the lack of effect of applied SOD could be due to the tissue possessing very high endogenous levels, in which case the effects of superoxide anions need to be unmasked by applying an inhibitor of SOD, such as diethyldithiocarbamate (Martin *et al.*, 1994; Paisley & Martin, 1996).

11.5 Agents that potentiate or inhibit NO activity

Because the target of neuronally released NO is guanylate cyclase, this enzyme and phosphodiesterases that degrade cGMP can be the sites of attention:

$$\text{GTP} \xrightarrow{\textit{guanylate cyclase}} \text{cGMP} \xrightarrow{\textit{phosphodiesterase}} \text{GMP} \qquad 11.3$$

Methylene blue inhibits guanylate cyclase (Murad *et al.*, 1978), and should produce a parallel inhibition of the effects of applied NO and nerve stimulation, as it does, for example, in the rabbit urethra and rabbit and human corpus cavernosum (Ignarro *et al.*, 1990; Andersson *et al.*, 1991; Dokita *et al.*, 1991; Rajfer *et al.*, 1992). Similarly in vascular tissues methylene blue should block both endothelium-dependent relaxations and responses to applied NO or NO donors (Kirkeby *et al.*, 1993). An analogue of cGMP, 8-bromo-cGMP, is lipid soluble and can therefore pass through cell membranes. It can be used to show what the effect on a tissue is of elevating its intracellular cGMP levels.

Soluble guanylate cyclase can be inhibited by 1*H*-(1,2,4)oxadiazolo(4,3-a)quinoxalin-1-one (ODQ) (Garthwaite *et al.*, 1995; Moro *et al.*, 1996). In the rabbit anococcygeus muscle, in which the NANC inhibitory neuromuscular transmission is mediated by NO (Graham & Sneddon, 1993; Kasakov, Cellek & Moncada, 1995), ODQ causes a concentration-dependent inhibition of responses to EFS and SNP, and prevents the elevation of intracellular cGMP (Cellek, Kasakov & Moncada, 1996). ODQ seems to be specific for guanylate cyclase and has no effect on cAMP levels (Cellek *et al.*, 1996), and has the potential to become a widely used tool.

6-Anilino-5,8-quinolenedione (LY 83583) was originally developed as an inhibitor of guanylate cyclase, and it does indeed block this enzyme and prevent cGMP production in smooth muscle (Jin *et al.*, 1993). However, it also generates superoxide anions, and although it can attenuate the response to applied NO or nerve stimulation this effect can be prevented by superoxide dismutase (Barbier & Lefebvre, 1992; Martin *et al.*, 1994). In the guinea-pig trachea LY 83583 has been shown to inhibit nitrergic transmission (Kannan & Johnson, 1995), but this could be due to generation of superoxide anions, or even to a direct effect of inhibition of NOS activity (Luo, Das & Vincent, 1995). Thus, LY 83583 would be a difficult drug to work with because of its multiple actions.

The compound 2-*o*-propoxyphenyl-8-azopurine-6-one, otherwise known as zaprinast or M&B22948, is a selective inhibitor of the phosphodiesterase that cleaves cGMP (phosphodiesterase V). In addition to raising the basal levels of cGMP (Bowman & Drummond, 1984) it markedly enhances accumulation of cGMP and relaxation evoked by neural stimulation in the bovine retractor penis muscle (Bowman & Drummond, 1984), rabbit urethra (Dokita *et al.*, 1991) and guinea-pig pulmonary arteries (Liu *et al.*, 1992a,b). It also causes parallel potentiation of responses to nerve stimulation and NO in human and rabbit isolated corpus cavernosum (Bush *et al.*, 1992a; Rajfer *et al.*, 1992) and cat gastric fundus (Barbier & Lefebvre, 1995).

Either reduced haemoglobin (deoxygenated haemoglobin) or oxyhaemoglobin will bind NO and effectively causes its inactivation (Stark, Bauer & Szurszewski, 1991). In both these compounds the iron is in its ferrous state (Fe^{2+}), i.e. haemoglobin–Fe(II) or haemoglobin–Fe(II)–O_2. Methaemoglobin, in which the iron is oxidized to its ferric (Fe^{3+}) state, does not interact with NO. Atmospheric O_2 will cause the oxidation of haemoglobin to form methaemoglobin. Oxyhaemoglobin can be prepared from erythrocyte haemolysates or by reducing methaemoglobin, which is the predominant component of commercially available haemoglobins (Martin *et al.*, 1985). Incubation of a tissue with haemoglobin should attenuate or abolish responses to applied NO in the same way as it attenuates or abolishes the neurogenic response if NO is the neurotransmitter or the endothelium-dependent relaxation if NO is the EDRF. Several studies have shown this to be the case using haemoglobin obtained from erythrocyte haemolysates, or recombinant human haemoglobin (Ignarro *et al.*, 1987a,b,1988a; Bult *et al.*, 1990; Ignarro *et al.*, 1990; Stark *et al.*, 1991; Keef *et al.*, 1993; Chakder, Rosenthal & Rattan, 1995; Conklin *et al.*, 1995; Murray *et al.*, 1995; Rattan, Rosenthal & Chakder, 1995). Methaemoglobin does not affect nitrergic transmission, for example as in the bovine retractor penis or rat anococcygeus muscle (Bowman & Gillespie, 1982).

Hydroxocobalamin is also known as vitamin B_{12a}, and because it is readily nitrosylated it should be able to trap free NO (Rajanayagam, Li & Rand, 1993; Rand & Li, 1993; Jenkinson, Reid & Rand, 1995; Rochelle *et al.*, 1995). Using a superfusion cascade system to examine nitrergic transmission in the canine ileocolonic junction, it has been shown that hydroxocobalamin blocks responses to applied NO in parallel with those to neural stimulation (Boeckxstaens *et al.*, 1995; De Man *et al.*, 1995). In some tissues, notably the rat gastric fundus and anococcygeus muscle, hydroxocobalamin has little effect against NANC relaxations, while having severe effects against applied NO (Li & Rand, 1993; Rajanayagam *et al.*, 1993; Jenkinson *et al.*, 1995). Although these results are negative, they do not preclude a transmitter role for NO (see Chapter 5). Thus it would be unwise to rely on hydroxocobalamin alone in the evaluation of putative nitrergic transmission. In the isolated rat aorta, ACh-stimulated release of EDRF is blocked simultaneously with inhibition of relaxant responses to applied NO (Rajanayagam *et al.*, 1993).

The imidazolineoxyl *N*-oxide derivative 2-(4-carboxyphenyl)-4,4,5,5-tetraethylimidazoline-1-oxyl-3-oxide (carboxy-PTIO) is a stable free radical that reacts with NO. It inhibits endothelium-dependent relaxations in the rabbit aorta by removing NO (Akaike *et al.*, 1993), and it causes concentration-dependent inhibition of responses to EFS in the bovine retractor penis muscle, also by scavenging NO (Paisley & Martin, 1996). The mechanism of action of carboxy-PTIO does not appear to involve superoxide anions, or anything other than a direct interaction with NO (Akaike *et al.*, 1993; Paisley & Martin, 1996), and therefore is another promising tool.

Hydroquinone is a scavenger of free radicals (Mittal & Murad, 1977) that generates superoxide anions (Chakder & Rattan, 1992; Paisley & Martin, 1996), and may be a useful tool for inhibiting responses to applied NO and EFS. For example, in the opossum internal anal sphincter and canine ileocolonic junction inhibitory responses due to applied NO and EFS are both suppressed by hydroquinone, the effects of which are similarly reversed by SOD (Chakder & Rattan, 1992; De Man *et al.*, 1995). Like hydroxocobalamin, hydroquinone should not be totally relied upon, because in several tissues, including anococcygeus, trachea and stomach, it may attenuate responses to applied NO while having little or no effect on responses to putatively nitrergic nerve stimulation (Gillespie & Sheng, 1990; Hobbs *et al.*, 1991; Gibson *et al.*, 1992, 1995). For example, in the bovine retractor penis muscle, hydroquinone has little effect against responses to EFS, but when endogenous SOD activity is blocked by diethyldithiocarbamate it becomes a most efficacious inhibitor (Paisley & Martin, 1996) (see Fig. 5.2 in Chapter 5). In the absence of SOD inhibition, hydroquinone has been used more in endothelial systems where it is a useful tool for inhibiting responses to EDRF and applied NO (Alonso *et al.*, 1992b,1993).

Pyrogallol is a potent generator of superoxide anions (Marklund & Marklund, 1974). Superoxide anion radicals react with the NO radical thereby inactivating NO. In blood vessels, such as rabbit ear artery or cerebral arteries, or bovine pulmonary vessels, pyrogallol markedly inhibits the endothelium-dependent relaxation evoked by ACh (Ignarro *et al.*, 1987a,b,1988a; Girard *et al.*, 1995). Pyrogallol should inhibit the activity of NO when applied exogenously and when released from nerves, as indeed it does in the rat anococcygeus muscle (Gillespie & Sheng, 1990) and in the canine ileocolonic junction in a superfusion cascade bioassay (Boeckxstaens *et al.*, 1994,1995). However, again one must be cautious, because in some tissues pyrogallol may inhibit the activity of exogenous NO but not responses to nitrergic nerve stimulation (Knudsen, Svane & Tottrup, 1992). It has been suggested that apparently incongruous effects of superoxide anion generators, such as hydroquinone, pyrogallol and LY 83583, are due to the endogenous tissue levels of SOD being so high that these agents have little effect against nitrergic transmission, but when endogenous SOD is blocked their true actions become unmasked (Martin *et al.*, 1994; Paisley & Martin, 1996).

As already discussed, SOD removes superoxide radicals, which in many tissues are a major determinant of NO inactivation. The two main isoforms of SOD are extracellular superoxide dismutase type C (EC-SOD C), which is secreted by cells and binds to heparan sulphate proteoglycans in the glycocalyx of many cell types (including endothelial cells) and in the connective tissue matrix; the other is the cytosolic ZnCu form of SOD, which has no affinity for heparan sulphate and does not bind to cell surfaces. Both types can be used to mop up superoxide anions *in vitro* (Abrahamsson *et al.*, 1992), but the ZnCu isoform is the more commonly used. SOD produces a parallel potentiation of endothelium-dependent vasorelaxant responses evoked by ACh and applied NO (Gryglewski *et al.*, 1986; Rubanyi & Vanhoutte, 1986; Ignarro *et al.*, 1987a,b, 1988b; Girard *et al.*, 1995; Fukuto & Chaudhuri, 1995). In many cases SOD causes a parallel potentiation of responses to applied NO and neurogenic responses, including those in the canine ileocolonic junction, porcine cerebral artery, rat gastric fundus and rat colon (Bult *et al.*, 1990; Hata *et al.*, 1990; Boeckxstaens *et al.*, 1991a; Chen & Lee, 1993).

Diethyldithiocarbamate is an inhibitor of ZnCu-SOD and, in bovine coronary arteries, it has been shown to abolish the ACh-evoked endothelium-dependent relaxation in parallel with inhibition of relaxant responses to applied NO and NO donors (Mugge *et al.*, 1991; Omar *et al.*, 1991). Inhibition of SOD by diethyldithiocarbamate also leads to the increased production of superoxide anions by endothelial and smooth muscle cells (Omar *et al.*, 1991), but has no effect on the release of nitrogen oxides, measured using chemiluminescence (Mugge *et al.*, 1991).

Recently a synthetic flavone, 6,7-dimethoxy-8-methyl-3′,4′,5-trihydroxyflavone, has been developed that protects NO from superoxide anions. In rabbit ear artery it markedly potentiates the relaxant responses to ACh (Girard *et al.*, 1995). It seems to be such an efficient scavenger that in the presence of excess superoxide anions, generated by pyrogallol, this flavone will still allow endothelium-dependent responses to be maintained (Girard *et al.*, 1995).

11.6 Summary

The experimental approaches to determine whether NO is released either from neurones or endothelial cells are broadly similar. In both cases NO has to be shown to be synthesized and released from a specific source, then the released NO has to be shown to act on its target in a predictable way. NOS itself is the object of many investigations aimed at determining the release of NO. The binding of L-arginine to NOS is stereo-specific and its enantiomer D-arginine is not a substrate. L-NMMA, LNNA and L-NAME are competitive antagonists of the L-arginine-binding site, and they can inhibit or abolish neurogenic responses and endothelium-dependent responses that are due to synthesis and release of NO: application of L-arginine, but not D-arginine, should reverse the antagonism. Other types of NOS antagonist are available, including DPI, 7-nitroindazole and aminoguanidine. DPI blocks the NADPH-binding site on NOS, thereby inhibiting its flavoprotein function, while the latter two are nitrogen-containing compounds that interact with the arginine-binding site. Inhibition of constitutive NOS (types I and III) can also be accomplished by inhibiting calmodulin activity with agents such as calmidazolium, W-7 or fendiline. The organoselenium compound ebselen is a thiol-modifying agent that inhibits NOS activity, especially type III NOS. Finally, agents that bind haem groups can also inhibit NOS activity, and include such compounds as NO itself, CO, miconozole and cyanide. Of all the types of NOS inhibitor, the arginine analogues are the most useful for determining whether or not NOS is involved in an observed pharmacological or physiological response. NO can be applied to a tissue in an organ bath as a gas dissolved in water or as an acidified solution of sodium nitrite. Alternatively NO donors can be used that release NO from their molecular structure. There are several NO donor compounds available, with different characteristics and time courses of NO release. Other pharmacologically useful agents include those that interfere with the mechanisms of inactivation of NO, and include compounds that entrap NO (oxyhaemoglobin, hydroxocobalamin and carboxy-PTIO), inhibitors of NO inactivation (SOD), promoters of NO inactivation (hydroquinone, pyrogallol and diethyldithiocarbamate). Soluble guanylate cyclase is often a useful focus in determining NO activity, and inhibitors of this enzyme (LY 83583, ODQ and methylene blue) are readily available. cGMP analogues and inhibitors of cGMP degradation (SK&F 9631 and zaprinast or M&B22948) may also be meaningfully employed.

11.7 Selected references

Archer, S. (1993). Measurement of nitric oxide in biological models. *FASEB Journal*, **7**, 349–60.

Bauer, J. A., Booth, B. P. & Fung, H. L. (1995). Nitric oxide donors: biochemical pharmacology and therapeutics. *Advances in Pharmacology*, **34**, 361–81.

Feelisch, M. (1991). The biochemical pathways of nitric oxide formation from nitrovasodilators: appropriate choice of exogenous NO donors and aspects of preparation and handling of aqueous NO solutions. *Journal of Cardiovascular Pharmacology*, **17** (Suppl 3), S25–33.

Fukuto, J. M. & Chaudhuri, G. (1995). Inhibition of constitutive and inducible nitric oxide synthase: potential selective inhibition. *Annual Reviews of Pharmacology and Toxicology*, **35**, 165–94.

Lancaster, J. R. (1994). Simulation of the diffusion and reaction of endogenously produced nitric oxide. *Proceedings of the National Academy of Sciences USA*, **91**, 8137–41.

12

Microscopy

12.1 Introduction

The localization of the enzyme responsible for the synthesis of NO has played an important role in furthering our knowledge of the physiological activity of this unusual signal molecule. At both the light microscope (LM) level and the electron microscope (EM) level two techniques have been widely employed. The first is histochemical localization of NADPH-diaphorase activity, and the second is immunocytochemical localization of NOS. In this chapter the relationship between NADPH-diaphorase and NOS is discussed, as is the rationale behind, and problems associated with, the use of staining for diaphorase activity in order to localize NOS, and presumptive sites of NO production. Microscopic visualization of guanylate cyclase and its product, cGMP, in relation to being the main substrate of NO activity, and in relation to sites of NO release, is also discussed.

12.2 Tissue fixation and the relationship between NOS and NADPH-diaphorase

It is crucial to understand that the term 'diaphorase' has no physiological meaning, and is merely a generic term for any enzyme that is capable of catalysing the oxidation of either NADH or NADPH by any artificial electron acceptor, such as a dye, ferricyanide, quinone or tetrazolium salt. Furthermore, there are three main types of diaphorase (see Pearse, 1972): those that oxidize NADH (NADH-diaphorases); those that oxidize NADPH (NADPH-diaphorases); and one that oxidizes either NADH or NADPH, called DT-diaphorase. This name derives from the older names for NAD and NADP, i.e. diphosphopyridine nucleotide (DPN) and triphosphopyridine nucleotide (TPN), respectively. Diaphorases may also be described according to other attributes: some diaphorases are mitochondrial, some extramitochondrial; some are cytosolic (soluble), and some particulate (insoluble).

The first evidence that NOS could also be a NADPH-diaphorase was provided by Hope *et al.* (1991), who showed that in the rat brain the extracted diaphorase has the same molecular weight as type I NOS. They also showed that the two enzymes could be co-purified to homogeneity, and that both immunoprecipitated with the same antibody, directed against the diaphorase. At almost the same time Dawson *et al.* (1991), in addition to showing extensive co-localization of NADPH-diaphorase activity and NOS-immunoreactivity in regions of the brain, adrenal medullary and myenteric neurones, also showed that a human kidney cell line transfected with cDNA for NOS expressed NOS-immunoreactivity and NADPH-diaphorase activity simultaneously. Since then, similar data have been obtained from insect neurones too (Muller, 1994).

When neuronal NADPH-diaphorase was initially characterized (Hope & Vincent, 1989), it was established that NADH could not act as a substrate, thus excluding the enzyme from being NADH-diaphorase or DT-diaphorase. Furthermore, the enzyme was not inhibited by dicoumarol, and tetrazolium salts could act as electron acceptors, again excluding DT-diaphorase (Pearse, 1972). The significance of these data is that neuronal NOS activity is not related to either NADH-diaphorase or DT-diaphorase.

There is not always a close relationship between NOS and NADPH-diaphorase. In contrast to the situation in endothelial cells, in which the NOS and NADPH-diaphorase activities are both particulate (Tracey *et al.*, 1993), it has been determined from biochemical studies that neuronal NADPH-diaphorase activity (in a neuroblastoma cell line) is restricted mostly to the particulate fraction, while the NOS activity is mostly found in the soluble fraction (Tracey *et al.*, 1993). Also, quiescent macrophages possess substantial diaphorase activity, but little or no NOS activity. When stimulated, the NADPH-diaphorase and NOS activities both increase, but without any correlation between the two (Tracey *et al.*, 1993). These results indicate that neuronal type I NOS and neuronal NADPH-diaphorase are not necessarily a shared entity, and nor are macrophage type II NOS and macrophage NADPH-diaphorase activity. Thus, NADPH-diaphorase does not categorically correspond to NOS. This having been said, it has been shown that purified type I, type II and type III NOS all possess NADPH-diaphorase activity (Förstermann *et al.*, 1991b; Hope *et al.*, 1991; Tracey *et al.*, 1993).

When the molecular structures of the three NOS isoforms are considered – and it is appreciated that they are all types of cytochrome P_{450} reductase with a highly conserved sequence homology especially towards the C-terminal, which represents the site of NADPH activity (Chapter 2) – it is only to be expected that they are all NADPH-diaphorases. Nevertheless, the caveat remains that not all NADPH-diaphorases are a type of NOS, and the two terms are not synonymous. A common mistake is to assume that co-localization of NOS-immunoreactivity with NADPH-diaphorase activity can be taken to mean that the two enzymes are one and the same: this is not necessarily so.

The NADPH-diaphorase that is also NOS (whatever isoform) is resistant, at least partially, to fixation by formaldehyde, whether used *per se* or as paraformaldehyde. Only when the tissue has been fixed can NADPH-diaphorase staining be used to demonstrate NOS (Nakos & Gossrau, 1994; Worl *et al.*, 1994b). The use of fixatives to unmask NADPH-diaphorase activity that reflects NOS may be species dependent since this seems to be the case for mammals and insects (Muller, 1994; Muller & Bicker, 1994) but not for anurans (Gabriel, 1991) where the converse situation holds.

Results from fractionation studies have shown that in fixed tissues NADPH-diaphorase is cytosolic and is co-localized with NOS in neurones (Matsumoto *et al.*, 1993). However, in unfixed tissues the larger proportion of diaphorase activity is particulate, and does not correspond to NOS. Paraformaldehyde fixation abolishes particulate non-NOS NADPH-diaphorase activity, and maximally reduces soluble, NOS-related activity by approximately 50% (Matsumoto *et al.*, 1993). Interestingly, it has also been shown that while paraformaldehyde fixation abolishes NADPH-diaphorase staining in hippocampal pyramidal cells, which contain type III NOS rather than type I NOS, glutaraldehyde fixation does not (Dinerman *et al.*, 1994). However, the way in which fixation conditions influence diaphorase staining has been emphasized by Spessert and Layes (1994) who found that increasing the power of the fixative solution of paraformaldehyde, by adding glutaraldehyde or periodate, decreased the intensity of the diaphorase staining, but did not affect the pattern of diaphorase staining.

Many histological studies have shown that there is a 1:1 correspondence, or very nearly so, between the localization of NOS and of NADPH-diaphorase (Bredt *et al.*, 1991a; Dawson *et al.*, 1991; Afework *et al.*, 1992; Hassall *et al.*, 1992; Saffrey *et al.*, 1992; Ward *et al.*, 1992d; Young *et al.*, 1992; Luth *et al.*, 1994), and it is noteworthy that all these authors used formaldehyde fixation of the tissue prior to staining for NADPH-diaphorase activity. On the other hand, several studies have shown that there is not necessarily a 1:1 correspondence between NADPH-diaphorase and type I NOS localization in neural tissues (Belai *et al.*, 1992; Schmidt *et al.*, 1992a; Loesch, Belai & Burnstock, 1993; Tay, Moules & Burnstock, 1994). The lack of correspondence is most apparent in the staining of sacral spinal cord and sacral autonomic pathways where large discrepancies can exist, with up to approximately 20% of NOS-immunoreactive neurones being NADPH-diaphorase negative, and 25% of NADPH-diaphorase-positive neurones being NOS-immunonegative (Vizzard *et al.*, 1994; Pullen & Humphreys, 1995). A possible reason for discrepancy is that some fixation-resistant NADPH-diaphorase is not NOS. Another possible reason is more technical, and may reflect that fact that the antibodies used to label the NOS diffuse less ably than the substrates for the diaphorase reaction, giving the diaphorase activity an apparently broader distribution. Thus, unless proven otherwise, it would not be safe to assume that NADPH-diaphorase activity is an accurate marker of neuronal NOS.

When using NADPH-diaphorase to mark the distribution of NOS, some form of

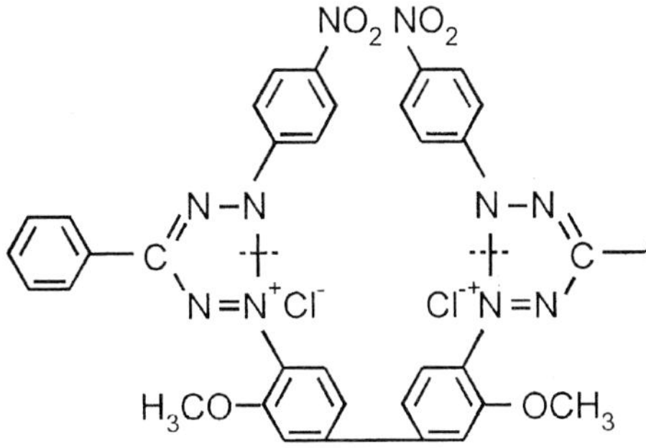

Fig. 12.1 Nitroblue tetrazolium (2,2′-di-*p*-nitrophenyl-5,5′-diphenyl-3,3′-[3,3′-dimethoxy-4,4′-biphenylene]ditetrazolium chloride). In the reduction of nitroblue tetrazolium to form nitroblue diformazan, the N–N bonds are cleaved (dotted line), and a total of $4H^+$ is consumed, with the concomitant generation of 2HCl.

fixation of the tissue is essential. This is because although all known forms of NOS are NADPH-diaphorases, only a small proportion of NADPH-diaphorases are NOS (Tracey *et al.*, 1993), and most of those that are not NOS are destroyed by aldehyde fixation.

12.3 Histochemical localization of NADPH-diaphorase

The development of staining techniques for visualizing NADPH-diaphorase reactivity is closely linked to the development of staining techniques for dehydrogenases involved in various metabolic pathways. In accomplishing this, tetrazolium salts have been used extensively (Burstone, 1962; Pearse, 1972). These are pale or colourless compounds, which when reduced form darkly coloured, insoluble formazans.

Of several tetrazolium compounds that have been developed (Burstone, 1962; Pearse, 1972), the one that is the most widely used for LM work, and to some extent EM work, is nitroblue tetrazolium (NBT), which produces a blue–purple formazan. NBT is a ditetrazolium (Fig. 12.1), which could account for its protein substantivity (see Section 12.4). It is fairly readily reduced to its formazan derivative, and has a higher redox potential than the system from which it accepts electrons, which also makes it a preferential electron acceptor over atmospheric oxygen. The formazan produced is insoluble in water, and appears to be amorphous rather than crystalline, and is resistant to oxidation (back to NBT) by atmospheric oxygen. Hence NBT has proven to be a valuable tool.

12.4 Substantivity

Substantivity is a term that is perhaps more at home in the dyeing industry, where it relates to the affinity of a dyestuff for textiles (Vickerstaff, 1950).

In histochemical terms it relates to the affinity of a reactant or reaction product, involved in stain production, to tissue or cellular components. Some tetrazolium salts bind non-specifically to protein through electrostatic interaction and, when applied to a tissue section cannot be removed by thorough washing. This is possibly an undesirable quality because if it becomes reduced the formazan deposit will be laid down at this site; thus even though the diaphorase might be cytosolic, the formazan product would appear to be associated with membranous structures, giving the impression that the enzyme itself was membrane bound. On the other hand, the substantivity of, for example, NBT for lipoprotein can be advantageous in that, because the tetrazolium is held to the protein, the tendency towards crystallization is reduced and an amorphous stain, rather than a crystalline stain, is produced (Pearse, 1972). In contrast, the non-substantive monotetrazole, which yields a more simple formazan, also yields a grossly crystalline staining pattern. The substantivity of the reaction product is also an important consideration: one that has substantivity for a particular cellular component could easily yield a false impression of the localization of the enzyme.

At the EM level the formazan reaction product is closely associated with membranous intracellular structures (Hope & Vincent, 1989), leading to the conclusion that this diaphorase is membrane bound. However, if the substantivity of NBT is taken into account (Pearse, 1972) the ultrastructural localization of the formazan must be regarded with caution.

12.5 NADPH-diaphorase: light microscopy

Two general approaches have been used to localize NADPH-diaphorase histochemically: one may be regarded as indirect, and is historically older, and the other as more direct.

In the indirect method, NADP is used as a co-enzyme for a dehydrogenase that is activated by its preferred substrate. During this process the NADP is reduced to NADPH, which then goes on to be a substrate for NADPH-diaphorase. The general scheme of the reaction is:

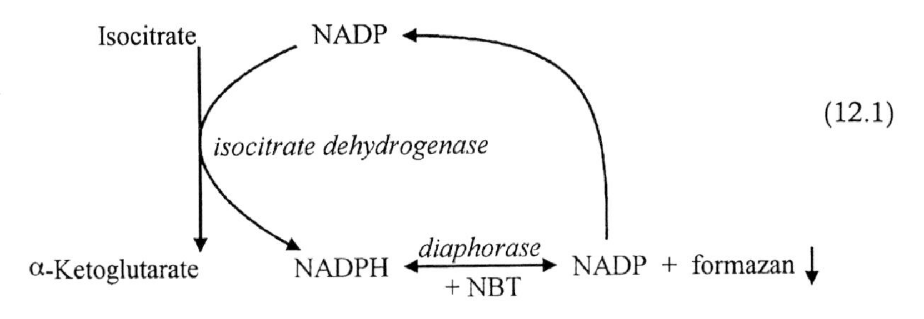

In this example isocitrate is shown as a substrate for isocitrate dehydrogenase, but other substrates for other dehydrogenases could be used, e.g. citrate, glucose-6-phosphate or malate (Farber, Sternberg & Dunlap, 1956; Vincent, Staines & Fibiger, 1983; Dail, Galloway & Bordegaray, 1993; Sancesario *et al.*, 1993). This method relies upon the presence of the specific dehydrogenase. Farber *et al.* (1956) found a similar pattern of staining in the rat kidney regardless of substrate, but Hess, Scarpelli & Pearse (1958) found that with the direct method the diaphorase staining pattern was non-specific, and more widely spread.

The more direct method uses reduced NADP (i.e. NADPH) as the substrate for the diaphorase (Hess *et al.*, 1958) (Fig. 12.2). Thus the general scheme becomes:

$$\text{NADPH} + \text{tetrazolium} \xleftrightarrow{\textit{diaphorase}} \text{NADP} + \text{formazan}\downarrow \qquad (12.2)$$

The equilibrium of the chemical reactions in scheme 12.1 is such that formazan production is favoured because the NADP formed by the diaphorase is removed from the system by the activity of the dehydrogenase. In scheme 12.2 the equilibrium lies sufficiently to the right, but is also pushed there by supplying high concentrations of NADPH. Both sets of reactions favour formazan production also because the formazan itself is insoluble, and therefore effectively unavailable once it has dissociated from the enzyme.

In practice the second (direct) method is the more widely used, although several variations on this theme exist. One variation is to use malate as a substrate for the cytosolic malic enzyme, which utilizes NADP as a co-enzyme in the production of pyruvate, and is thus used to stimulate regeneration of NADPH:

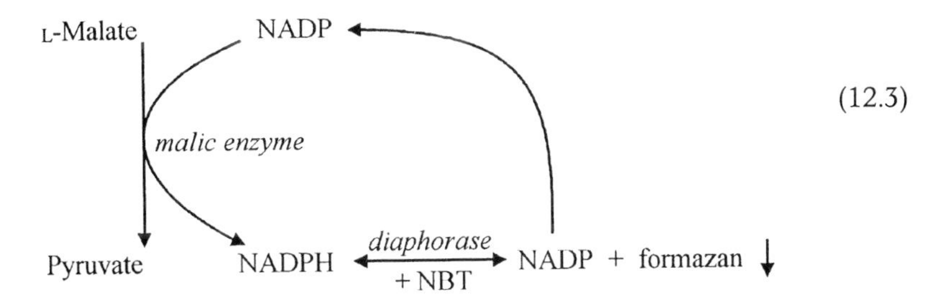

A modification of the indirect method for visualizing NADPH-diaphorase activity that would appear to be beneficial, but which does not seem to have been adopted, is to include a dye such as methylene blue in the reaction mixture, at a catalytic concentration (Farber & Louviere, 1956). With methylene blue present the reaction time is substantially reduced (thereby reducing the amount of non-specific tetrazolium reduction), and the quality of staining appears to improve. It is thought that the dye acts as an electron carrier between the diaphorase and the tetrazolium

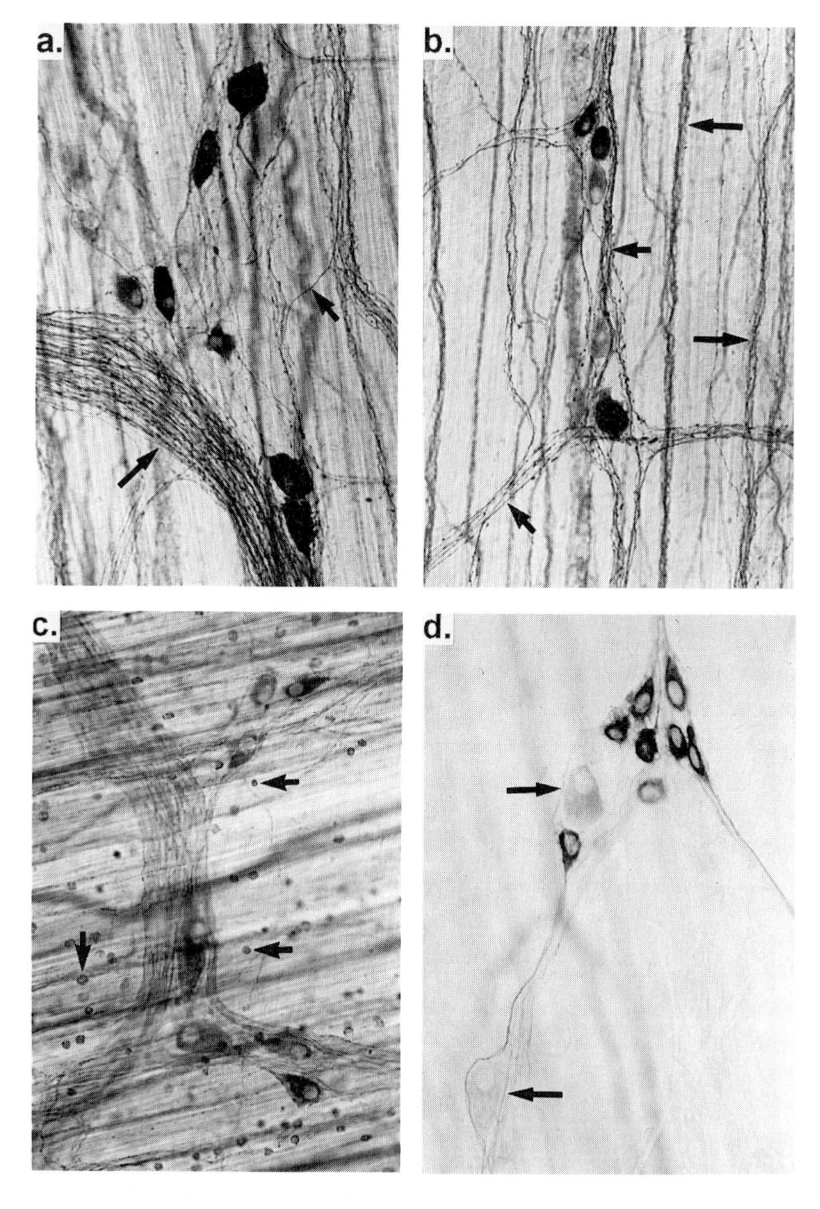

Fig. 12.2 NADPH-diaphorase staining of whole-mount preparations of rat caecum – light microscope. The staining was carried out according to Hoyle *et al.* (1996) (Chapter 16, Table 16.1, column 3), using NADPH and nitroblue tetrazolium as the substrates. (a) Myenteric plexus with heavily stained nerve cell bodies, and typical varicose autonomic nerve fibres arranged in bundles (larger arrow) or as single interconnecting fibres (smaller arrow). (b) Myenteric plexus with heavily stained and lightly stained nerve cell bodies, nerve fibres arranged in interconnecting bundles (smaller arrows) and in bundles embedded within the circular muscle layer (larger arrows). In panels

salt, which makes it possible to stain for enzymes that do not interact with tetrazolium:

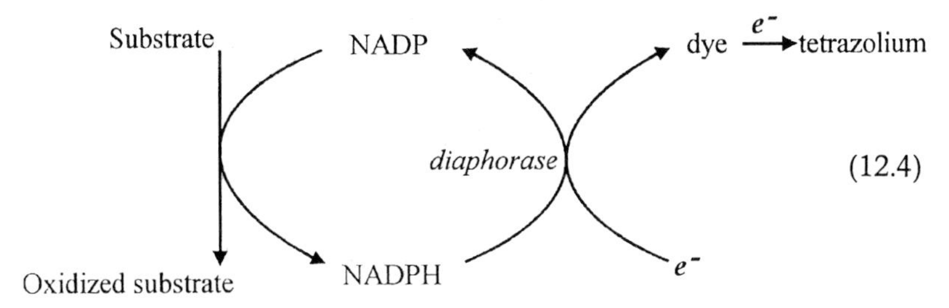

(12.4)

Whether or not addition of low concentrations of methylene blue improves the direct method of diaphorase staining remains to be investigated.

At the concentrations that tend to be used for histochemical staining, NBT is only just soluble. For this reason some recipes call for the use of Triton X-100, a detergent, to aid solvation. Triton X-100 is also used, particularly in whole-mount preparations or thick sections, to aid the penetration of the reaction mixture into cells. Empirically, this detergent is beneficial, and in a systematic study the improvement in staining gained by incorporating Triton X-100 into the reaction mixture has been shown to be related to a catalytic effect on formazan production (Fang *et al.*, 1994).

12.6 NADPH-diaphorase: electron microscopy

NADPH-diaphorase activity has also been examined at the ultrastructural level using EM. However, there are several problems associated with this technique, not least of which is that the most commonly used tetrazolium salt,

---

Caption for Fig. 12.2 (*cont.*) (a) and (b) the circular muscle orientation is approximately from top to bottom. (c) Myenteric plexus, with heavily stained nerve cell bodies, but in this preparation there are many monocytes/macrophages that have infiltrated the muscle layer, appearing as small dots (arrows). In many of those that are in the plane of focus the reaction product appears to be in cytoplasmic granules surrounding a very small nucleus. The axis of the circular muscle is running approximately from left to right. (d) Submucous plexus, with many small heavily stained nerve cell bodies and some larger nerve cell bodies that are only lightly stained (arrows). Note the good resolution that is obtainable, and the fine definition of neural processes. Close observation of many nerve cell bodies that are not too heavily stained reveals a greater density of reaction product in perinuclear cytoplasm as opposed to peripheral cytoplasm. The longer arrows represent 45 μm in (a), (b) and (c), and 30 μm in (d).

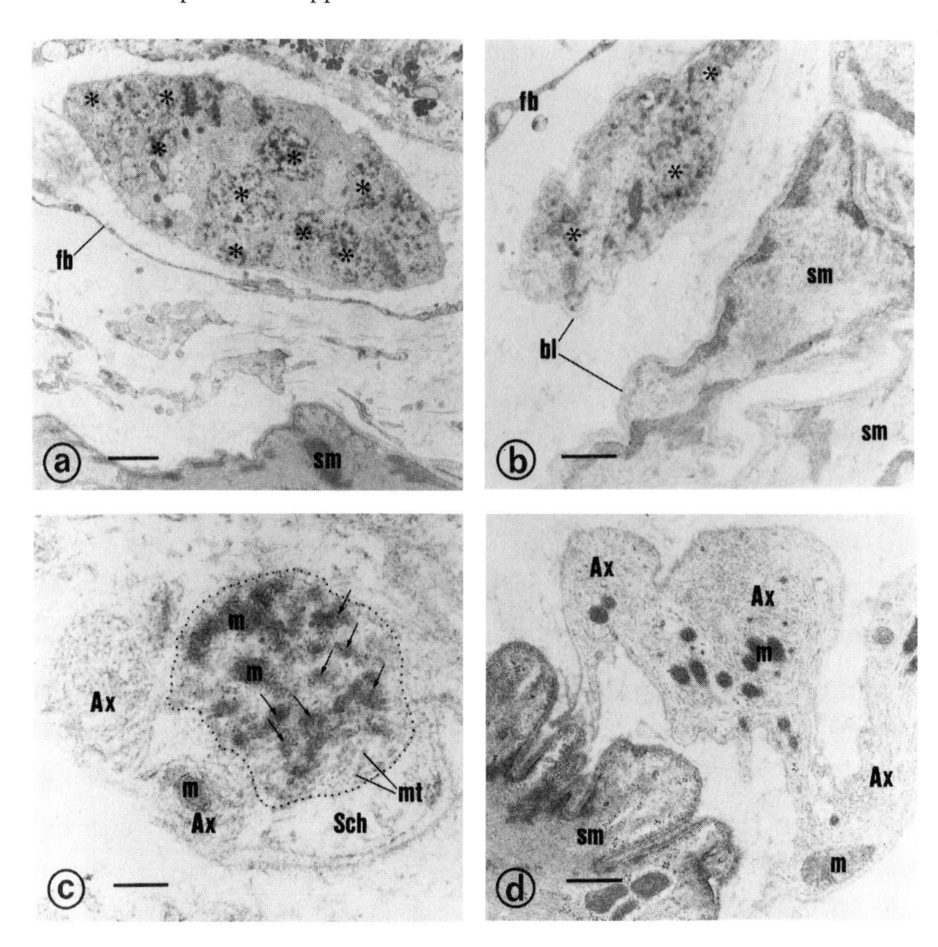

Fig. 12.3 NADPH-diaphorase staining in the innervation of the rat basilar artery – electron microscope. NADPH and nitroblue tetrazolium were used as the substrates. (a) and (b) Many axon profiles in a nerve bundle are NADPH-diaphorase positive (asterisks): bl, basal lamina; fb, fibroblast process; sm, smooth muscle. Scale bar = 1 μm in (a) and 0.5 μm in (b). (c) Higher magnification of a nerve bundle with a positive axon (dotted line outlines the axolemma) containing the nitroblue diformazan reaction product (arrows) and two negative axons (Ax): m, mitochondria; mt, microtubules; Sch, Schwann cell process. Scale bar = 0.2 μm. (d) A control preparation in which NADPH was omitted from the staining procedure, showing the lack of formazan deposits. Scale bar = 0.5 μm. Note that the nitroblue diformazan precipitate is not very electron-dense, and that it has a patchy distribution, appearing typically as a mottled grey cloud. (Reproduced with permission from Loesch, Belai & Burnstock, 1994.)

'NBT, produces a formazan that is not very electron-dense (Fig. 12.3). A second important problem is related to the substantivity of NBT. Being a ditetrazolium, and carrying two positive charges, it has greater substantivity than a monotetrazolium. Additionally, the greater the charge of the tetrazolium, the slower its rate of penetration into the tissue is likely to be. Finally, the lipid solubility of the tetrazolium salt can be responsible for artefactual results. Although NBT formazans are lipophobic, they can diffuse into lipid-rich areas, and secondary formazan deposits can be laid down at the lipid–water interface, thus giving a false impression of the site of diaphorase activity (Altman, Hoyer & Andersen, 1979).

Over 20 years ago a tetrazolium salt was synthesized, specifically with the purpose in mind for the use with EM. This salt, 2-(2′-benzothiazolyl)-5-styryl-3-(4′-phthalhydrazidyl)tetrazolium chloride (BSPT), is non-osmiophilic, lipophobic, singly charged and produces an osmiophilic formazan (Kalina *et al.*, 1972). The fact that the tetrazolium is non-osmiophilic while the formazan is osmiophilic means that the reaction product will become electron dense when associated with the heavy metal ion of osmium. Osmium tetroxide (OsO_4) is routinely used as a fixative during the preparation of tissue sections for EM, and it becomes chelated by the BSPT–formazan (Kalina *et al.*, 1972). BSPT can also be used at the LM level. The usual colour of the BSPT–formazan is purple–brown, but better staining may be obtained by complexing it with nickel, in which case the formazan appears green (Altman *et al.*, 1979).

In rat brain neurones, using BSPT to reveal ultrastructural sites of NADPH-diaphorase activity, the reaction product tends to be located on membranes, especially those of the endoplasmic reticulum and nuclear envelope, and mitochondria, but not Golgi apparatus (although it may appear within the cisternae), plasmalemma or synaptic vesicles (Wolf, Wurdig & Schunzel, 1992; Calka, Wolf & Brosz, 1994). The formazan precipitate appears clearly on both sides of the endoplasmic reticulum, not as a diffuse cloud; however, there is a loss of resolution of ribosomes (Wolf *et al.*, 1992; Calka *et al.*, 1994). Furthermore, in cells that appear negative at the LM level, formazan deposits can be found at the EM level, which distribute sparsely among mitochondria and on membranes of the endoplasmic reticulum (Wolf *et al.*, 1992).

Despite the advantages of BSPT over NBT, and the fact that it is readily commercially available, BSPT has not been used extensively for EM work; NBT has been, but with difficulty. The NBT–formazan deposit tends to appear as a light-grey cloud (Loesch *et al.*, 1993; Loesch, Belai & Burnstock, 1994), or with a characteristic mottled texture (Gabbott & Bacon, 1993) (Fig. 12.3). At present there is little information available on the neuronal ultrastructural localization of NADPH-diaphorase as revealed by using NBT: it can be found in nerve cell bodies (cyta), dendrites and axons (Gabbott & Bacon, 1993; Loesch *et al.*, 1994); although the deposits may appear to be in the cytoplasm, there does not always seem to be any relationship with cellular organelles. Thus, the true localization of the diaphorase remains equivocal.

In endothelial cells, in general, at the EM level NADPH-diaphorase, which would represent both type I and type III NOS activity, is not necessarily associated with membranous structures (Gabbott & Bacon, 1993; Loesch *et al.*, 1993; Tomimoto *et al.*, 1994), and could therefore be interpreted as being cytoplasmic. However, to be critical, the areas surrounding the 'cytoplasmic' formazan are generally rich in rough endoplasmic reticulum and mitochondria (Gabbott & Bacon, 1993). Thus, until the problems associated with the substantivity are taken into account, the subcellular distribution of NADPH-diaphorase activity, as determined by using NBT, has to be treated with caution.

12.7 Immunocytochemical localization of NOS-immunoreactivity and co-localization with NADPH-diaphorase activity

12.7.1 NOS-immunoreactivity: light microscopy

Appendix I lists species and tissues in which NOS has been localized by immunocytochemical methods.

In immunocytochemical (or immunohistochemical) techniques a specific component is identified *in situ* by means of an antigen–antibody interaction. The most popular methods are indirect, derived from methods established by Coons, Leduc and Connolly (1955), in which the primary antibody, which recognizes the desired component, is identified by a secondary antibody that has been raised against the immunoglobulin of the animal species in which the primary antibody was raised. The secondary antibody is tagged in some way to render it visible. Commonly a fluorescent marker is conjugated to the secondary antibody, such as fluorescein isothiocyanate (FITC) or tetrarhodamine isothiocyanate (TRITC). FITC fluoresces green when excited by ultraviolet light with a wavelength of approximately 490 nm, and TRITC fluoresces red at around 530 nm. Thus, with a microscope that is equipped to provide the appropriate illumination, structures that contain the primary antigen fluoresce, while the unlabelled ground tissue does not, providing a high-contrast image (Fig. 12.4a).

Variations on this theme include the use of a biotinylated secondary antibody and a complementary third layer of FITC-conjugated streptavidin. Originally methods were designed, and are still very much in use, that used avidin rather than streptavidin. Avidin and streptavidin have a tremendous affinity for biotin, reacting very quickly and forming non-covalent bonds together that are among the strongest known in nature (Coggi, Dell'Orto & Viale, 1986). Four molecules of avidin or streptavidin bind to one molecule of biotin, and several moieties of avidin become conjugated to the secondary antibody, hence the avidin–biotin complex provides an amplification system. Because of this signal amplification, higher dilutions (lower concentrations) of the primary antibody can be used, which, especially if it is a polyclonal antibody, means that unwanted antibodies that might also be present are diluted out, and the degree of non-specific staining is reduced. Another benefit is that because the primary antibody is used at higher dilutions resources are conserved. The advantage of streptavidin over avidin is that it binds fewer tissue structures non-specifically.

Instead of using fluorescence to visualize the antibody–antigen complex, the system can be linked to enzymes whose reactivity can be used to generate a chromogenic reaction product. In the original method a peroxidase–anti-peroxidase (PAP) complex was used (Graham & Karnowsky, 1966). Essentially the primary antibody is linked to the third-layer PAP complex by a second layer of unconjugated anti-rabbit immunoglobulin antibody. The method relies upon antibodies having two reactive sites (Fab portions of the IgG antibody molecule), one of which is bound to the primary antibody, the other of which is free to bind to the PAP IgG (also raised in rabbit). Colour is developed by activating the peroxidase, supplying substrates such as 3,3′-diaminobenzidine (DAB) and hydrogen peroxide, resulting in an insoluble, dark-brown, osmiophilic precipitate, suitable for LM. Because the reaction product is insoluble the tissue sections can be dehydrated, cleared and mounted in resin, and thus made permanent. Because it is osmiophilic the reaction product can be rendered electron dense, and suitable for EM work.

A popular modification of the PAP method uses an avidin-conjugated PAP complex as a third layer to complement the secondary antibody which is biotinylated (Hsu, Raine & Fanger, 1981). This has come to be called the ABC method (avidin–biotin complex).

The PAP and ABC methods provide a low level of background staining (although this can be variable), good resolution and high sensitivity. There is little to choose between the fluorescence methods and enzyme-linked methods when only one antigen is being visualized. However, the fluorescence method is a better starting point when co-localization studies, either with another antibody or with diaphorase activity, are planned. For example, if two primary antibodies are raised in different species one can end up being labelled with FITC and the other with TRITC, so their distributions can be examined using the same section.

12.7.2 Co-localization of NADPH-diaphorase activity and NOS-immunoreactivity

Because the diaphorase reaction produces a dark product and the immunocytochemistry a fluorescent product, it should be possible to perform the

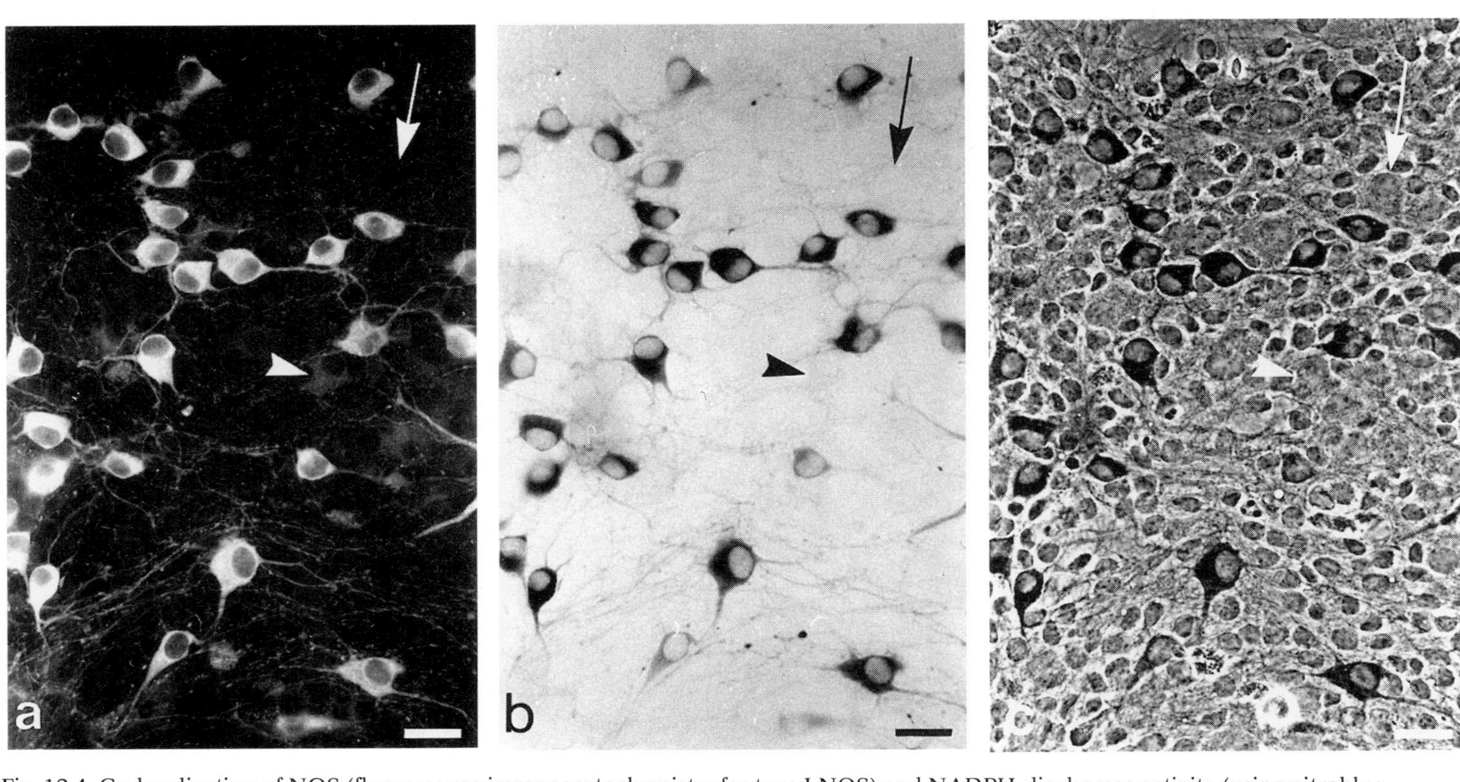

Fig. 12.4 Co-localization of NOS (fluorescence immunocytochemistry for type I NOS) and NADPH-disphorase activity (using nitroblue tetrazolium) in neurones of the guinea-pig myenteric plexus grown in culture. Fixed cultures were processed for NADPH-diaphorase activity (scheme 12.3) and then for NOS immunoreactivity. (a) NOS-immunoreactivity in many neurones. (b) The same field, showing the NADPH-diaphorase activity. (c) Phase-contrast micrograph of this field. The arrow indicates a neurone that is NOS- and NADPH-diaphorase-negative. The chevron indicates a neurone that has a low level of immunoreactivity and NADPH-diaphorase activity. Scale bars = 25 μm. (Reproduced with permission from Saffrey *et al.*, 1992.)

two staining techniques using a single preparation, and thus be able to visualize the co-localization of the two products (Fig. 12.4). In practice the production of formazan quenches the fluorescence, and this may (Belai *et al.*, 1992; Saffrey *et al.*, 1992; Afework, Tomlinson & Burnstock, 1994; Tay *et al.*, 1994) or may not (Ward *et al.*, 1992d; Kitchener & Diamond, 1993; Schemann, Schaaf & Mader, 1995) be tolerable. If the two staining techniques are combined, the diaphorase reaction can be carried out either before (Hassall *et al.*, 1992; Saffrey *et al.*, 1992; Alonso *et al.*, 1994; Tay & Moules, 1994; Tay *et al.*, 1994) or after (Belai *et al.*, 1992; Afework *et al.*, 1994) the immunological procedure, or even between application of the primary and secondary antibodies (personal observations). It does not seem to matter, but generally the immunofluorescence will be suppressed. The formazan will quench immunofluorescence whether the primary antibody is directed against NOS or any other neuropeptide (Hoyle & Burnstock, 1989; Tay *et al.*, 1994).

Immunocytochemical staining using the PAP method, in which DAB is used to produce a brown reaction product, has to be used with care in conjunction with NADPH-diaphorase staining (Alonso *et al.*, 1992a; Kitchener & Diamond, 1993; Gabbott & Bacon, 1994a,b). When the PAP immunoreaction and DAB development are performed before the diaphorase staining there can be an accretion of formazan at the peroxidase site, yielding a false localization of the diaphorase (Kitchener & Diamond, 1993). On the other hand, if the diaphorase staining is performed first, the density of the formazan can obscure visualization of the DAB reaction product. For this reason it may be beneficial to use 3-amino-9-ethylcarbazole to develop the PAP immunolocalization, which produces a less dense, red precipitate (Christensen & Fang, 1994).

Some authors have managed to obtain a successful double stain, using either PAP or immunofluorescence, by keeping the incubation period with NBT short, around 10 min (Hassall *et al.*, 1992; Saffrey *et al.*, 1992; Christensen & Fang, 1994), thereby resulting in only a light labelling of diaphorase activity, and less interference with the immunochrome. If double staining does not work there are two further options. The first is to use thin sections (approximately 5 μm), and to stain adjacent sections with each procedure (Dawson *et al.*, 1991; Kitchener & Diamond, 1993). The second method is to perform the immunostaining first, photograph the areas of interest, remove the coverslip, wash the preparation and then stain again, this time for diaphorase activity (Ward *et al.*, 1992d; Kitchener & Diamond, 1993; Valtschanoff *et al.*, 1993a; Schemann *et al.*, 1995). A variation on the second method is to perform the diaphorase staining first, followed later by the immunostaining. However, in the author's opinion, this way must still run the risk of interference with the fluorescence by formazan. A further variation is to perform a PAP immunolocalization, photograph the preparation, then remove the reaction product by incubation with dimethylformamide

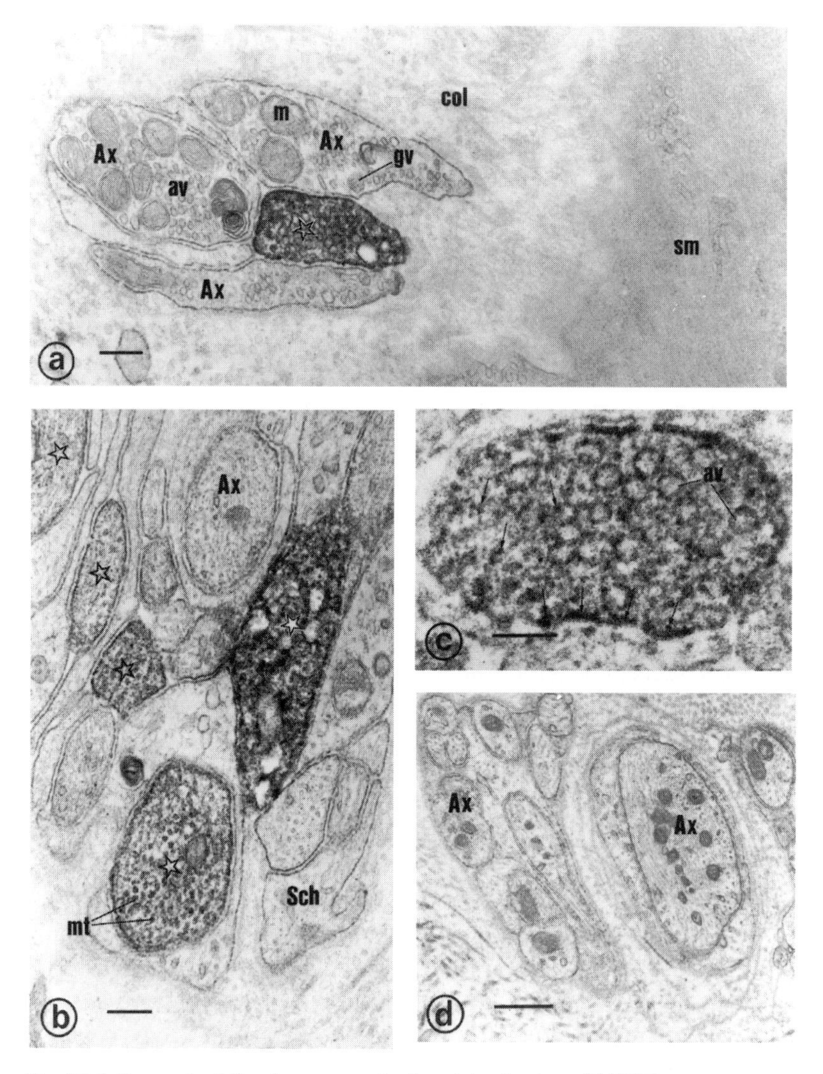

Fig. 12.5 Pre-embedding immunocytochemistry for type I NOS in perivascular nerves of the rat basilar artery. (a) A NOS-positive axon (star) and three NOS-negative axons (Ax) in close apposition to a smooth mucle cell (sm): av, small agranular vesicles; col, collagen fibres; gv, large granular vesicle; m, mitochondria; Schwann cell. Scale bar = 0.2 μm. (b) Several axon profiles in a nerve bundle are NOS positive (stars). Labelled microtubules are seen in an intervaricosity region (mt). Scale bar = 0.1 μm. (c) A neuronal varicosity showing immunoprecipitate attached to the membrane of synaptic vesicles (av) as well as the axoplasmic surface of the axolemma (arrows). There is also NOS-immunoreactivity free in the axoplasm. Scale bar = 0.1 μm. (d) Control preparation, in which the immunostaining procedure was carried out using anti-NOS serum that had been preadsorbed with purified NOS. No immunoprecipitate can be seen. Scale bar = 0.5 μm. (Reproduced with permission from Loesch, Belai & Burnstock, 1994.)

(100%, several days, 60 °C) before staining for diaphorase activity (Valtschanoff *et al.*, 1993b).

12.7.3 NOS-immunoreactivity: electron microscopy

For examining immunoreactivity at the EM level two general approaches are used. The first is called pre-embedding staining, and the second is called postembedding staining. In the pre-embedding method the tissue is initially lightly fixed, for example with paraformaldehyde and glutaraldehyde, fairly thick sections are cut (up to several hundred microns) and these are stained immunocytochemically as for LM on sections. The sections are then fixed with OsO_4 (osmicated), contrasted with uranyl acetate and lead citrate, infiltrated with resin and cut to make thin or ultrathin sections. In other words the immunostaining is carried out before the tissue is embedded in resin (Fig. 12.5).

For postembedding staining the tissue is fixed and embedded in resin for cutting thin or ultrathin sections, which are then stained using a standard immunocytochemical procedure and contrasted. Postembedding staining is necessary for the successful use of immunogold, but it does have two drawbacks. The first is relatively minor: the embedding procedure causes a loss of antigenicity, and therefore the primary antibody has to be used at a high concentration. The second is that the actual subcellular site of immunoreactivity cannot be visualized because there is a loss of resolution of the tissue. In particular membranes and membranous structures can be difficult to define.

There are several versions of methods that employ gold labelling, the simplest of which is to use a secondary antibody conjugated to colloidal gold particles. After processing, the gold particles, indicating the localization of the primary antibody and antigen, appear as black dots with a very smooth outline (Fig. 12.6). The secondary antibody can be conjugated to gold particles with a known diameter (usually in the range of 1–15 nm); thus, double labelling can be carried out using primary antibodies raised in different species and secondary antibodies conjugated to different sizes of gold particle. Immunogold methods have been used successfully at the LM level, and are particularly good when combined with a silver enhancement. However, they tend not to be generally employed because they offer few advantages, if any, over fluorescence or PAP methods.

For EM work the immunogold technique is usually used for postembedding staining. This is because colloidal-gold-labelled antibodies do not penetrate tissues at all well. One advantage of immunogold labelling over PAP or ABC methods is that there is no diffusion of the reaction product away from the site of the primary antibody–antigen complex, so the gold particles should accurately indicate the disposition of the primary antigen.

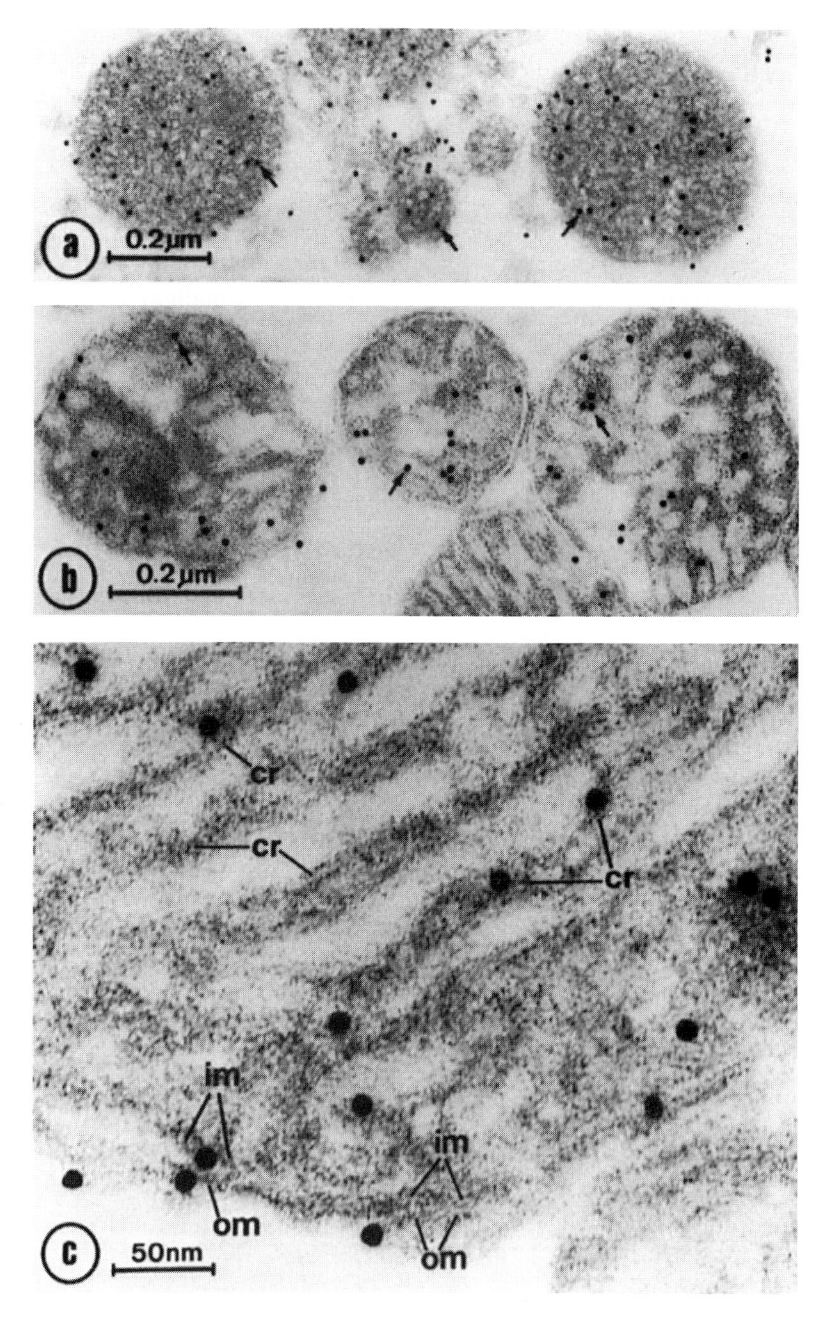

Fig. 12.6 Postembedding immunocytochemistry for type III NOS in mitochondrial fractions prepared from rat liver and brain. (a) Liver, (b) and (c) brain. The black dots are gold particles (10 nm diameter) conjugated to the anti-rabbit IgG antiserum, used to localize the anti-NOS antibody, and are found predominantly within the mitochondria (a) and (b), associated with (c) mitochondrial outer and inner membranes (om and im, respectively) and cristae (cr). (Reproduced with permission from Bates *et al.*, 1995.)

12.8 Localization of the guanylate cyclase system

Because NO stimulates guanylate cyclase, thereby resulting in elevated intracellular levels of cGMP, methods that localize this enzyme or its product have been used to identify those cells that are possible targets of NO. A NO donor, such as SNP, can also be used to stimulate guanylate cyclase. For example, in the brain, when SNP is perfused through the cerebral circulation, followed immediately by fixative, cGMP visualized immunocytochemically using a specific antibody can be found in the same general regions as NOS, but in a different population of cells (Southam & Garthwaite, 1993). NOS could be found in nerve terminals, and cGMP in adjacent nerve cell bodies, or *vice versa* (Southam & Garthwaite, 1993). These results imply that NO acts in cells that are local to its source, rather than in the same cell as that in which it is produced. Thus, NO functions as an intercellular signal molecule as opposed to an intracellular signal molecule and, as such, it can be considered to be a paracrine substance.

A paracrine role for NO has also been suggested in the periphery. In the lung, soluble guanylate cyclase, identified using a monoclonal antibody, is present only in bronchial smooth muscle cells, and not in respiratory epithelium. In contrast, NOS is localized in the epithelium (Rengasamy, Xue & Johns, 1994). The likely function is that NO is synthesized and released from the epithelial cells to act on the local smooth muscle cells, where it causes bronchodilatation.

Using a broadly similar approach the sites of activity of neuronally released NO have been determined in the canine proximal colon (Shuttleworth *et al.*, 1993b). In this organ a substantial component of the NANC inhibitory neuromuscular transmission is due to NO (Thornbury *et al.*, 1991; Ward *et al.*, 1992b,c; Keef *et al.*, 1994). In the experiments aimed at localizing the target cells of NO by immunolocalization of cGMP, phosphodiesterase inhibitors, M&B22948 and isobutylmethylxanthine were used. These compounds retard the degradation of cGMP, allowing it to accumulate as it is being synthesized, providing higher concentrations for the antibody to recognize. EFS of the intramural nerves *in vitro*, in isolated strips of colonic wall, results in accumulation of cGMP predominantly in smooth muscle cells (Shuttleworth *et al.*, 1993b). Also, when NO is applied as an aqueous solution the smooth muscle cells respond with increased cGMP production. Furthermore, in the myenteric plexus some neurones accumulate cGMP in response to these challenges. In fact, 94% of nerve cell bodies in the myenteric plexus that had elevated cGMP-immunoreactivity in response to exogenous NO did not contain NOS, and not one cell that had elevated levels of cGMP as a result of EFS contained NOS (Shuttleworth *et al.*, 1993b). In addition, in this same study, interstitial cells of Cajal, lying at the junction of the circular muscle coat and the submucous plexus, which receive innervation from the *plexus submucosus exter-*

nus, or Henle's plexus (Hoyle & Burnstock, 1989), also accumulated cGMP. This implies that these cells are functionally innervated by enteric neurones that utilize NO as a neurotransmitter (Shuttleworth *et al.*, 1993b), in addition to providing evidence for ganglionic and neuromuscular transmission by NO in intrinsic enteric neurones. The localization of cGMP has also been used to identify the smooth muscle cells of enteric arterioles and venules as targets of NO (Shuttleworth *et al.*, 1993b).

In rat myenteric plexus and dorsal root ganglia, cGMP localized immunologically is found predominantly in glial cells, with some in a relatively small number of neurones close to NOS-containing nerve fibres (Aoki *et al.*, 1993b). In the rat central nervous system, cGMP accumulation can be evoked by application of SNP, which releases NO from its molecular structure, which activates the soluble guanylate cyclase. Elevated levels of cGMP are found in nerve fibres and astrocytes, rather than nerve cell bodies (de Vente & Steinbusch, 1992). The increased levels of cGMP in peripheral and central glial cells might indicate an as yet undetermined function of these cells that is controlled locally by NO.

12.9 Summary

The localization of the enzyme responsible for the synthesis of NO has played an important role in furthering our knowledge of the physiological activity of this unusual signal molecule. At both the light microscope level and electron microscope level two techniques have been widely employed. The first is histochemical localization of NADPH-diaphorase activity, and the second is immunocytochemical localization of NOS. There is not always a close relationship between NOS localization and NADPH-diaphorase staining. Neuronal type I NOS localization and neuronal NADPH-diaphorase activity are not necessarily a shared entity, and nor are macrophage type II NOS and macrophage NADPH-diaphorase activity. However, purified type I, type II and type III NOS all possess NADPH-diaphorase activity. The NADPH-diaphorase that is also NOS is resistant to fixation by formaldehyde, and only when the tissue has been fixed can NADPH-diaphorase staining be used to demonstrate NOS. Many histological studies have shown that there is a 1:1 correspondence, or very nearly so, between the localization of NOS and of NADPH-diaphorase in neural tissue but also several studies have shown that there is no 1:1 correspondence. The lack of correspondence is most apparent in the staining of sacral spinal cord and sacral autonomic pathways. NOS can be localized using immunocytochemical techniques that result in a fluorescent or dark-coloured stain. It should be possible to perform two staining techniques on a single preparation, and thus be able to visualize the co-localization of NADPH-diaphorase and NOS. There are practical problems associated with double staining whether immunofluorescence of a PAP technique is used to label the NOS. NADPH-diaphorase activity has been examined at the ultrastructural level using EM. However, there are several problems associated with this technique too, not least of which is that the most commonly used tetrazolium salt, NBT, produces a formazan that is not very electron dense. NBT also has high substantivity, and its lipid solubility can be responsible for artefactual results. Although NBT–formazans are lipophobic, they can diffuse into lipid-rich areas, and secondary formazan deposits can be laid down at the lipid–water interface, thus giving a false impression of the site of diaphorase activity. Another tetrazolium salt, BSPT, is non-osmiophilic, lipophobic, and produces an osmiophilic formazan, and should therefore be useful for EM work. The localization of NOS can be identified at an ultrastructural level using pre- or postembedding EM methods. In determining the relationship between cells that contain NOS and cells that can be a target for released NO, immunocalization of soluble guanylate cyclase may be useful.

12.10 Selected references

Coggi, G., Dell'Orto, P. & Viale, G. (1986). Avidin-biotin methods. In *Immunocytochemistry: Modern Methods and Applications*, second edn., eds. J. M. Polak & S. Van Noorden, pp. 54–70. Bristol: Wright.

Pearse, A. G. E. (1972). *Histochemistry: Theoretical and Applied*, third edn., Edinburgh: Churchill Livingstone.

Pullen, A. H. & Humphreys, P. (1995). Diversity in localization of nitric oxide synthase antigen and NADPH-diaphorase histochemical staining in sacral somatic motor neurones of the cat. *Neuroscience Letters*, **196**, 33–6.

Tracey, W. R., Nakane, M., Pollock, J. S. & Förstermann, U. (1993). Nitric oxide synthases in neuronal cells, macrophages and endothelium are NADPH diaphorases, but represent only a fraction of total cellular NADPH diaphorase activity. *Biochemical and Biophysical Research Communications*, **195**, 1035–40.

Section 4

Protocols and techniques

13

Measurement of nitric oxide

13.1 Introduction

It is ironic that many of the very properties of NO that enable it to carry out its diverse functions also present considerable problems when attempting its detection and quantitation. This can readily be deduced from the literature. Among the thousands of publications investigating the synthesis of NO or its physiological effects, the percentage that includes estimates of NO levels is extremely low. A range of techniques have been developed based on different principles. While each technique has certain individual advantages, all of the techniques have disadvantages that limit their application. These disadvantages range from lack of sensitivity or specificity to interference by factors commonly present in biological systems. The choice of technique is therefore heavily dependent upon the application. In this chapter, detailed protocols will only be provided for those techniques that do not require specialized equipment and can therefore be attempted by most laboratories. However, the scientific basis behind all the major techniques available will be described together with selected references demonstrating their use. Should a particular application justify the expense (and commitment of time) of obtaining specialized equipment, then it is always advisable to consult with researchers who already have experience of its use. Most manufacturers are working closely with scientific institutions and readily provide this information. In any study requiring the measurement of NOS activity, the radiochemical assay based on the conversion of arginine to citrulline (see Chapter 14) should be considered seriously as an alternative to measuring the production of NO. For a detailed description of the measurement of NO in biological systems and the problems associated with sample preparation, readers should refer to an excellent review by Archer (1993).

13.2 Standard solutions of NO

For quantitation of NO it is obviously necessary to be able to calibrate the technique using some form of standard NO solution and this is not as simple as it might seem. The types of standards that can be used also depend upon the technique to be used.

13.2.1 Authentic NO

The ideal solution for calibrating assays are those prepared by serial dilution of a saturated solution of NO gas in water. The preparation of saturated NO solutions requires both meticulous and stringent conditions to ensure purity and stability and to limit the hazards involved. Commercial NO gas may need to be purified to remove other gaseous oxides of nitrogen present as contaminants and this should only be carried out by experienced personnel (Feelisch, 1991). The water used for solution must be of high quality without any oxidizing impurities. Oxygen must be excluded at all stages of preparation, by gassing with argon or helium, to prevent nitrite/nitrate formation (Greenberg, Wilcox & Rubanyi, 1990; Feelisch, 1991; Archer, 1993). As a gas, NO is extremely toxic. On exposure to air, NO forms nitrogen dioxide which can be lethal at levels as low as 200 ppm. Thus, solutions should be prepared in a fume hood and steps should be taken to remove vented NO by a series of gas scrubbers (Archer, 1993). The simplest procedure for the preparation of saturated NO solutions has been described by Ishii *et al.* (1991). Throughout the procedure distilled water is used that has been passed through a Milli-Q water system, degassed by applying a vacuum for 30 min and purged with 100% argon for 30 min. Then 10 ml of the water can be saturated with NO by applying NO gas (commercially available) to the water in a small sealed flask via a 21-gauge needle at a rate of 100 ml min^{-1} for 20 s. Excess NO is vented by a second needle. By this method the resultant solution is assumed to be 2 mM (Ishii *et al.*, 1991). However, other values are reported in the literature depending on the solubility constant that has been taken for NO in water at 20 °C. Serial dilutions can be made using degassed, argon-purged water. Solutions prepared in this way are not very stable and should be prepared fresh for each experiment. The chemiluminescence technique can be used to measure NO in the form of a gas. Gas mixtures of NO at various concentrations are commercially available and can be used directly as standards with this technique (Chung & Fung, 1990; Archer, 1993).

Due to the difficulty in preparing authentic NO solutions, many studies use alternative standards that produce known amounts of NO in aqueous solution.

13.2.2 NO-producing standards

In the case of the nitrite/nitrate assay, standard solutions consist simply of sodium nitrite at a range of concentrations. Nitrite solutions can also be used to produce known amounts of NO by reduction under acidic conditions according to the following equation:

$$2KNO_2 + 2KI + 2H_2SO_4 + 2K_2SO_4 \rightarrow 2NO + I_2 + 2H_2O + 4K_2SO_4 \quad 13.1$$

Thus a range of NO standards can be prepared by reacting known concentrations of potassium nitrite with a mixture of potassium iodide and sulphuric acid or, as an alternative, glacial acetic acid (Schmidt *et al.*, 1989b; Tsukahara *et al.*, 1994).

Such highly acidic solutions are unsuitable for the bioassays and, in many studies, the use of bioassays is limited to assessing a relative increase in NO production rather than an absolute determination of NO levels. However, apart from using authentic NO, it may be possible to use the NO generator *S*-nitroso-*N*-acetyl-D,L-penicillamine (SNAP) as an alternative. This compound decomposes to produce NO at a known constant rate in solution at 37 °C. SNAP solutions of known concentration have been used as NO calibration standards with an NO microelectrode (Ichimori *et al.*, 1994) and may be applicable to other systems. In the case of the spectrophotometric assay utilizing haemoglobin, NO standards are not required since the NO concentration can be calculated on the basis of the extinction coefficient for the reaction (see Section 13.4.2).

13.3 Bioassays

There are two assays available that use the biological properties of NO to detect the synthesis and release of NO. These bioassays played a major part in early studies attempting to determine the roles of NO as a physiological messenger and to identify the enzyme responsible for NO synthesis. To a certain extent, these methods have now been superseded by other more quantitative techniques. However, they are still used in the detection of NO and are useful in demonstrating that the levels of NO synthesized are sufficient to produce a physiological response.

13.3.1 Cascade superfusion bioassay

13.3.1.1 Scientific basis

The preparation that releases NO and a detector preparation are connected in a perfusion/superfusion cascade. The detector preparation consists of an isolated blood vessel from which the endothelium has been removed. Any NO released from the donor preparation superfuses onto the detector vessel resulting in relaxation of the vascular smooth muscle. The size of the response is monitored by measuring changes in the tension of the detector vessel and is proportional to the amount of NO being released.

13.3.1.2 Protocol

This bioassay has most often been used for the study of NO synthesized by endothelial cells and the donor preparations can consist of isolated blood vessels with an intact endothelium (Rubanyi & Vanhoutte, 1986; Ignarro, Buga & Chaudhuri, 1988a; Greenberg *et al.*, 1990) or endothelial cells grown on microcarrier beads (Myers, Guerra & Harrison, 1989; Buga *et al.*, 1991). However, it should be noted that the synthesis and release of NO following electrical stimulation of NOS-containing nerves can equally be studied by this technique (see Chapter 15). Isolated blood vessel segments need to be examined for leakage of the perfusate and any small holes or branches ligated. The vessel segments are fitted at either end with polyethylene tubing and perfused through the lumen with oxygenated Krebs' solution at 37 °C at a constant flow rate (2–3.5 ml min^{-1}). The vessel preparations are then submerged in chambers containing oxygenated Krebs' solution at 37 °C. Endothelial cells grown on microcarrier beads can be packed in small columns and perfused. The perfusate from the donor preparation is allowed to superfuse the detector vessel downstream. The detector vessel (endothelium denuded) is connected to a force transducer for recording of isometric tension. For measurement of relaxation responses the vessel needs to be precontracted. After equilibration, NO synthesis in the donor preparation can be stimulated. In the case of isolated vessels, stimulation can be achieved by adding agonists such as ACh, that act on endothelial receptors, or Ca^{2+} ionophores to the perfusing solution. It needs to be established that the agent used does not have a direct effect on the detector vessel. Stimulation of NO synthesis by shear stress can also be investigated by subjecting endothelial cells grown on microcarrier beads to changes in flow. Several detector vessels can be connected in series. Due to the short half-life of NO, the transit time between the donor and the detector preparations should be kept as short as possible. Performing the experiments under reduced oxygen tension and in the presence of SOD can reduce the rate of inactivation of NO released (Rubanyi & Vanhoutte, 1986).

13.3.1.3 Advantages and disadvantages

The main advantage to this technique is that it detects the synthesis of NO under physiological conditions and, by its very nature, it demonstrates that the amount of NO produced is sufficient to produce a physiological response. Technically, estimates of the levels of NO released can be obtained by comparison of the size of the responses with those to authentic NO or NO donors such as SNAP. However, at best this bioassay is only regarded as semi-quantitative. Since the response in the detector vessel could occur by the release of a variety of vasoactive agents synthesized in the donor preparation in addition to NO (see Chapter 3), NOS inhibitors, inhibitors of prostaglandin synthesis and specific receptor antagonists need to be used to increase the specificity of the technique.

13.3.1.4 Applications

The cascade/superfusion bioassay has a significant part to play in studies of cardiovascular disease. Different combinations of control and pathological specimens as the donor and the detector vessel may be used. In this way the sensitivity to stimulation, the capacity for NO synthesis and the responsiveness to NO can be examined in the same system. Unless the laboratory is already routinely involved in vascular pharmacology, the cascade/superfusion system has limited application as an assay technique nowadays. However, the procedure is still being used as a detection system and to distinguish between authentic NO and the various agents that have been proposed as candidates for EDRF (Feelisch *et al.*, 1994).

13.3.2 Guanylate cyclase bioassay

13.3.2.1 Scientific basis

NO activates soluble guanylate cyclase resulting in increased production of cGMP. The measurement of the increase in cGMP levels can therefore be used to detect and measure NO synthesis.

13.3.2.2 Protocol

The protocol for the guanylate cyclase bioassay has been described in detail (Ishii *et al.*, 1991) and has been used for the detection of NO synthesis in a variety of preparations (Förstermann *et al.*, 1990; Gorsky *et al.*, 1990; Ishii *et al.*, 1991; Pollock *et al.*, 1991; Schmidt *et al.*, 1991; Sheng *et al.*, 1992). Rat foetal lung fibroblast cells (RFL-6) are used as the source of guanylate cyclase. The cells are grown to confluence in tissue culture microwell plates in Hams F-12 nutrient mixture supplemented with 15% uninactivated foetal calf serum. After removal of the tissue culture medium, the cells are washed twice with Locke's buffer containing 0.3 mM 3-isobutyl-1-methylxanthine (IBMX). Locke's buffer consists of 10 mM HEPES buffer, pH 7.4, containing 154 mM NaCl, 5.6 mM KCl, 2.0 mM $CaCl_2$, 1.0 mM $MgCl_2$, 3.6 mM $NaHCO_3$ and 5.6 mM glucose. IBMX is a phosphodiesterase inhibitor that prevents the degradation of cGMP. The cells are then equilibrated with the above IBMX-containing Locke's buffer for 20 min at 37 °C. Then, 3–5 min before assay, SOD (final activity 100 U ml^{-1}) is added to the wells to remove superoxide anions. The sample is then added to the wells and incubated for 3 min at 37 °C. The reaction is stopped by aspirating the incubation medium and adding ice-cold 50 mM sodium acetate buffer, pH 4.0. The final samples can be frozen in liquid nitrogen and stored at −70 °C. The cGMP content of the samples is determined by radioimmunoassay. This is a routine assay for which kits are available commercially.

Using authentic NO or possibly SNAP solutions, the amount of NO that produces a certain level of increase in cGMP can be calibrated. Blanks need to be prepared to determine the basal levels of cGMP in the RFL-6 cells. By this method, levels of NO in the conditioned medium, e.g. that released from endothelial cells or from activated macrophages in culture, can be determined. In addition, preparations of NOS can be added directly to the well and NOS activity can be measured. In this case it would be necessary to add all the factors that may be required for optimal NOS activity (e.g. Ca^{2+}, calmodulin, FAD, BH_4, arginine; see Chapter 14). The sensitivity of the assay is largely governed by the size of the wells. The use of 48-well plates enables smaller volumes to be used and theoretically lowers the limit of detection to a concentration as low as 1 nM (equivalent to 100 fmol in 100 μl of sample) (Ishii *et al.*, 1991). NOS inhibitors can be used in the preparation of the sample to establish that the cGMP response is specifically due to the production of NO.

13.3.2.3 Advantages and disadvantages

This bioassay is extremely sensitive and can readily be performed in most laboratories, the only major equipment required being a gamma counter. It can be used to measure both the release of NO and NOS activity. However, the assay is extremely susceptible to interference by haemoglobin, which can completely inhibit the activation of guanylate cyclase by NO (Knowles & Moncada, 1994). As a consequence, this technique cannot be used to measure NOS activity in crude enzyme preparations unless contaminating haemoglobin is removed. This can be achieved with phosphocellulose (Förstermann *et al.*, 1990). The major disadvantage to this assay is that considerable interassay variability has been reported, making it difficult to pool data from different experimental days (Ishii *et al.*, 1991).

13.3.2.4 Applications

The guanylate cyclase bioassay provides a relatively simple and sensitive method for detecting and measuring NO release and NOS activity, particularly in purified preparations (Förstermann *et al.*, 1990; Gorsky *et al.*, 1990; Ishii *et al.*,

1991; Pollock *et al.*, 1991; Schmidt *et al.*, 1991; Sheng *et al.*, 1992). It is not particularly suited to kinetic experiments or to the routine assay of crude preparations of NOS.

13.4 Spectrophotometric assays

There are two different spectrophotometric assays available for the determination of NO production based on very different principles. Depending on the application, one of these assays will probably provide the best starting point for attempting to measure NO.

13.4.1 Nitrite/nitrate assay

13.4.1.1 Scientific basis

In oxygenated solutions, NO ultimately forms nitrite and nitrate as stable end-products according to the following reactions (Hevel & Marletta, 1994):

$$2{\cdot}NO + O_2 \rightarrow 2{\cdot}NO_2 \tag{13.2}$$

$${\cdot}NO + {\cdot}NO_2 \rightarrow N_2O_3 \text{ (in water)} \rightarrow 2NO_2^- \tag{13.3}$$

$$2{\cdot}NO_2 \rightarrow N_2O_4 \text{ (in water)} \rightarrow NO_2^- + NO_3^- \tag{13.4}$$

NO also reacts directly with superoxide anions to produce nitrate. Nitrate can be converted back to nitrite by enzymic reaction. Thus the determination of nitrite levels can provide an indirect method for estimating NO production (Archer, 1993; Hevel & Marletta, 1994).

13.4.1.2 Protocol A – Griess reaction

A very simple protocol for measuring total nitrite/nitrate levels has been described in which the whole process, including the conversion of nitrate to nitrite, can be carried out using the same disposable semi-microcuvette (Hevel & Marletta, 1994). The Griess reagent needs to be prepared fresh by mixing 0.1% naphthylethylenediamine dihydrochloride with 1% sulphanilamide in 5% phosphoric acid in equal proportions. The two solutions can be stored separately as stocks at 4 °C for at least 4 weeks. The stocks and Griess reagent need to be protected from light.

First, 20 μl of a solution containing nitrate reductase (60 mU) and 25 μM NADPH are added to 210 μl of sample in a cuvette and incubated for 30 min at room temperature. This reduces all the nitrate to nitrite. NADPH interferes with the assay so it is removed by adding 20 μl of a solution containing L-glutamate dehydrogenase (200 mU), 100 mM NH_4Cl and 4 mM α-ketoglutarate (freshly prepared). After 10 min at room temperature, 250 μl of Griess reagent is added and the absorbance of the reaction product is read at 543 nm after 5 min at 37 °C. Nitrite can be measured on its own by omitting the enzyme solutions and nitrate levels can be determined by subtraction of this measurement from the total nitrite/nitrate levels measured. Nitrate reductase (*Aspergillus niger*) and L-glutamate dehydrogenase (*Candida utilis*) are both available commercially. Blanks consist of the media in which the sample was prepared taken through the entire procedure. Standards simply consist of known concentrations of nitrite and nitrate prepared in the appropriate media. The limit of detection is in the micromolar range (Archer, 1993). Large numbers of samples can readily be processed by carrying out the entire procedure using a 96-well microplate commonly used for ELISA and reading the absorbance with a microplate reader (Amano & Noda, 1995). The assay can technically be used with whole-cell systems and as a discontinuous assay of NO production by crude or purified preparations of NOS. Major interference can occur with certain tissue culture media which not only have high levels of nitrate but may also contain phenol red, which absorbs in the same region as the product of the Griess reaction (Hevel & Marletta, 1994). Therefore, such media should be avoided in the assay. Haemoglobin and other haem proteins can also interfere spectrophotometrically. Haemoglobin can be removed prior to assay with phosphocellulose (Förstermann *et al.*, 1990) or after the formation of nitrite/nitrate by filtration with 10 000 M_r cut-off filters (Misko *et al.*, 1993b). Control experiments need to be performed to confirm that the nitrite/nitrate measured is actually due to the production of NO. This usually involves the use of arginine analogues to inhibit NOS. In this particular assay, it should be noted that arginine analogues containing nitro groups (L-NAME and L-NOARG) should *not* be used since they interfere with the Griess reaction (Greenberg *et al.*, 1995b). The addition of carboxy-PTIO can enhance the levels of nitrite produced. This radical scavenger favours the reaction converting NO to nitrite and limits the reactions of NO with other groups such as thiols or haem, which do not necessarily result in the formation of nitrite (Amano & Noda, 1995).

13.4.1.3 Protocol B – fluorimetric assay

For those laboratories equipped with a spectrofluorimeter, there is an alternative method for measuring nitrite/nitrate levels that is 50–100 times more sensitive than the Griess reaction (Misko *et al.*, 1993b). This is based on the reaction of nitrite with 2,3-diaminonaphthalene under acidic conditions to produce the fluorescent product 1-(*H*)-naphthotriazole. The 2,3-diaminonaphthalene reagent (0.05 mg ml^{-1} in 0.62 M HCl) needs to be freshly prepared and protected from light. Then, 10 µl of this reagent is added to 100 µl of the sample and incubated for 10 min at 20 °C after which the reaction is terminated by the addition of 5 µl of 2.8 N NaOH. Fluorescence is measured using excitation and emission wavelengths of 365 nm and 450 nm, respectively. Nitrate levels can be measured after conversion to nitrite as described in Protocol A (Section 13.4.1.2). Similarly the protocol can readily be carried out using 96-well plates for batch analysis. The limit of detection is at a nitrite concentration of 10 nM (Misko *et al.*, 1993b). The same standards and controls are required as for Protocol A and the reaction is subject to the same types of interference that occur with the Griess reaction.

13.4.1.4 Advantages and disadvantages

Both methods described here are quick and simple and can be used to process a large number of samples. Careful control experiments can usually overcome the problem that the assays provide an indirect rather than direct indication of the NO that is actually being produced. These methods are the only ones available in which the presence of oxygen or superoxide anions in the sample and the instability of NO are not a problem. The major disadvantage is the lack of sensitivity. The level of detection is generally not sufficient to measure the low levels of NO synthesis by type I or type III NOS unless they are present as purified and concentrated preparations.

13.4.1.5 Applications

The lack of sensitivity severely restricts the application of the nitrite/nitrate assays described here. However, both have been used to measure the high levels of nitrite/nitrate produced by type II NOS after induction in culture systems (Gross & Levi, 1992; Misko *et al.*, 1993a; Amano & Noda, 1995; Nakane *et al.*, 1995) and to assess NOS activity in purified preparations (Misko *et al.*, 1993a; Sup, Gordon Green & Grant, 1994). The fluorimetric assay is also sufficiently sensitive to measure nitrite levels in plasma (Misko *et al.*, 1993b). The measurement of plasma nitrite levels represents the main way in which clinical studies can attempt to assess changes in NO production *in vivo*. Nitrate levels should not be measured in this context due to interference by the dietary intake of nitrates (Leone *et al.*, 1994a). More sophisticated techniques, such as high-performance capillary electrophoresis with capillary ion analysis, are available for the measurement of plasma nitrite (Leone *et al.*, 1994a) but these require specialized equipment and the fluorimetric assay provides a viable alternative.

13.4.2 Haemoglobin method

13.4.2.1 Scientific basis

NO reacts directly with the oxygenated ferrous form of haemoglobin (HbO_2) to yield the ferric form, methaemoglobin (metHb), according to the following equation:

$$\text{Haemoglobin–Fe(II)–}O_2 + NO \rightarrow \text{Haemoglobin–Fe(III)} + NO_3^- \quad (13.5)$$

This reaction results in a shift in the absorbance spectrum for haemoglobin which can be measured spectrophotometrically. The concentration of NO can be calculated on the basis of the extinction coefficient for the reaction.

13.4.2.2 Protocol

The protocol for the haemoglobin method and its application for the measurement of NO has been reviewed in detail (Noack, Kubitzek & Kojda, 1992; Hevel & Marletta, 1994; Murphy & Noack, 1994). Original methods measured the change in the *difference* in absorbance between 401 and 411 nm. This requires the use of a double-beam dual-wavelength spectrophotometer. However, the absorbance at 411 nm should not change since it is the isosbestic point for the reaction. The method described here has therefore been simplified to measuring absorbance at 401 nm only, which can be achieved with most standard spectrophotometers (Hevel & Marletta, 1994).

HbO_2 is available commercially ($>95\%$) and a stock solution can be prepared as a 25 mg 2 ml^{-1} solution ($\approx 300\ \mu M$) in 100 mM HEPES buffer, pH 7.5 and stored in aliquots at −80 °C (Hevel & Marletta, 1994). Alternatively, HbO_2 can be freshly prepared from haemoglobin or metHb by adding 1–2 mg of sodium hydrosulphite powder to 25 mg ml^{-1} haemoglobin in buffer and blowing oxygen gently into the flask. The colour of the solution changes two times ending up as a bright, light red when the conversion to HbO_2 is complete. The resulting HbO_2 solution is then desalted by Sephadex G-25 chromatography. The main band is collected (≈ 1 mM HbO_2) stored on ice and protected from light (Murphy & Noack, 1994). The final concentration of the HbO_2 used in the assay should not be greater than 10 μM so that the baseline absorbance is not too high. However, concentrations greater than 2 μM are recommended to ensure it is not depleted during the assay. The actual concentration of the HbO_2 solution (in μM) can be calculated from its absorbance spectrum based on the following equation, where A is the absorbance at the wavelength indicated (nm):

$$[HbO_2] = [1.013(A_{576}) - 0.3269(A_{630}) - 0.7353(A_{560})] \times 10^2 \quad (13.6)$$

Other haem proteins such as myoglobin can equally be used for the assay (Gross & Levi, 1992). The procedure then varies depending on the nature of the preparation to be assayed.

13.4.2.2.1 Release of NO from intact cell or tissue preparations

The preparation is incubated in the appropriate physiological buffer containing HbO_2. Aliquots of the incubation solution are sampled at regular intervals to measure the change in absorbance at 401 nm (i.e. the production of metHb) over time. Absorbance readings should be corrected for the change in absorbance of HbO_2 in the absence of tissue over the same period of time since auto-oxidation can occur (Noack *et al.*, 1992).

13.4.2.2.2 Perfusion experiments

If the spectrophotometer is equipped with a flow-through cell, continuous monitoring of NO release from a perfused preparation is technically possible. The preparations, e.g. endothelial cells grown on microcarrier beads or isolated heart preparations, should be perfused with oxygenated physiological solutions containing HbO_2. Cold HbO_2 stock solution should be added continuously to the oxygenated buffer just before the preparation to avoid the excessive oxidation of HbO_2 that can occur under these conditions. The absorbance of the perfusate can then be monitored at 401 nm. It is important that the concentration of HbO_2 in the perfusate should be kept constant throughout (Noack *et al.*, 1992).

13.4.2.2.3 Assay of NOS activity *in vitro*

The synthesis of NO by *in vitro* preparations of NOS can be measured within the spectrophotometer cell at 37 °C if the cell holder can be heated. A typical procedure that is used (Hevel & Marletta, 1994) is to preincubate the following components (final concentration) in the cell at 37 °C:

100 mM HEPES buffer pH 7.5
1 mM arginine
1 mM $CaCl_2$ and 1 μM calmodulin (if required)
4–8 μM HbO_2
100 μM NADPH
4 μM FAD
$\leq$12 μM BH_4.

The reaction is started by the addition of NOS and monitored continuously at

401 nm (Hevel & Marletta, 1994). BH_4 interferes strongly with the assay at greater concentrations. SOD (100 U) can be added to prevent interference by superoxide anions. Non-oxygenated haemoglobin absorbs in the region under study. In addition, NO can undergo a different type of reaction with non-oxygenated haemoglobin to form nitrosylhaemoglobin. Therefore, any haemoglobin should be removed from the NOS preparation. In all the above procedures media containing phenol red should not be used as this absorbs at 401 nm.

13.4.2.3 Calculation of NO levels

The extinction coefficient for the absorbance change at 401 nm [A_{401}(metHb)$-A_{401}$(HbO_2)] needs to be calibrated (Noack *et al.*, 1992; Hevel & Marletta, 1994; Murphy & Noack, 1994). A value for $\Delta\varepsilon_{401}$ of 60 000 M^{-1} cm^{-1} has been reported to be the best approximation (Hevel & Marletta, 1994); however, this can change with the HbO_2 preparation and the conditions used. The method is extremely sensitive and can measure NO levels in the low nanomolar range depending on the sensitivity of the spectrophotometer used. Calibration of HbO_2 is achieved by adding excess potassium ferricyanide to a range of known concentrations of HbO_2. This results in the complete conversion to metHb. The change in absorbance at 401 nm can then be plotted against HbO_2 concentration, the graph should be linear and go through zero (Noack *et al.*, 1992). The change in absorbance that would occur in the conversion of 1 M HbO_2 using a 1-cm pathlength cell can then be extrapolated to calculate NO concentration.

13.4.2.4 Advantages and disadvantages

This technique provides the most sensitive method available that does not require highly specialized apparatus for the measurement of NO release. It is relatively specific, although peroxynitrite, in addition to NO, will cause absorbance changes (Schmidt, Klatt & Mayer, 1994). Since HbO_2 is present when the NO is being produced, loss of free NO through other reactions is restricted. Because the detection reaction is continuous, this is also the only method described so far that can be used to monitor continuously the synthesis of NO. Interference by factors that absorb can largely be overcome. However, interference by BH_4 may mean that NOS activity has to be measured at suboptimal conditions.

13.4.2.5 Applications

This method can be used both to measure NO released from intact cells (K. Schmidt *et al.*, 1992) and NOS activity in *in vitro* preparations (McCall *et al.*,

1991; Gross & Levi, 1992). It is not very convenient for processing large numbers of samples and, for crude enzyme preparations, the radiochemical assay (Chapter 14) is still to be preferred. The method is particularly suited to kinetic analysis since NO synthesis can be continuously monitored (Abu-Soud *et al.*, 1994a) and for investigating whether the action of NOS inhibitors is reversible (Frey *et al.*, 1994; Narayanan *et al.*, 1995).

13.5 Methods requiring specialized apparatus

13.5.1 Chemiluminescence

13.5.1.1 Scientific basis

In the gaseous phase, NO reacts with ozone according to the following reaction:

$$NO + O_3 \rightarrow NO_2\cdot + O_2 \qquad (13.7)$$

$$NO_2\cdot \rightarrow NO_2 + h\nu \qquad (13.8)$$

where $h\nu$ represents energy in the form of light.

The light produced in the reaction is proportional to the amount of NO and can be measured using a luminometer.

13.5.1.2 General procedure

The original equipment for this procedure was designed to measure levels of NO as a pollutant in air. In view of the great interest in NO in biological systems, manufacturers have modified the equipment such that it can now be used to measure NO in aqueous samples. This involves as additional step whereby NO in aqueous solution is driven into the gas phase by a process of 'stripping'. Stripping is achieved by subjecting the sample to a constant flow of nitrogen gas under vacuum. The NO gas is then transferred into a chamber for reaction with ozone that is generated by the machine. The chamber is connected to a photomultiplier tube for the measurement of the light produced (Menon *et al.*, 1989; Archer, 1993). Standards can consist of NO gas, authentic NO solutions or nitrite solutions reduced under acidic conditions (see Section 13.2). Fewer than 10 pmoles of NO can be detected by this method (Archer & Cowan, 1991). This procedure can easily be adapted to provide a very sensitive measure of total nitrite/nitrate levels. Nitrate can be converted to nitrite (see Section 13.4.1). After reaction with potassium iodide in glacial acetic acid or sulphuric acid, nitrite is reduced to NO which can be measured as already described (Schmidt *et al.*, 1989b, 1990).

13.5.1.3 Advantages and disadvantages

This method provides a sensitive quantitative technique that is accurate and highly specific for NO. However, its very specificity creates limitations with regard to how well the levels of free NO measured actually reflect the amount of NO being produced. This is exemplified by the fact that the production of NO by a purified preparation of type I NOS is not detected by chemiluminescence unless SOD is present to remove superoxide anions (Schmidt *et al.*, 1995). Thus, detected levels of free NO are likely to underestimate NO synthesis and careful sample preparation is required to reduce the loss of free NO. This can partly be overcome by sacrificing specificity and using the procedure to measure nitrite instead of free NO (Schmidt *et al.*, 1989b, 1990). This reintroduces a requirement for many of the control experiments described in Section 13.3 to ensure that the nitrite measured has been derived from NO.

13.5.1.4 Applications

This sophisticated technique will continue to play an important role in any application that requires free NO to be identified specifically. It has been used to detect NO release from endothelial cells (Myers, Guerra & Harrison, 1989; Archer & Cowan, 1991) and from nitrovasodilators (Chung & Fung, 1990; Brien *et al.*, 1991). Furthermore, it has the advantage that it can be used both to measure free NO and very low levels of nitrate/nitrite. Initial difficulties in applying this technique to biological specimens are being overcome as the equipment is still being refined to suit its specific application to the biological study of NO, such that sample preparation and the time of processing can be kept to a minimum. However, for the routine analysis of NOS activity, it is likely that other techniques (see Chapter 14) are more practical.

13.5.2 NO electrodes

Scientific basis 13.5.2.1

When subjected to an applied potential in aqueous solution, NO is electrochemically active (Ichimori *et al.*, 1994):

$$NO + 2H_2O \rightarrow NO_3^- + 4H^+ + 3e^- \qquad (13.9)$$

The electrons produced are proportional to NO concentration and can be measured as changes in current.

13.5.2.2 General procedure

NO is measured using a working electrode as a probe. Several different types of working electrode have been made. Those that are commercially available work on the same principle as oxygen electrodes. Relative specificity for NO is achieved by coating the electrode with a membrane through which NO can diffuse but which is impermeable to ions or other larger molecules that are electrochemically active (Archer, 1993; Tsukahara, Gordienko & Goligorsky, 1993; Ichimori *et al.*, 1994). By using a positive voltage, it is also possible to discriminate against oxygen (Ichimori *et al.*, 1994). The electrode can be calibrated with authentic NO solutions, SNAP or nitrite solutions reduced under acidic conditions (Malinski *et al.*, 1993b; Tsukahara *et al.*, 1993; Ichimori *et al.*, 1994). The probes are now being manufactured such that they are relatively robust and can be inserted directly into tissue to measure NO synthesis *in situ*. In addition, the diameters of the electrodes are small such that NO can be measured in small volumes of solution. Measurements are made simply as changes in current by connecting the output from the electrode to a chart recorder or computer. NO levels are calculated from standard curves produced at the time of the experiment. Commercial electrodes are able to measure NO at concentrations in the low nanomolar range. Very high levels of sensitivity have been reported for a porphyrin-based electrode with a tip diameter of approximately 0.5 μm (Malinski & Taha, 1992). This electrode was made in the laboratory and used a porphyrin coating to catalyse the electrochemical oxidation of NO. The porphyrin coat was covered with a negatively charged film to prevent interference by nitrite ions. Using this electrode, it was possible to detect NO release from a single endothelial cell (Malinski & Taha, 1992). This electrode will also respond to catecholamines although with three orders of magnitude less sensitivity.

13.5.2.3 Advantages and disadvantages

Very little sample preparation is required, NO can even be measured in tissue samples *in situ*. The response time of the electrode is in the order of 1 s (Ichimori *et al.*, 1994), meaning that loss of free NO through its reaction with other factors present in the sample is limited by their rates of reaction. Even so, it has been reported that SOD needs to be added to a purified preparation of type I NOS in order to detect NO synthesis with an NO-sensitive electrode (Schmidt *et al.*, 1995). Because the electrode is inserted directly into the sample where NO is produced, NO synthesis can be monitored in real time and the progress of the reaction can be monitored over time. For reliable operation at high sensitivities, care has to be taken for proper earthing and shielding of the system from electrical interference. In addition, mechanical disturbance, e.g. changes in perfusion flow (Ichimori *et al.*, 1994) or even the careful introduction of the substrate to initiate the NOS reaction, can result in signal interference (personal observation).

13.5.2.4 Applications

NO-sensitive electrodes have only recently been manufactured and the equipment is still being improved and modified to suit individual applications. Studies have already been carried out in which NO release from endothelial cells in culture and from isolated blood vessels has been detected and quantified (Tsukahara *et al.*, 1993; Ichimori *et al.*, 1994; Tsukahara *et al.*, 1994). The fact that NO can be detected in tissue specimens or cell cultures *in situ* and that NO synthesis can be monitored continuously over time means that this approach may be highly versatile and has considerable potential in a wide variety of applications. To date, the number of publications reporting the use of commercially available NO sensors is limited. Review of future literature in the field is required to establish if NO sensors can supersede the other techniques described in this chapter for the detection and quantitation of NO.

13.5.3 Electron paramagnetic resonance spectroscopy

13.5.3.1 Scientific basis

When radicals are exposed to energy and a magnetic field, unpaired electrons are promoted to higher energy levels. Relaxation of the stimulated electron to the ground state produces a spectrum which can be measured by electron paramagnetic resonance (EPR) spectroscopy (Archer, 1993; Henry *et al.*, 1993).

13.5.3.2 General procedure

Although NO is a free radical and paramagnetic, free NO cannot be detected by conventional EPR spectroscopy because its response is too rapid. Therefore, NO is detected using spin-trapping techniques in which NO is allowed to interact with a suitable compound (the spin-trap) such that it forms a more stable adduct that is still detectable by EPR spectroscopy. NO is distinguished from other free radicals by the type of spectrum that is obtained. The first spin-trap to be used was haemoglobin, which reacts with NO to form nitrosylhaemoglobin:

$$\text{Haemoglobin–Fe(II)} + \text{NO} \rightarrow \text{Haemoglobin–Fe(II)–NO} \qquad (13.10)$$

The presence of the characteristic spectrum of nitrosylhaemoglobin can then be used to detect NO and the magnitude of the signal is proportional to the amount of NO present. In this way, the synthesis of NO was detected *in vivo* in rats with septic shock and in macrophages in culture following activation by cytokines (see Henry *et al.*, 1993 for review). In addition, the failure to detect nitrosylhaemoglobin after stimulation of endothelial cells has been used as evidence that EDRF is not free NO but rather an entity derived from NO (Greenberg *et al.*, 1990; Chapter 3). One problem with using haemoglobin as a spin-trap is that NO can undergo an alternative reaction with oxygenated haemoglobin (Eq. 13.5, Section 13.4.2.1) which would compete with the formation of nitrosylhaemoglobin *in vivo*. Thus an increased signal with EPR spectroscopy could result not only as a result of increased NO production but also as a consequence of increased oxygen demand and a reduction in the proportion of haemoglobin present as HbO_2 (Murphy & Noack, 1994).

EPR spectroscopy of nitrosylhaemoglobin has also to be carried out at very low temperatures and other spin-traps which form spin adducts that can be detected at room temperature have been investigated. These include nitrones and nitrosocompounds, which have also been used for the detection of other oxygen-derived free radicals. However, there are problems with these spin-traps in obtaining unequivocal identification of NO (see Archer, 1993; Henry *et al.*, 1993 for review). Research is continuing to develop new spin-traps that are more specific to NO and less susceptible to interference. The complex of Fe^{2+} with diethyldithiocarbamate has been proposed as a promising spin-trap since the adduct is more stable in the presence of oxygen (Henry *et al.*, 1993; Zweier, Wang & Kuppusamy, 1995). However, this complex is relatively insoluble in water. The complex of Fe^{2+} with *N*-methyl-D-glucamine dithiocarbamate has been suggested as an alternative spin-trap for use under physiological conditions since the adduct is both stable in the presence of oxygen and soluble in water (Komarov *et al.*, 1993; Zweier *et al.*, 1995).

13.5.3.3 Advantages and disadvantages

EPR spectroscopy requires expensive equipment and considerable expertise to perform. It is not particularly sensitive and is not generally used to quantify NO (Archer, 1993). It does, however, represent a powerful technique with which to identify NO and to detect its presence under physiological conditions.

13.5.3.4 Applications

EPR spectroscopy has played, and will continue to play, an important role in determining the molecular targets for NO in biological systems (Henry *et al.*, 1993; Chapter 10). In addition, by performing experiments in the presence of spin-traps, EPR spectroscopy can now be used to monitor the production of NO *in vivo* and in isolated tissue *in situ*. In this way, the production of NO has been demonstrated in isolated heart preparations following ischaemia (Zweier *et al.*, 1995). Similarly, NO production can be detected in the brain after cerebral ischaemia in control mice but not in mice in which the gene for type I NOS has been disrupted (Mullins *et al.*, 1995). Remarkably, direct monitoring of the blood circulation in the tail in the presence of a spin-trap has enabled NO production to be detected in real time and *in vivo* in conscious mice (Komarov *et al.*, 1993).

13.6 Selected reference

Archer, S. (1993). Measurement of nitric oxide in biological models. *FASEB Journal*, **7**, 349–60.

14

Measurement of nitric oxide synthase activity

14.1 Introduction

Technically, NOS activity can be measured by measuring the amount of NO produced using any of the methods described in Chapter 13. In practice, this is not the approach that is used by the vast majority of studies. This is due to the fact that the techniques for measuring NO require expensive specialized equipment and/or are very susceptible to interference, not sufficiently quantitative or are relatively insensitive (see Chapter 13). Fortunately, there is an alternative approach which has none of these disadvantages and, provided the proper controls are used, can give reliable quantitative data on relatively small amounts of sample. The basis of this approach is that, during the synthesis of NO from arginine, citrulline is produced in equivalent amounts. Using [^{3}H]arginine or [^{14}C]arginine, it is possible to measure very small amounts of radioactive citrulline produced in a radiochemical assay. In this chapter, the protocols for preparing the enzyme sample and solutions and the assay procedure will be described together with modifications that can be made and problems that can arise. It should be noted that the standard procedure described here has been based on personal experience and derived from numerous descriptions of the assay in the literature, all of which differ with regard to individual components and concentrations used. The majority of these differences are minor and unlikely to be of significance. However, if it is intended to use the assay routinely or on a particular tissue, then the procedure should only be regarded as a good starting point. Additional control experiments should be done in order to optimize conditions for each individual application.

14.2 Sample preparation

For many applications, NOS activity is measured in crude preparations and purification of the enzyme is not necessary. It should be remembered that assay of tissue homogenates does not distinguish which particular cell type within a tissue is expressing NOS. This requires the parallel use of microscopical techniques (see Chapter 16) for the localization of NOS. To a certain extent, the different isoforms of NOS can be distinguished in crude preparations by altering the preparative procedures and/or the conditions for the assay. In the majority of tissues, type I and type II NOS are predominantly cytosolic and are thus present in the soluble rather than the particulate fraction. There are exceptions to this, notably type I NOS in fast-twitch skeletal muscle fibres (Brenman *et al.*, 1995), and distinction on the basis of subcellular fractionation alone should not be regarded as absolute. Type II NOS can be distinguished on the basis that its activity is Ca^{2+}-independent (see Section 14.5.3). It has been estimated that 95% of type III NOS activity in endothelial cells is membrane bound and is thus localized predominantly in the particulate fraction (Förstermann *et al.*, 1991b). There are numerous studies in the literature that have investigated NOS activity using a variety of different preparative procedures on a range of different tissues and the following references are given as examples (Bredt & Snyder, 1990; Mitchell *et al.*, 1991; Pollock *et al.*, 1991; Salter, Knowles & Moncada, 1991; Schmidt *et al.*, 1991; Bush, Gonzalez & Ignarro, 1992b; Hiki *et al.*, 1992; Sheng *et al.*, 1992; Carlberg, 1994; Förstermann *et al.*, 1994; Knowles & Moncada, 1994; Lowenstein & Snyder, 1994; Lincoln & Messersmith, 1995; Nakane *et al.*, 1995). Crude preparations can be used for screening NOS activity in a variety of tissues and for investigating whether changes in NOS activity have occurred in a variety of experimental and pathological conditions.

14.2.1 Crude enzyme preparations

14.2.1.1 Homogenizing medium

This is comprised of 10 mM HEPES buffer (pH 7.4) containing:

0.32 M sucrose
1 mM dithiothreitol
0.1 mM EDTA
10 $\mu g\ ml^{-1}$ leupeptin
10 $\mu g\ ml^{-1}$ soybean trypsin inhibitor
2 $\mu g\ ml^{-1}$ aprotonin
[± detergent e.g. 0.1% (v/v) Triton X-100 or 20 mM CHAPS].

Concentrations of HEPES up to 100 mM may be used. Dithiothreitol is a reducing agent that prevents the formation of disulphide bonds. Leupeptin, soybean trypsin inhibitor and aprotonin are protease inhibitors. Some methods use phenylmethylsulphonyl fluoride (up to 100 $\mu g\ ml^{-1}$) as an additional protease inhibitor. This is relatively insoluble and can be dissolved by sonication. Detergent should only be added to the homogenizing medium when it is needed to solubilize membrane-bound NOS activity. In all the procedures described in the following sections the homogenizing medium should be kept at 4 °C and tissue preparations kept cold and stored on ice until assay. Once tissues have been homogenized, they should be taken through the assay procedure as quickly as possible since loss of NOS activity can occur fairly rapidly with time.

14.2.1.2 Whole homogenates

This method involves minimal sample preparation and can be used when no distinction between soluble or membrane-bound NOS activity is required. Samples are homogenized with 10–20 volumes of the homogenizing medium using a motor-driven homogenizer. Sonication is not recommended since there can be some loss of NOS activity as a result. The whole homogenate can then be used for the assay. If the samples are cells in culture, then homogenization is not always appropriate. Cells can be removed in the homogenizing medium by scraping. Disruption of the cells can then be achieved by repeated rapid freezing and thawing.

14.2.1.3 Soluble NOS activity

If only soluble NOS activity is to be measured then the homogenizing medium must *not* contain detergent and the homogenate has to be centrifuged to remove the particulate fraction. For complete removal of membrane-bound activity, the sample should be centrifuged at 100 000 g for 1 h. In practice, a lower speed and shorter time for centrifugation is usually adequate and more convenient. The supernatant after centrifugation at 12 000 g for 10 min from a homogenate of endothelial cells (containing membrane-bound type III NOS only), prepared in the absence of detergent, contains minimal NOS activity (personal observation). However, if it is important to an application that a contribution from membrane-bound NOS is totally excluded then control experiments using 100 000 g for 1 h should be carried out to validate the results. After centrifugation the supernatant is removed and used for assay.

14.2.1.4 Membrane-bound NOS activity

This usually applies to the measurement of type III NOS activity in endothelial cells. If purified endothelial cell preparations are to be used then the cells can be homogenized in the homogenizing medium containing detergent. The assay can then be performed using the whole homogenate or the supernatant after centrifugation. It may be necessary to exclude a contribution from soluble NOS, for example in tissue samples that might also contain type I NOS or where type II NOS has been induced. In this case, the samples should be homogenized in medium *without* detergent. After centrifugation, the pellet can then be re-homogenized in medium containing detergent. Following further centrifugation, the supernatant can be used for assay.

14.2.2 Removal of endogenous arginine

The tissue concentration of arginine may be approximately 100 μM and cell culture media may contain 0.5–1 mM arginine (Knowles & Moncada, 1994). Since the basis of the NOS assay is to measure the conversion of radioactive arginine to radioactive citrulline, the higher the levels of endogenous, non-radioactive arginine the lower the sensitivity of the assay. Arginine can be removed from crude supernatants prepared as described previously by passing them through a Dowex ion-exchange column (for preparation see Section 14.3.1. Arginine binds to the column leaving the eluant essentially arginine free. In practice, NOS activity and changes in NOS activity can be measured in many tissues without removing endogenous arginine. However, it should be noted that values for NOS activity obtained without removal of, or correction for, endogenous levels of arginine will be relative rather than absolute. This is because the specific radioactivity of the [^{3}H]arginine or [^{14}C]-arginine will be altered once it is incubated with sample containing non-radioactive arginine (Knowles & Moncada, 1994).

14.2.3 Preparations of individual isoforms of NOS

Some investigations, such as the study of the action of inhibitors, require standard preparations of individual isoforms of NOS. This requires a large amount of source material. The most common sources for the different isoforms of NOS are (Förstermann *et al.*, 1994; Nakane *et al.*, 1995):

- type I NOS – rat brain;
- type II NOS – Murine RAW 264.7 macrophages cultured until confluent and then activated for up to 18 h with 100–300 ng ml^{-1} lipopolysaccharide from *Escherichia coli* and 3 U ml^{-1} INF-γ;
- type III NOS – bovine aortic endothelial cells cultured until confluent.

Type I and type II NOS are prepared by homogenizing in 50 mM Tris-HCl buffer, pH 7.5, containing: 1 mM EDTA; 5 mM β-mercaptoethanol; 10 μg ml^{-1} pepstatin; 10 μg ml^{-1} aprotonin; 10 μg ml^{-1} leupeptin; and 100 μg ml^{-1} phenylmethylsulphonyl fluoride and centrifuging at 100 000 *g* for 1 h. The resulting supernatant can then be used for study. For type III NOS, the pellet resulting from centrifugation should be resuspended in this buffer additionally containing 20 mM CHAPS. Experiments can be performed using the supernatant obtained after further centrifugation at 100 000 g for 1 h.

Purification of the individual isoforms of NOS is only generally necessary when detailed kinetic or mechanistic analyses are to be performed. For type I and type III NOS, this involves affinity chromatography with 2′,5′-ADP Sepharose followed by fast protein liquid chromatography. In the case of type II NOS, BH_4 needs to be present during the purification steps which involve sequential 2′,5′-ADP Sepharose and DEAE-agarose chromatography. For a very detailed description of the purification steps for all the isoforms of NOS, readers should refer to Förstermann *et al.* (1994). Once NOS has been purified, the samples can be concentrated by ultrafiltration and aliquots frozen in liquid nitrogen and stored at −80 °C.

14.3 NOS assay: standard solutions and chemicals

14.3.1 Dowex ion-exchange resin

The Dowex ion-exchange resin should be prepared in advance of the assay. This is used to remove [^{14}C]arginine or [^{3}H]arginine from the samples after the assay so that the radioactivity measured reflects the amount of [^{14}C]citrulline or [^{3}H]citrulline produced. Dowex 50X8-200 ion-exchange resin is the type that should be used. This is supplied in the hydrogen form and needs to be converted to the sodium form. This is achieved by repeated washing of the resin with 1 M NaOH until the effluent appears clean and clear. The resin is then washed repeatedly with double-distilled deionized water until the pH returns to that of the water. The resin can then be stored and used for chromatography either in water or in 50 mM HEPES buffer pH 7.4 containing 1 mM EDTA.

Ion-exchange columns can be simply prepared by adding the resin to 1-ml disposable syringes plugged with glass wool. Samples are added to the top of the columns which are then rinsed with 3 × 1-ml portions of water or HEPES buffer. The combined eluants can be collected directly in scintillation vials for counting radioactivity. When processing large numbers of samples, it is more convenient to add the resin (1:1 mixture with water or buffer) batchwise to each sample. After vortexing, the resin quickly settles and the supernatant can be transferred to vials for counting. To assess the effectiveness of the chromatography, known amounts of [^{3}H]arginine can be added and the eluant tested for the efficiency of its removal. Similarly, standard samples of [^{3}H]citrulline can be used to determine that there has not been significant loss of citrulline during the separation process (Salter *et al.*, 1991; Hevel & Marletta, 1994; Lincoln & Messersmith, 1995).

It is possible to recycle the ion-exchange resin once it has been used. Radioactivity that is bound to the column can be removed by repeated washes with 0.5 M NH_4OH. The eluant should be checked for radioactivity until it approaches background levels. The resin can then be prepared as already described and used again. In practice, there is a limit to the number of times that the resin can be reused. If blank values become too high then the resin should be replaced by fresh stock.

14.3.2 [^{14}C]Arginine or [^{3}H]arginine substrate solution

The substrate for the NOS assay is either [^{14}C]arginine or [^{3}H]arginine both of which are readily available commercially as solutions. Provided the specific activity is high enough (approximately 1.11–2.22 TBq $mmol^{-1}$ or 30–60 Ci $mmol^{-1}$), [^{3}H]arginine is sufficient for most applications. [^{14}C]Arginine can be more expensive but it is less susceptible to self-decomposition. Although [^{3}H]arginine is the isotopic form that will be stated throughout the remainder of this chapter, [^{14}C]arginine can equally be used in all cases. Commercial preparations are not 100% pure and decomposition or bacterial contamination can occur over time. As a consequence, significant radioactivity is present in reagent blanks carried through the entire assay procedure. This radioactivity is not due to citrulline and the levels can increase with time of storage of the stock. The levels of radioactivity in the blanks largely determines the sensitivity of the assay. If they are too high, the [^{3}H]arginine should be purified before use. This can be achieved by adding the [^{3}H]arginine to a Dowex ion-exchange column (see Section 14.3.1). After 4 × 1-ml washes with water, the [^{3}H]arginine can be eluted with 0.5 M NH_4OH (4 × 1-ml). The combined eluants should then be freeze-dried. The purified [^{3}H]arginine can then be redissolved in 2% ethanol using sterile procedures. It is often convenient to store the [^{3}H]arginine in aliquots sufficient for each assay run to prevent contamination of the stock.

In order to have a high enough arginine concentration in the assay, direct use of the commercial [^{3}H]arginine can be prohibitively expensive. In addition, it may be necessary to use an arginine concentration above that which is supplied. In order to increase the arginine concentration, [^{3}H]arginine can be mixed with non-radioactive arginine. It is necessary to know the relative proportions of radioactive and non-radioactive arginine in the working solution so that the change in specific activity of the [^{3}H]arginine can be calculated. The reduction in specific activity that results will reduce the sensitivity of the assay. However, in most cases, this is compensated for by the increase in NOS activity that occurs with higher concentrations of arginine (see Section 14.5.2.2). In addition, the blanks will be kept at a reasonably low level. In practice, a working solution with a specific activity in the range 11.1–111 G Bq $mmol^{-1}$ or 0.3–3 Ci $mmol^{-1}$ is sufficient for most applications

(Hevel & Marletta, 1994; Lincoln & Messersmith, 1995). The actual concentration of arginine to use is best determined by experiment.

14.3.3 Incubating media

The standard incubating medium A that is recommended is 50 mM HEPES, 1 mM EDTA buffer (pH 7.4) containing:

1 mM dithiothreitol
1.25 mM $CaCl_2$
10 μg ml^{-1} calmodulin
1 mM NADPH
10 μM FAD
100 μM BH_4.

Modifications can be made to incubating medium A depending on the application (see Section 14.6). In order to ensure that the activity being measured is due to NOS and to characterize this activity, incubating media should also be prepared containing the appropriate inhibitors, e.g.:

incubating medium B: as A above but also containing 1 mM L-NAME;
incubating medium C: as A above but also containing 2 mM EGTA;
incubating medium D: as A above but also containing 200 μM trifluoperazine (TFP).

Unfortunately, NADPH, FAD and BH_4 are not stable in solution and so the incubating media should be used as fresh as possible and cannot be stored. For convenience, 100 mM HEPES, 2 mM EGTA buffer, pH 7.4 (2 × conc.) can be stored as a stock. The remaining constituents can be preweighed and stored at −20 °C. On the day of the assay, the constituents can be made up in water at concentrations higher than required. The constituents can then be added to the 2 × conc. buffer stock together with additional water in the right proportions to give the final concentrations shown above. When adding the inhibitors in incubating media B–D, care should be taken not to alter the concentrations of the other constituents.

14.4 NOS assay: standard procedure

1 Prepare enzyme sample – keep on ice.
2 Prepare incubating media – keep on ice.
3 Prepare [^{3}H]arginine substrate solution, i.e. mix with non-radioactive arginine.

4 Add 25 µl of the enzyme sample to each tube. It is recommended that each measurement should be carried out in triplicate.
5 Add 25 µl of the homogenizing medium to three tubes to act as reagent blanks.
6 Add 100 µl of the respective incubating media to each sample or blank. The effects of inhibitors should be assessed in the same assay run.
7 Equilibrate at 37 °C in a water bath.
8 Start the reaction by adding 25 µl of [^{3}H]arginine substrate solution at 15–30-s intervals to each tube and mix.
9 Incubate at 37 °C for 10 min.
10 Stop the reaction by the addition of 2–3 ml of the Dowex ion-exchange resin (1:1 mixture in water or buffer) at 15–30 s intervals and vortex mixing. (If columns are to be used, the reaction can be stopped by the addition of excess L-NAME and EGTA in HEPES buffer and the samples processed as described in section 14.3.1).
11 Allow the resin to settle and transfer the supernatant to scintillation vials with a Pasteur pipette. Take care not to get any resin in the pipette.
12 Add any scintillation fluid that is designed for aqueous samples, e.g. 5 ml Optiphase 'HiSafe 3' and mix well.
13 Count the radioactivity on a liquid scintillation counter.

There are alternatives to this procedure which have been used routinely. If the amount of sample is very small, then the volumes used can be scaled down proportionately. The reaction can be started by the addition of the enzyme rather than the [^{3}H]arginine. Alternatively, NADPH can be prepared as a separate solution and used to initiate the reaction. If the co-factors, FAD and BH_4 are not to be added to the assay then this last modification has the advantage that the incubation media can be prepared well in advance of the assay (Salter *et al.*, 1991; Hevel & Marletta, 1994; Lincoln & Messersmith, 1995).

14.5 Control experiments and blanks

14.5.1 Blanks

However well the [^{3}H]arginine has been purified, some radioactivity that is not due to the production of [^{3}H]citrulline will be measurable. In addition, a small percentage of the [^{3}H]arginine may not be removed by the ion-exchange resin. Because of this, it is extremely important to correct the radioactivity measured in the samples for the radioactivity present in the appropriate blanks. In many studies, blanks consist of samples incubated in media containing 1 mM L-NAME and 2 mM EGTA (see Salter *et al.*, 1991 as an example). In most cases this may be adequate; however, it is recommended that two different blanks should be used during each assay run. This is particularly important when measuring NOS activity for the first time in a new tissue. The first blank consists of 25 μl of the homogenizing medium taken through the entire procedure. The second blank consists of 25 μl of the sample incubated with incubation medium B (containing 1 mM L-NAME) and taken through the entire procedure. Since L-NAME inhibits all the isoforms of NOS, the radioactivity measured in the reagent blanks should be the same as the radioactivity measured in the sample in the presence of L-NAME. If they are not the same then this provides a strong indication that there has been interference with the assay (see Section 14.7). Once it has been established that this is not the case then a single blank may be sufficient. For routine analysis of different tissues, L-NAME has been selected as the NOS inhibitor. A variety of arginine analogues are available that inhibit NOS. L-NMMA and ADMA can be metabolized by certain tissues (see Chapter 10). In addition L-NOARG is much less soluble than its methyl ester L-NAME.

14.5.2 General control experiments

14.5.2.1 Linearity with time and tissue concentration

It might be assumed that increasing the time of incubation or tissue concentration would increase the sensitivity of the assay. However, the activity of enzymes can rapidly reach a plateau with time and with increasing tissue concentrations, particularly in crude preparations. If it is intended to use the NOS assay to determine whether changes in NOS activity have occurred in a given situation then it is very important to ensure that this plateau has not been reached (Hevel & Marletta, 1994). Incubation times of 30 min or longer are not appropriate for many crude enzyme preparations despite what may be read in the literature. Therefore, whenever performing the NOS assay on a new tissue for the first time, control experiments should be carried out in which NOS activity is measured after different incubation times and using different concentrations of tissue. NOS activity can then be plotted against time and concentration. The optimum time and concentration can then be selected from the linear portion of the graph.

14.5.2.2 [^{3}H]arginine substrate concentration

For kinetic analysis of NOS activity, it is necessary to carry out the NOS assay using different concentrations of the [^{3}H]arginine substrate. If it is only intended to assess whether NOS activity is present in a tissue then the concentration of arginine used may not be critical. However, if the aim of the study is to

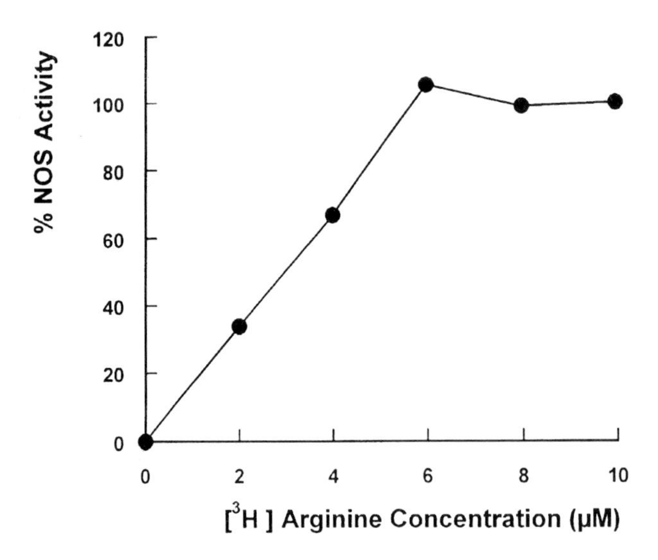

Fig. 14.1 Effect of arginine concentration on NOS activity in a preparation from the rat ileum containing the myenteric plexus and smooth muscle layers. [3H]Arginine substrate solutions consisted of a range of concentrations with the same specific activity (3.61 Ci mmol^{-1}). Activity at each concentration was corrected for its own blank (homogenate omitted). Results have been expressed as a percentage of the activity under standard conditions (10 μM arginine), measured concurrently. Data consist of a representative experiment, which was repeated, and have been given as the mean of triplicate determinations. (Reproduced from Lincoln & Messersmith, 1995; *Journal of Neuroscience Methods*, **59**, 191–7, with permission of the publishers, Elsevier, Amsterdam.)

measure whether changes in NOS activity have occurred then it is important to establish that the concentration of arginine used is not rate limiting. If it is rate limiting then any increase in NOS activity may not be detectable. In order to assess this, [3H]arginine solutions should be prepared with a range of concentrations but the same specific activity. The assay should be performed using these substrate solutions on the same tissue sample. An example of this type of experiment is given in Fig. 14.1. It can be seen that, in this particular preparation, concentrations less than 6 μM were rate limiting whereas a concentration of 10 μM was not.

14.5.3 Characterization of NOS activity by inhibitors

The total NOS activity in a sample can be calculated as *radioactivity (dpm) in the sample incubated in medium A minus dpm in the reagent blank.* This can then be converted to nmole h^{-1} g^{-1} tissue by simple calculation.

Alternatively, total NOS activity can also be calculated as *radioactivity (dpm) in the sample incubated in medium A minus dpm in the sample incubated in medium B, i.e. the difference in activity in the presence and absence of L-NAME.*

If there is no interference with the assay these two values should be the same.

It is often useful to characterize the NOS activity further by assessing whether it is Ca^{2+}- or calmodulin-dependent, using EGTA as a Ca^{2+} chelator or a calmodulin antagonist such as TFP. This can then be used to determine if activity is due to type II NOS (Ca^{2+}-independent) or to types I and III NOS (Ca^{2+}-dependent).

Ca^{2+}-dependent NOS activity can be calculated as *radioactivity (dpm) in the sample incubated in medium A minus dpm in the sample incubated in medium C, i.e. the difference in activity in the presence and absence of 2 mM EGTA.*

If there is a mixture of Ca^{2+}-dependent and Ca^{2+}-independent activity in the same sample then the total activity and the Ca^{2+}-dependent activity will not be the same.

Ca^{2+}-independent NOS activity can be calculated as *total NOS activity minus Ca^{2+}-dependent NOS activity.*

A similar process can be used to determine calmodulin dependence. Omission of calmodulin from the incubating media is not sufficient since crude preparations contain significant levels of calmodulin from the cell cytosol. It should be noted that type II NOS still binds calmodulin even though its activity is not dependent on Ca^{2+}. Thus, calmodulin antagonists can inhibit type II NOS activity although higher concentrations of antagonists are often required and inhibition is not always complete.

An example of the effects of these inhibitors on NOS activity measured in the soluble fraction of a myenteric plexus/smooth muscle preparation from the rat ileum is given in Fig. 14.2. From these data it can be seen that all of the NOS activity was Ca^{2+}-dependent and thus not due to type II NOS. All of the NOS activity was also inhibited by the calmodulin antagonist TFP. The fact that the activity was measured in the soluble fraction also provides an indication that the NOS activity was probably due to type I NOS, which is cytosolic, rather than type III NOS, which is membrane-bound (Lincoln & Messersmith, 1995).

14.6 Modifications of the assay

The composition of the standard incubating medium A may appear to be unnecessarily complex. It is often assumed that the addition of all the components is only necessary for purified enzyme preparations since crude preparations will still contain many of the components from the cell cytosol and the co-factors will be bound to the NOS. However, before omitting any of the components it is

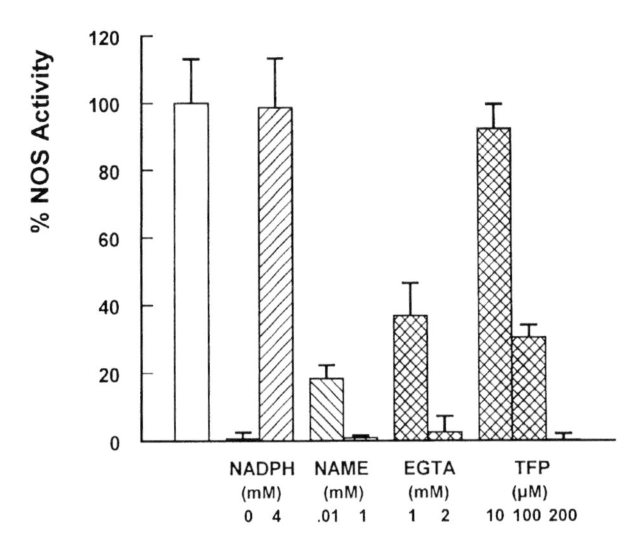

Fig. 14.2 Effects of the co-substrate, NADPH, and of the inhibitors, L-NAME, EGTA and TFP, on NOS activity in a myenteric plexus/smooth muscle preparation from the rat ileum. The clear bar represents control activity measured under standard conditions (see text for the composition of the standard incubating medium A). The incubating media were altered as indicated under each of the hatched bars. Results are expressed as a percentage of the control activity, measured concurrently, and have been given as the mean ± SEM of 3–5 experiments each consisting of triplicate determinations. (Reproduced from Lincoln & Messersmith, 1995; *Journal of Neuroscience Methods*, **59**, 191–7, with permission of the publishers, Elsevier, Amsterdam.)

advisable to check whether this influences the NOS activity even when assaying a crude enzyme preparation. This point is illustrated well in Figs. 14.2–14.4 which provide data obtained from experiments on a crude soluble fraction of a myenteric plexus/smooth muscle preparation from the rat ileum. From these data, it can be seen that it is an absolute requirement for NADPH and Ca^{2+} to be added in the incubation medium. In the absence of either one of these components NOS activity was not detectable (Figs. 14.2 and 14.3). It was also possible to demonstrate that, at a concentration of 1 mM, the co-substrate NADPH was not rate limiting since a higher concentration of NADPH did not result in any increase in NOS activity. Interestingly, increasing the Ca^{2+} concentration above a certain level resulted in inhibition of NOS activity (Fig. 14.3), indicating that it is possible to add too much as well as too little of a particular component. It should be noted that the concentration of Ca^{2+} added does not represent the final free Ca^{2+} concentration since the non-specific divalent cation chelator EDTA is present in both the homogenizing and incubating buffers. In order to determine the precise free Ca^{2+}

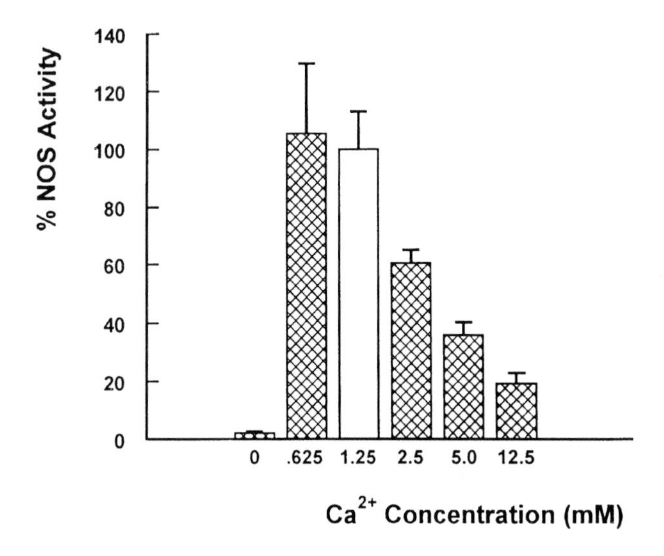

Fig. 14.3 Effects of Ca^{2+} concentration on NOS activity in a myenteric plexus/smooth muscle preparation from the rat ileum. The clear bar represents control activity measured under standard conditions (Ca^{2+} concentration, 1.25 mM). Ca^{2+} concentrations were altered as indicated under the hatched bars. Note that these concentrations represent those added exogenously and not the final free concentration in the incubation medium which contains a non-specific chelator, EDTA. Results are expressed as a percentage of the control activity, measured concurrently, and have been given as the mean ± SEM of 3–4 experiments each consisting of triplicate determinations. (Reproduced from Lincoln & Messersmith, 1995; *Journal of Neuroscience Methods*, **59**, 191–7, with permission of the publishers, Elsevier, Amsterdam.)

concentration present it is necessary to calibrate the solutions with a Ca^{2+} electrode. In the case of crude preparations of type II NOS, it may be necessary to add 1 mM magnesium diacetate to the incubating medium. The activity of crude but not purified preparations of type II NOS is dependent on the presence of Mg^{2+} (Hevel & Marletta, 1994).

Many studies do not add BH_4, FAD or FMN to the incubating medium when assaying NOS activity in crude preparations. However, it can be seen from Fig. 14.4 that maximum NOS activity might not be measured if these co-factors are not added. The NOS activity of this particular preparation measured in the absence of added co-factors was only 10% of the activity measured in the presence of added BH_4 and FAD. The addition of FMN did not increase NOS activity further. Individually, BH_4 and FAD increased NOS activity, but not to the level observed when they were present together (Lincoln & Messersmith, 1995). Unfortunately,

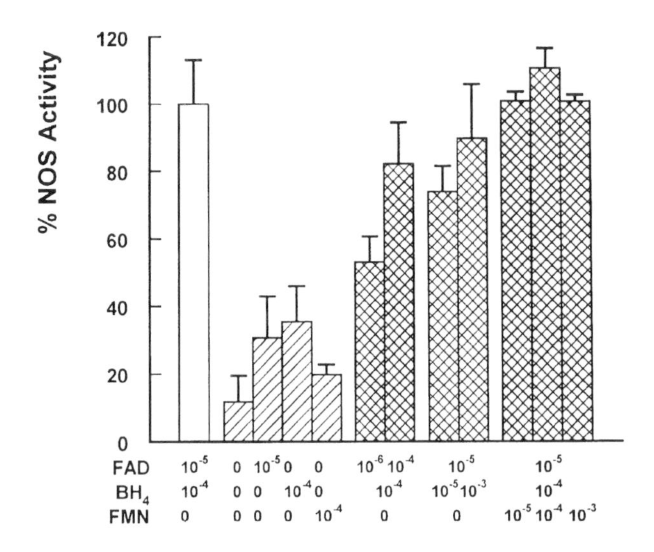

Fig. 14.4 Effects of the co-factors, FAD, BH_4 and FMN, on NOS activity in a myenteric plexus/smooth muscle preparation from the rat ileum. The clear bar represents control activity under standard conditions (see text for the composition of the standard incubating medium A). Co-factors were investigated either singly (single hatching) or in combination (cross hatching). The concentrations (M) used have been indicated under each bar. The composition of all the other components of the incubating medium were kept constant. Results have been expressed as a percentage of the control activity, measured concurrently, and have been given as the mean ± SEM of 3–5 experiments each consisting of triplicate determinations. (Reproduced from Lincoln & Messersmith, 1995; *Journal of Neuroscience Methods*, **59**, 191–7, with permission of the publishers, Elsevier, Amsterdam.)

the requirement for added co-factors may vary between different tissues, particularly in the periphery. Therefore the optimal conditions for a given tissue may need to be assessed individually.

Measuring NOS activity under different defined conditions may also be important when investigating changes in NO synthesis in pathological conditions. As an example, BH_4 synthesis can be reduced in diabetes mellitus (see Chapter 9). Thus, pathological changes in NO synthesis may be reflected by a change in the requirement of NOS for added BH_4 as well as by changes in NOS activity under optimal conditions.

14.7 Problems and interference with the NOS assay

The procedures described here should be able to provide good indications of NOS activity in the majority of tissues where NOS is known to be present.

Initial problems are likely to be solved by carrying out the various control experiments suggested in order to establish the optimal conditions for the particular tissue under investigation. There is no question that the speed with which the assay is carried out and the precision with which the solutions are made largely determine the variability of the results. It is very important to keep all solutions and enzyme preparations cold throughout the initial stages. With practice, the assay can provide highly reproducible and accurate results and is very sensitive. One of the main disadvantages of the radiochemical assay is in kinetic experiments since it is discontinuous, i.e. the reaction cannot be monitored continuously over time. In such cases, using specific NO electrodes to monitor the reaction might be considered as an alternative, provided interference with the NO detection can be excluded (see Chapter 13).

There are aspects of the NOS assay described here that need to be viewed with some caution and which may be of particular importance in certain tissues. Firstly, in crude preparations, it cannot be assumed that NOS is the only enzyme present that is involved in the metabolism of arginine or citrulline. Secondly, the presence of radioactivity produced by the samples cannot always be assumed to be due to the production of citrulline exclusively. In both the liver and the kidney, there are pathways which can synthesize arginine from citrulline. Thus, in these tissues, measurements of [^{3}H]citrulline produced are likely to underestimate the NOS activity that is actually present. This can largely be overcome by carrying out the NOS assay in the presence of excess (1 mM) non-radioactive citrulline (Knowles & Moncada, 1994). Several tissues also contain arginase, which converts arginine to ornithine. Ornithine can be subsequently metabolized by ornithine aminotransferase or by ornithine decarboxylase, which forms the first step in a pathway which synthesizes polyamines such as spermine or putrescine. If arginase is present in significant amounts then it will not only compete with NOS for the [^{3}H]arginine substrate, but it could also result in the synthesis of a variety of radioactive products in addition to citrulline. The presence of such interference can often be detected by using more than one blank, as described in Section 14.5.1. If NOS activity assessed by its sensitivity to L-NAME is significantly less than that assessed by correction with the reagent blank, then it is probable that radioactive products in addition to citrulline are present. This effect can be considerable. In the outer medulla of the rat kidney, activity that is sensitive to L-NAME can be less than 30% of the activity measured by correction with the reagent blanks (unpublished observations). In many cases, this problem can be overcome by adding valine (50–60 mM) to the incubating media as an inhibitor of arginase activity (Knowles & Moncada, 1994). However, from personal experience, this may not be sufficient in crude homogenates of the cortex and outer medulla of the kidney, which contain particularly high levels of arginase, ornithine decarboxylase and ornithine aminotransferase. An alternative approach with particularly problematic tissues may be to subject the eluant from the ion-exchange resin to high-performance liquid chromatography such that the citrulline can be identified unequivocally (Carlberg, 1994).

14.8 Selected references

Knowles, R. G. & Moncada, S. (1994). Nitric oxide synthases in mammals. *Biochemical Journal*, **298**, 249–58.

Lincoln, J. & Messersmith, W. A. (1995). Conditions required for the measurement of nitric oxide synthase activity in a myenteric plexus/smooth muscle preparation from the rat ileum. *Journal of Neuroscience Methods*, **59**, 191–7.

15

Pharmacology

15.1 Introduction

In this chapter general protocols for examining the transmitter status or endothelium-derived relaxing factor status of NO *in vitro* and *in vivo* are described. There is no substitute for experience, and it is not intended that this chapter comprehensively describes all common and less common preparations; rather, it describes the principles of NO pharmacology. Many types of apparatus for recording activity in several isolated organs and tissues have been comprehensively described by Hugo Sachs Elektronik (1995). The principles for neural and endothelial preparations are the same, and follow the scheme:

1 Determine that the neural response is non-adrenergic, non-cholinergic (NANC), and confirm that the response is indeed neurogenic. In vascular preparations, determine that the response is endothelium dependent.
2 Determine that activation of NOS is taking place by showing that inhibition of NOS abolishes the response. Demonstrate that inhibition of NOS by an arginine analogue can be reversed or prevented by application of excess L-arginine, but not D-arginine.
3 Show that application of NO or NO donors mimics the response.
4 Show that the response can be severely attenuated or abolished by haemoglobin.
5 Show that raising the level of superoxide anions, by using superoxide anion generators or inhibitors of SOD, inhibits the response.
6 Show the involvement of soluble guanylate cyclase and cGMP in the effector mechanism by blocking guanylate cyclase activity (and blocking the response), or by inhibiting cGMP metabolism with phosphodiesterase inhibitors (and potentiating the response).

Steps 1–3 are crucial, steps 4–6 less so. Step 6 will depend upon the observed response being mediated via guanylate cyclase and cGMP, this should not be assumed to be the case, although in smooth muscle preparations this will almost certainly be so.

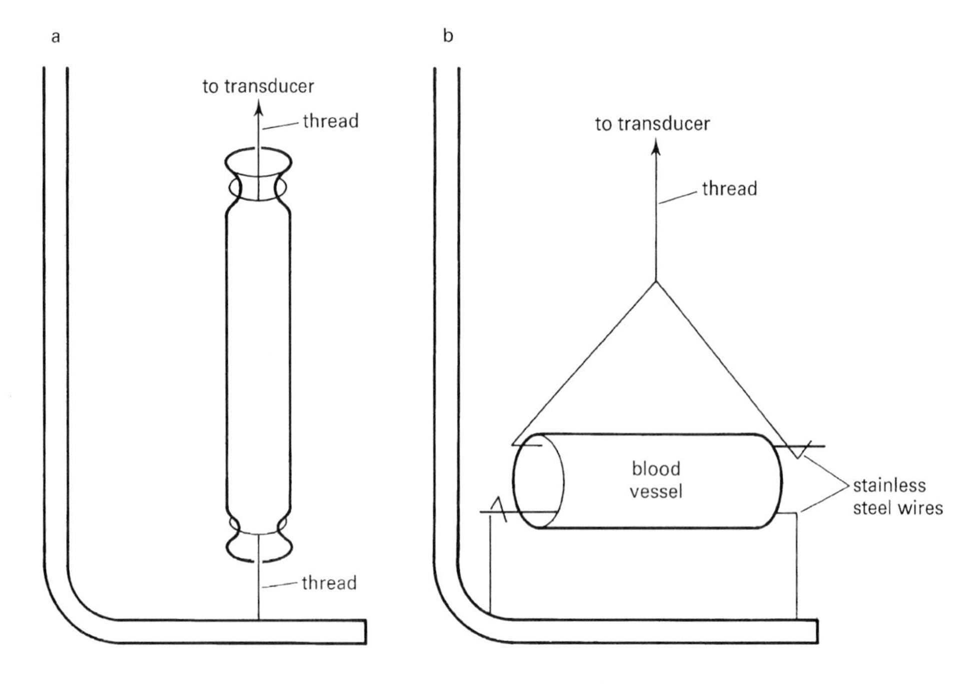

Fig. 15.1 Diagram of typical ways for mounting isolated smooth muscle preparations in organ-baths. (a) The preparation could be any smooth muscle that has the majority of its cells running in the vertical axis, such as longitudinal muscle from the gastrointestinal tract, or a circular muscle dissected circumferentially, or even a blood vessel such as the portal vein, or a helical strip preparation of a blood vessel. The whole assembly would be immersed in an organ bath containing a physiological saline solution (e.g. Krebs' solution). The transducer would be either a force-displacement transducer, for measuring the tension in the preparation under near isometric conditions, or an auxotonic or isotonic transducer, in which case the change in muscle length is measured. (b) For measuring the radial tension in a blood vessel wall the dissected vessel is slid over two wires (tungsten or stainless steel), and the transducer would measure the isometric tension. Stimulating electrodes are not shown: in preparations in (a) a pair of Pt-wire ring electrodes, approximately 3 mm in diameter and 10 mm apart and through which the preparation is threaded, work well. Other common configurations are parallel Pt-wires approximately 20 mm long and 5 mm apart placed vertically on each side of the preparation. For vessels in (b) the electrodes would be parallel wires (Pt) placed horizontally, one on each side of the preparation. (Modified from Kennedy, 1984.)

15.2 Neurogenic responses (step 1)

Simple isolated smooth muscle preparations are mounted in standard organ baths of a suitable size (2–10 ml), maintained in standard bicarbonate-buffered physiological saline solution (Krebs' or Tyrode solution), gassed with

95% O_2/5% CO_2. The mechanical activity of the preparation is recorded, and suitable electrodes are present to allow electrical field stimulation (Fig. 15.1a). If a means of electrical field stimulation is not available, nicotine (1–100 μM) or the nicotinic receptor agonist 1,1-dimethyl-4-phenylpiperazinium (DMPP, 100 μM) may be used. Many isolated smooth muscles have little intrinsic tone and, because nitrergic transmission is often inhibitory, it may be necessary to raise the tone in order to be able to observe more easily any relaxant responses. The choice of contractile agent varies from tissue to tissue, and the advice here is to consult previously published papers. Autonomically innervated organs usually contain intramural neurones the utilize a variety of transmitter substances. In the first instance it is usually beneficial to block cholinergic transmission with atropine (0.3–1.0 μM), and to block sympathetic transmission with guanethidine (2–4 μM), or block α- and β-adrenoceptors with phentolamine and propranolol (both 0.1–1.0 μM), respectively. All these agents are inexpensive, and it is most convenient to include them in the Krebs' solution, rather than add them to the organ bath. In the presence of such agents the residual transmission processes are NANC. Responses to EFS should be constructed using a range of stimulation frequencies, from 0.5 to 32 Hz is common (e.g. Figs. 11.1 and 11.2 in Chapter 11), with train durations sufficiently long enough to allow a maximum response to be observed at any given frequency. Pulse durations should be kept short (0.1–1.0 ms), and the stimulus strength (voltage) should be either supramaximal or adjusted to give a known level of response. If nicotine or DMPP is being used, a concentration–response relationship is desirable but, in practice, this is seldom constructed, largely because nicotinic receptors readily desensitize, and it may be difficult to obtain reproducible responses to a given concentration.

When EFS is applied it must be confirmed, at least in some pilot experiments, that the stimulation parameters are specific for the neural elements, and that direct stimulation of the smooth muscle is not occurring. Tetrodotoxin (TTX, 0.1–1.0 μM) will block action potential conduction, and is the simplest way to confirm neural stimulation. Similarly, if nicotine or DMPP is being used TTX will confirm that responses are neurogenic.

15.3 Blood vessels: neural and endothelium-dependent responses (step 1)

There are two main approaches to the study of blood vessels. The first is to dissect short segments, a few millimetres long, and mount them in organ baths in a manner described by Bevan and Osher (1972) or in a more specialized apparatus, such as a myograph (Mulvany & Halpern, 1976), so that isometric tension of the muscle coat can be measured. The second is to perfuse long vessels segments *in vitro*, or to dissect out vascular beds or organs, and then perfuse them through a major artery, or to perfuse a vascular bed through a major artery *in situ* in an anaesthetized animal.

For vessels with an internal diameter greater than approximately 100 μm, two tungsten, stainless steel or platinum wires (diameter as small as 25 μm, but use stouter wires for larger vessels) can be passed through the lumen (Bevan & Osher, 1972). One is attached to a rigid support, and the other, via a silk thread, to an isometric force transducer (Fig. 15.1b). The assembly is immersed in an organ bath, typically 5 or 10 ml, containing physiological saline (Krebs' solution), gassed with 95% O_2/5% CO_2 (the O_2 provides a high pO_2, the CO_2 drives the bicarbonate buffering system, to keep the pH close to 7.3). Myographs similar to that described by Mulvany and Halpern (1976) have been developed for recording mechanical activity of small blood vessels with a diameter as small as 100 μm. The vessel segment is slid over two tungsten wire (40 μm) prongs, one of which is fixed, the other of which is attached to the armature of an isometric force transducer. The preparation is perfused with gassed physiological saline. The resting tension of the vessel segment is adjusted by means of a micrometer screw. In this apparatus, because the perfusate is gassed, there are no gas bubbles that rise through the chamber and which might hit the preparation or the silk thread, thereby causing artefactal 'blips' on the recording trace.

In the organ bath, and in the Mulvany apparatus a pair of platinum wire stimulating electrodes can be placed on each side of the preparation to facilitate EFS of the nerve terminals present. A particular feature of the electrodes used for stimulating blood vessels in the organ bath is that a high current density is required, thus the electrodes should be as close to the vessel as possible, and the electrodes themselves need to be of narrow diameter, 500 μm maximum. Parallel plate electrodes may work, but they are unlikely to allow an optimal stimulation.

For vascular beds a cannula is inserted into the main artery and perfusion pressure is recorded via a pressure transducer attached to a side-arm. The perfusate is a physiological saline, and often contains an inert protein such as bovine serum albumen (2 %), which raises the oncotic pressure and reduces the formation of oedema. This is especially important when the vascular bed is associated with an organ, for example the Langendorff heart preparation, or isolated perfused kidney, hepatic perfusion, or hind-limb perfusion. In preparations like the isolated mesenteric vascular bed, which is freed of its organ (the small intestine), inclusion of an osmotic agent in the saline may not be necessary. In isolated systems it is possible to apply EFS by using the cannula itself as one electrode, and by laying the preparation on a stainless steel mesh, which becomes the complementary electrode. With judicious experimental conditions it is even possible to stimulate sympathetic or sensory nerves, selectively (Ralevic *et al.*, 1996).

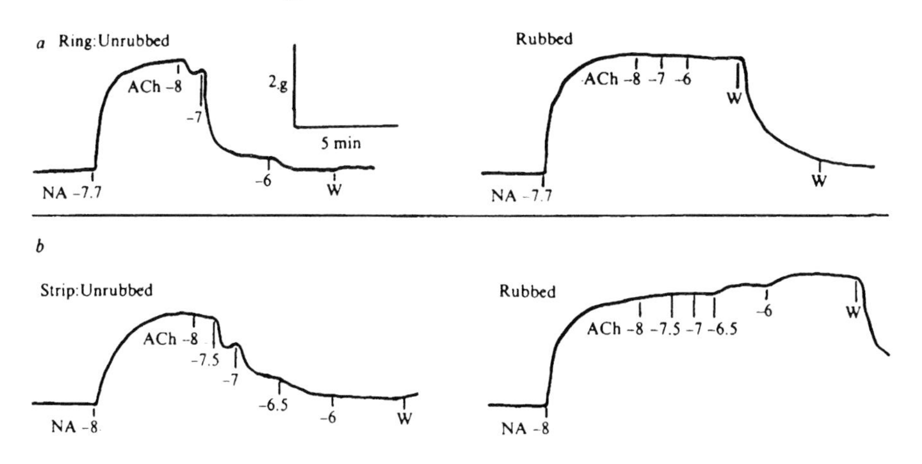

Fig. 15.2 Confirmation of endothelium-dependent responses in ring and strip preparations of isolated rabbit aorta. (a) A ring preparation is constricted with noradrenaline (NA, $10^{-7.7}$ M) to obtain a standard level of tone. Acetylcholine is added cumulatively (ACh, 10^{-8}, 10^{-7}, 10^{-6} M), and produces a concentration-dependent relaxation. The endothelium was removed by gently rubbing the intimal surface. The experiment was repeated, and the loss of reactivity to ACh confirms its removal. (b) Essentially the same as (a) except that the preparation was a helical strip rather than a ring segment. W indicates washout. (From Ralevic & Burnstock, 1993.)

As for other isolated smooth muscle preparations, it may be necessary to generate a controlled level of tone in the blood vessel in order to be able to evaluate relaxant responses. Often α-adrenoceptor agonists are used, and sometimes prostaglandins, histamine, 5-HT or some other agent might be used. As above, the advice is to consult the literature to obtain some idea about what agent might be useful.

In vascular preparations of any sort the integrity of the endothelium should be tested using an agent that is known to evoke an endothelium-dependent response. Most commonly ACh (10 nM and higher), bradykinin (1 nM upwards) and the Ca^{2+} ionophore A23187 (0.1 nM and upwards) are used (see Ralevic & Burnstock, 1993). If responses to these agents are present in control preparations but absent in 'endothelium denuded' preparations, then it is almost certain that the endothelium has been adequately removed (Fig. 15.2).

15.4 Removal of endothelial cells (step 1)

There are several methods for removing the endothelium from blood vessels, some are better applied to isolated vessels, usually ring segments being examined isometrically, and some more appropriate for whole beds or perfused vessels. Whatever method is used, it is necessary to confirm that the endothelial cells have been removed, and also that the smooth muscle of the preparation has not been damaged. Removal of endothelium can be verified histologically, or by the lack of response to an endothelial cell stimulant (e.g. ACh, bradykinin or A23187). Viability of the smooth muscle can be checked by application of non-endothelium-dependent agonists, such as papaverine or SNP, whose responses should be unaltered by the de-endothelializing procedure.

15.4.1 Isolated vessels

Mechanical methods are the most efficient for removing endothelium from short segments of isolated blood vessels or helical strip preparations; all these methods involve using an implement of some sort to rub off the endothelial cells. The method chosen is, by and large, a matter of personal preference, and it should not matter how the endothelium is removed as long as it is sufficiently damaged to prevent endothelium-dependent responses, and as long as the underlying smooth muscle is not damaged.

1. Large vessels such as the aorta and great veins (diameter >1 mm), and strip preparations of vessels.
   - i Rub the intimal surface with moistened cotton wool (Moore *et al.*, 1990).
   - ii Rub the intimal surface gently with a wooden stick (something like a cocktail stick, or the handle of a single-ended applicator) (Martin *et al.*, 1985).
   - iii Rub the intimal surface gently with filter paper (Molina, Hidalgo & Garcia de Boto, 1992).
2. Medium sized vessels (diameter >250 μm).
   - i Pass a cotton or silk thread through the lumen.
   - ii Gently press the isolated vessel segment with a Krebs-moistend finger – the correct pressure can be determined from experience.
3. Small vessels (<250 μm), too small to pass a thread through the lumen.
   - i Gently press the vessel with a Krebs-moistened finger.
   - ii Pass a fine tungsten wire through the lumen, and gently press the vessel.

15.4.2 Perfused vascular beds and long segments of vessels used for perfusion

Most of these methods involve using a chemical substance to destroy the endothelial cells, often a detergent. Some methods have been developed that effectively desiccate the endothelium, and some methods involve mechanical means. It is possible to use these methods *in situ* or *in vitro* before dissecting small segments for study. Optimal protocols need to be developed for individual preparations and the information given here can only be taken as a guide.

1. Detergents.
   i CHAPS – perfuse the vascular bed with CHAPS (30 s to 3 min, 3.0–4.7 mg ml^{-1}) (Moore *et al.*, 1990; Molina *et al.*, 1992).
   ii Sodium deoxycholate – perfuse the preparation with sodium deoxycholate (0.15–0.3%, 30 s) (Takenaga *et al.*, 1995; Ralevic & Burnstock, 1991).
   iii saponin – perfuse the preparation with saponin (50 mg ml^{-1}, 2 min) (Wiest, Trach & Dammgen, 1989).
   iv Triton X-100 – perfuse the preparation with Triton X-100 (0.1%, 2 min) (Furman & Sneddon, 1993).
2. Desiccation.
   i Air drying – pass a stream of air through the perfusion line (25 ml min^{-1}, 3 min) (Bjorling *et al.*, 1992).
   ii CO_2 drying – pass a stream of CO_2 through the perfusion line (Qiu, Henrion & Levy, 1994).
   iii Ethanol drying – perfuse the preparation with ethanol (96%, 30 s) (Peredo & Enero, 1993).
3. Mechanical.
   i Balloon catheterization angioplasty – a balloon catheter is guided into the vessels to be de-endothelialized, where it is inflated and gently pushed and pulled back and forth (three strokes), thereby damaging and removing the endothelial cells (Clowes, Reidy & Clowes, 1983). Balloon catheters (Fogarty embolectomy catheters) are available in a range of sizes, the smallest have a tip diameter of approximately 1 mm (Fogarty FG2), and the diameter of the maximally inflated balloon is approximately 4 mm.
   ii Use a combination of high perfusate flow and a stream of air bubbles, e.g. 2 min of high flow followed by 5 min of air bubbles, followed by 2 min of high flow (Ralevic *et al.*, 1989).

15.5 Inhibition of NOS (step 2)

Having established that the preparation has NANC neural responses or endothelium-dependent responses, the next step is to block NOS to provide the first line of evidence that NO is the transmitter or EDRF. The most commonly used

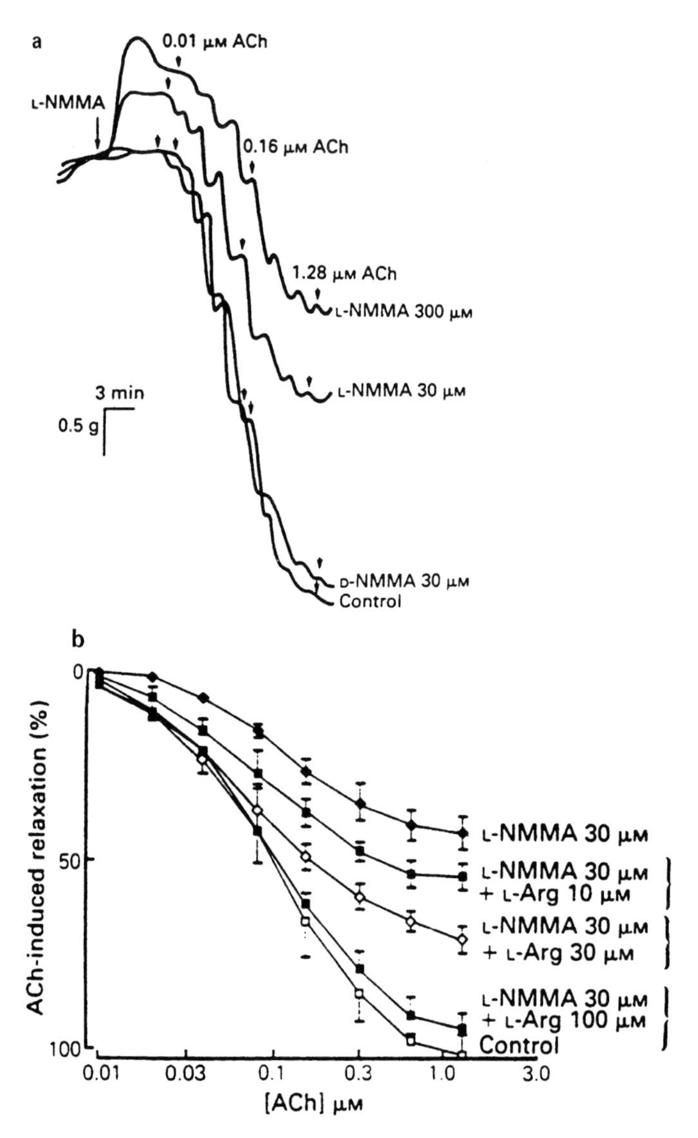

Fig. 15.3 Concentration dependency of inhibition of endothelium-dependent responses to ACh by L-NMMA in ring preparations of isolated rabbit aorta. (a) Four experimental traces have been overlaid, and show the effect of L-NMMA (30 and 300 μM) and D-NMMA (30 μM) on the relaxant responses to ACh (0.01–1.28 μM). (b) Concentration–response relationships for relaxant responses to ACh in the presence of L-NMMA and reversal of antagonism by L-arginine (10–100 μM). (From Ralevic & Burnstock, 1993.)

antagonists of NOS are the arginine derivatives: L-NMMA (100 μM), L-NNA (=L-NOARG, =NOLA, 10–100 μM) and L-NAME (10–100 μM). It is only necessary to test one of these, and the most potent, L-NAME, would be a good one to start with. None of these inhibitors needs a long incubation time, some experimenters use only 10 min. Equilibration of the antagonist can be checked by observing the responses to EFS. During incubation with the antagonist the responses to EFS will decline if they are mediated by NO, and they will have declined maximally when the antagonist has fully equilibrated. The maximum degree of antagonism will depend on the concentration of the antagonist. In neural preparations frequency–response curves should be constructed in the presence of various concentrations of the antagonist in order to determine a concentration that is sufficiently or maximally effective. In endothelial preparations full concentration–response curves to an endothelium-dependent agonist should be determined for the same reason (Fig. 15.3).

The arginine-binding site of the enzyme NOS is stereo-specific, thus the D-enantiomers of NMMA, NNA and NAME are inactive. It is not essential to test this. However, it is important to confirm that the antagonist is competing with L-arginine. This is accomplished by adding L-arginine to the organ bath, in the presence of the antagonist, at a concentration 5–10 times greater than that of the antagonist. This should cause full or partial reversal of the blockade. Application of D-arginine instead of L-arginine should have no effect, and would confirm the stereo specificity of the arginine binding-site.

There are other types of NOS antagonists available (see Chapter 11) but it is not necessary to use these compounds in preference to the competitive arginine-binding site antagonists.

15.6 Application of NO (step 3)

15.6.1 NO in solution

Preparation of NO in solution (sometimes referred to as authentic NO) needs to be carried out with great care in order to exclude O_2, and to prevent exposure of the experimenter to NO_2 or N_2O_4 that form when NO mixes with O_2. NO solutions should be prepared in a fume cupboard in a well-ventilated room, with exhaust NO being passed through a series of gas scrubbers that contain basic $KMnO_4$, which converts the NO to nitrite and nitrate, thus preventing the escape of NO into the atmosphere (Archer, 1993). NO can be obtained commercially as a pure gas (>99%), which is bubbled in a glass sampling bulb for 30 min through deionized distilled water that has previously been deoxygenated by bubbling with helium for 30 min. Archer (1993) recommends injecting N_2 with a gas-tight syringe through a rubber septum into the sampling bulb before withdrawing samples of NO solution to ensure that O_2 is excluded. At 20 °C and 1 atmosphere (101.3 kPa) the solubility of NO in H_2O is 0.056 g.l^{-1} (Aylward & Findlay, 1971), which is approximately a concentration of 1.5 mM. Dilutions of this stock solution are made using deoxygenated H_2O immediately before use, and are kept on ice. These are applied to the organ bath in the usual way.

15.6.2 Acidification of sodium nitrite ($NaNO_2$) to make solutions of NO

Dissolve $NaNO_2$ in distilled H_2O to make a solution of 1 M. Dilute in 0.1 M HCl (0.1% v/v concentrated HCl/distilled H_2O, pH < 2) to produce a stock of 10 or 100 mM and leave to stand for 10–15 min on ice before further serially diluting in decades in HCl. The concentration of NO is nominally equivalent to half the concentration of the nitrite: at a low pH the nitrite is protonated and forms nitrous acid (HNO_2, $pK_a = 3.3$), which is unstable and degenerates to NO:

$$2HNO_2 \rightleftharpoons NO\cdot + NO_2 + H_2O \qquad (15.1)$$

The acidified nitrite is added to the organ bath in the usual manner. It is important to perform control experiments in which equal volumes of HCl are added to the organ bath. Addition of 0.5 ml of acid vehicle to a 10-ml organ bath containing Krebs' solution with 25 mM HCO_3^-, gassed with 95% O_2/5% CO_2, decreases the pH by 0.26 (Tam & Hillier, 1992).

15.6.3 NO donor compounds

These compounds release NO from their intramolecular structure, and are another way of applying NO to the tissue. They are not strictly an equivalent to NO in solution because they may need to be taken into a cell and metabolized in order to release the NO (see Chapter 11), and subsequently stimulate guanylate cyclase.

The most commonly used NO donor for organ-bath work is SNP, followed by SIN-1 and SNAP. Added to the organ bath they should mimic nerve or endothelial cell stimulation and application of authentic NO. They can be used to examine the guanylate cyclase mechanism and to show, for example, that the observed effect of a NOS inhibitor is not due to a non-specific interaction with guanylate cyclase or phosphodiesterase. In a typical experiment, a compound such as L-NAME would shift an EFS frequency–response curve to the right, or even flatten it completely, but it would not have any effect on the concentration–response relationship for SNP. This would show that L-NAME is having its effect on responses to EFS by inhibiting NOS activity, rather than inhibiting the NO effector further downstream, at the level of guanylate cyclase or beyond.

All the common NO donors are water soluble and it is merely a question of constructing concentration–response relationships to determine their activity.

15.7 Preparation of oxyhaemoglobin (step 4)

Oxyhaemoglobin is applied to the organ bath (10 μM) and allowed to equilibrate for several minutes ($\approx$10 min) before being tested in the experimental system. It can be prepared from a haemolysate of blood, or it can be prepared from commercially obtained haemoglobin. An important point is that haemoglobin remains extracellular so, although it should prevent responses mediated by endogenous NO or applied authentic NO, it may not affect responses mediated by NO donors if the donor releases NO intracellularly.

15.7.1 Blood haemolysate

The methods commonly used are based on those described by Bowman and Gillespie (1982).

1 Whole blood is obtained from terminally anaesthetized rats (sodium pentobarbitone, 65 mg kg^{-1}, i.p.) via a cannula inserted into the carotid artery. The blood (2 × 3 ml) is collected into centrifuge tubes that contain heparin (50 IU).
2 Centrifuge the samples (1000 *g*, 20 min, 4 °C). Aspirate and discard the plasma and buffy coat.
3 Wash the red blood cells (RBCs) in isotonic phosphate buffer (pH 7.4) twice. Resuspend the RBCs in isotonic phosphate buffer added to a volume of 3 ml.
4 Transfer 1 ml of the RBC suspension to a polycarbonate tube (40 ml) and add 19 ml of hypotonic phosphate buffer (20 mosmol l^{-1}, pH 7.4).
5 Mix (the RBCs will lyse) and then centrifuge (20 000 *g*, 30–40 min, pH 7.4, 4 °C). The supernatant is the RBC haemolysate.
6 The haemolysate can be dialysed in order to remove low molecular weight impurities, but this step is not usually necessary. Put 15 ml of haemolysate into Visking tubing, in 5 l distilled water, continually stirred, 4 °C, overnight.
7 Assay for haemoglobin content using standard spectrophotometry.

Variations:

1 Step 1 – obtain blood from dogs by venipuncture (Keef *et al.*, 1993; Stark, Bauer & Szurszewski, 1991), or obtain postexpiry date human blood from a blood bank.
2 Step 4 – lyse 1 ml of RBC suspension with 4 ml of hypotonic buffer (Stark *et al.*, 1991), or lyse the RBCs with distilled water 1 : 1 (Keef *et al.*, 1993).

In this haemolysate the haemoglobin is predominantly in the oxyhaemoglobin form, and it should be bright red. It can be stored frozen (–20 °C) for a day or so. When the haemolysate goes off it turns brownish red due to the formation of the oxidized derivative, methaemoglobin. (Methaemoglobin will not bind NO.)

15.7.2 Reduction of commercial haemoglobin

Haemoglobin can be obtained commercially, and used to make solutions of known concentration, but the predominant form will probably be methaemo-globin not oxyhaemoglobin. The methaemoglobin can be reduced to oxyhaemo-globin by treating it with sodium dithionite (Martin *et al.*, 1985).

1 Make solution of haemoglobin (1 mM) in distilled water.
2 Add tenfold molar excess sodium dithionite ($Na_2S_2O_4$), mix.
3 Remove dithionite and low molecular weight impurities by dialysing against 100 volumes of distilled water (e.g. 10 ml haemoglobin solution in Visking tubing, in 1 l distilled water) for 2 h at 4 °C.

15.8 Manipulation of superoxide anions (step 5)

Hydroquinone, pyrogallol and diethyldithiocarbamate all increase the level of superoxide anions in the tissue, and therefore should inhibit NO-mediated responses. These compounds should be added to the organ bath or perfusate and allowed a period of time to equilibrate. If possible the time course of equilibration should be followed (as already described), but otherwise an equilibration period of 30 min should be adequate.

6,7-Dimethoxy-8-methyl-3′,4′,5-trihydroxyflavone and SOD should both protect NO against degradation, and therefore should potentiate the NO-mediated responses. This has the practical implication that it is necessary to test the compounds against submaximal responses, which can be potentiated, rather than against maximal responses which might not be. Again, these compounds are added to the organ bath or perfusate, and allowed time to equilibrate.

Following equilibration, construct stimulus–response curves. It may be necessary to use various concentrations to determine how the preparation behaves, and to determine that the chosen concentration is optimal. Also, it is important to test these agents against a control substance that is known not to involve NO production or even the guanylate cyclase mechanism.

15.9 Manipulation of the soluble guanylate cyclase system (step 6)

Essentially the same as Section 15.8, the guanylate cyclase inhibitor or phosphodiesterase inhibitor should be applied and allowed sufficient time to equilibrate. Experiments should be carried out to determine that the chosen concentration is optimal, and that it does not have undesired side-effects that could obscure the true result.

GMP mimetics have little value. For example, application of 8-bromo-cGMP will yield a result that shows what happens when cGMP levels are raised within the tissue, but it will not tell you whether or not cGMP is involved in the effector mechanism of the tissue being investigated.

15.10 Selected references

Archer, S. (1993). Measurement of nitric oxide in biological models. *FASEB Journal*, **7**, 349–60.

Hugo Sachs Elektronik (1995). *Research Equipment for Pharmacology and Physiology*, I.1 *Organ Baths and Apparatus for Isolated Organs and Tissues*. March-Hugstetten, Germany: Hugo Sachs Elektronik.

Ralevic, V. & Burnstock, G. (1993). *Neural–Endothelial Interactions in the Control of Local Vascular Tone*. Boca Raton, FL: CRC Press.

16

Microscopy

16.1 Introduction

It can hardly be imagined that anyone is going to be thrust into the deep end, and need to use an electron microscope to visualize the ultrastructural localization of NOS or NADPH-diaphorase without ever having used such a microscope before, or even having no experience of immunoelectron microscopy, unless they have adequate supervision. Thus, details of electron microscopy (EM) methods are not provided here. Any readers who find themselves in such a challenging situation are advised to seek help from a laboratory in which such procedures are successfully used. This chapter contains practical details of tissue fixation, histochemical localization of NADPH-diaphorase activity and immunocytochemical localization of NOS. Principles and variations of these techniques are discussed. It is assumed that the reader knows how to set up and use a microscope for the appropriate lighting conditions.

16.2 Tissue fixation

For the localization of NOS by visualizing sites of NADPH-diaphorase activity the tissue must be fixed. Most commonly segments of tissues or organs are immersed in 4% paraformaldehyde for periods of approximately 90 min to 2 h. For electron immunocytochemistry of tissues from laboratory animals a common method is to anaesthetize the animal very deeply, beyond possible recovery, and then perfuse fixative through a major blood vessel, such as the aorta, femoral artery or transcardially. The perfusion with fixative might be preceded by a short period of perfusion with saline that may contain an anticoagulant (heparin or citrate). The fixative would usually be a mixture of paraformaldehyde (4%) and glutaraldehyde (0.05–1%). Varndell and Polak (1986) have poignantly stated that

> good morphological preservation often precludes maximal retention of antigen immunoreactivity, and that compromise must be achieved to provided optimal reaction product deposition with acceptable ultrastructure.

Because it is such an efficient cross-linker glutaraldehyde can prevent the diffusion of antibodies through the tissue, and also it can cross-link enzymes and neuropeptides, thus decreasing their antigenicity. To overcome these problems the concentration of glutaraldehyde in the fixative can be kept to a low concentration (Valtschanoff *et al.*, 1993a; Bates *et al.*, 1995). If glutaraldehyde is used in the fixative prior to immunocytochemical work it will probably be necessary to wash the preparation with sodium borohydride (1% w/v in Tris buffer, pH 6–7.2, for 30–60 min) to block any free aldehyde groups that exist, and which could be available to cross-link with the primary antibody (Varndell & Polak, 1986; Valtschanoff *et al.*, 1993a). After perfusion for approximately 15 minutes tissues are removed and immersed in fixative for a further hour, to complete the fixation.

The advantage of perfusion fixation is that the fixative penetrates the tissue more rapidly and more deeply, because it is going through the vasculature under pressure, than it does by simple diffusion from the outside during immersion fixation. The more rapid penetration also reduces the time available for ultrastructural postmortem changes. Immersion-fixed material is preserved adequately for working with the light microscope, but it does not provide the best preservation of cellular substructure, especially of those cells that do not lie close to the surface of the specimen.

16.3 NADPH-diaphorase staining for light microscopy (LM)

Several methods for diaphorase staining are given in Table 16.1, and are representative of those found in the literature. As can be seen, there are several variations on a single theme: unfortunately a systematic study has not been performed in order to determine which one might be the best. The diaphorase staining reaction does not work at all well on tissues that have been fixed, infiltrated with paraffin wax, and subsequently dewaxed.

In practical terms the usual steps are to make up the solvent, add the tetrazolium salt and any other additives except for the NADPH, making sure that they are properly dissolved, then immediately before use the NADPH is added, and the staining solution is applied to the tissue.

16.3.1 Sections

1 Fix tissue by immersion in 4% v/v phosphate-buffered saline (PBS; 0.1 M, pH 7.4) for 60–120 min (depending on thickness of sample).
2 Wash thoroughly in PBS (three washes, 10 min each).
3 Immerse in PBS containing 7% (w/v) sucrose and 0.1% (w/v) sodium azide (PBS–sucrose–azide).

The sucrose increases the tonicity of the solution and acts as a cryoprotectant,

Table 16.1 *Recipes for histochemical staining of reduced nicotinamide adenine dinucleotide (NADPH) diaphorase activity*

Constituent	Recipe														
	1	2	3	4	5	6	7	8	9	10	11	12	13	14	15
Buffer[a] (pH)	7.4[b]	7.3	7.3	7.4	8.0	7.4	7.6	7.4	8.0[c]	8.0	—[d]	8.0	8.0	7.4	7.4
β-NADPH (mg ml^{-1}) [1.2][e]	1.0	1.0	0.83	0.5	1.0	1.0	1.0	1.0	2.5	1.2	1.0				0.2
β-NADP (mg ml^{-1}) [1.3]												0.77	0.77	0.5	
NBT (mg ml^{-1}) [1.2]	0.23	0.5	0.5	0.1	0.15	0.2	0.25	0.2				0.16	0.16	0.5	1.0
NT (mg ml^{-1}) [1.7]									3.0[f]						
BSPT (mg ml^{-1}) [2.0]										2.4[g]	3.0[g]				
L-Malate (mg ml^{-1}) [5.6]								2.7				2.7	2.7		
L-Lactate (mg ml^{-1}) [8.9]														11.2	
L-LDH (U ml^{-1})														50[h]	
$MgCl_2$ (mg ml^{-1}) [10.5]															0.05
$NiCl_2 \cdot 6H_2O$ (mg ml^{-1}) [4.2]											0.12[i]				
Triton X-100 (v/v %)				0.3	0.3	0.3	0.5	0.1					0.4[j]		

Abbreviations: β-NADPH, reduced β-nicotinamide adenine dinucleotide phosphate (tetrasodium salt); β-NADP, β-nicotinamide adenine dinucleotide phosphate (sodium salt); NBT, nitroblue tetrazolium, 2,2′-di-*p*-nitrophenyl-5,5′-diphenyl-3,3′-[3,3′-dimethoxy-4,4′-diphenylene]-ditetrazolium chloride; NT, neotetrazolium, 2,2′,5,5′-tetraphenyl-3,3′-(*p*-diphenylene)-ditetrazolium chloride; BSPT, 2-(2′-benzothiazolyl)-5-styryl-3-(4′-phthalhydrazidyl) tetrazolium chloride; L-malate, L-malic acid, disodium salt; L-lactate, L-lactic acid, sodium salt; L-LDH, L-lactic dehdyrogenase (EC 1.1.1.27).

In this table the information taken from the original papers has been normalized to give component concentrations as mass per unit volume (mg ml^{-1}). The original versions might have given amounts to make up a specific volume, or might have given the concentration of components in terms of molarity. The factor used to convert between molarity and mg ml^{-1} is given in the first column, within square brackets.

[a]Buffer is either phosphate-buffered saline (PBS, 0.1 M), or Tris-HCl, unless indicated otherwise.
[b]Veronal buffer (0.3 M), staining performed in an atmosphere of N_2.
[c]Glycylglycine buffer (0.05 M) containing 30% w/v polyvinyl alcohol.
[d]HEPES buffer (140 mM), no pH given, containing 32% polyvinyl alcohol (grade G04/140).
[e]The number given within square brackets here and for other substrates is the factor used to convert the concentration from mg ml^{-1} to mM, e.g. β-NADPH, 0.83 mg ml^{-1} = (1.2 × 0.83) mM = 1.0 mM.
[f]Neotetrazolium was used because the authors state that NBT and β-NADPH react in the absence of enzyme.
[g]BSPT was dissolved in a minimum volume (50 μl) of dimethylformamide.
[h]The value of 50 U ml^{-1} for this enzyme is a 'guestimate', based on the data available.
[i]A range of 0.06–0.18 mg ml^{-1} (0.25–0.75 mM) was suggested, depending on the tissue.
[j]The detergent was Tween-80 rather than Triton X-100.
References (column number)

1. Burstone, 1962.
2. Pullen & Humphreys, 1995.
3. Hoyle *et al.*, 1996.
4. Tomimoto *et al.*, 1994.
5. Southam & Garthwaite, 1993.
6 Valtschanoff *et al.*, 1993a.
7. Gabbott & Bacon, 1993; Song, Brookes & Costa, 1993.
8. Belai *et al.*, 1992; Saffrey *et al.*, 1992; Hassall *et al.*, 1992.
9. Chayen & Bitensky, 1991.
10. Wolf, Wurdig & Schunzel, 1992; Calka, Wolf & Brosz, 1994.
11. Altman, Hoyer & Andersen, 1979.
12. Sancesario *et al.*, 1993.
13. Vincent, Staines & Fibiger, 1983.
14. Nachlas, Wayler & Seligano, 1958.
15. Pearse, 1972.

preventing tissue damage from water expansion during freezing. The azide is needed to inhibit bacterial growth. If the morphology of the sections is inadequate, the sucrose concentration can be increased up to 20%. The sample should remain in PBS–sucrose–azide overnight in a refrigerator, and can be stored in this solution at 4–8 °C for several months.

4 Blocks for cryostat sectioning are prepared. The tissue is trimmed, oriented, and stuck onto cork blocks with embedding medium (OCT, Tissue Tek) and frozen in melting isopentane (–172 °C) suspended in a polythene beaker in liquid nitrogen, in a Dewar flask.

   These blocks should be stored in liquid nitrogen (–192 °C) until needed.

5 Cut sections at 10–15 μm.

This is a good thickness for visualizing nerve fibres, although better resolution of cellular distribution of the diaphorase activity may be gained on thinner sections. Nevertheless, sections of this thickness provide an adequate resolution, and also allow nerve fibres to be followed for longer distances than they probably could be in thinner sections. The sections should be lifted from the cryostat blade onto glass microscope slides coated with, for example, gelatin or poly(L-lysine), which prevents them lifting off during subsequent processing. When the slides are coated is best to do so by submersion (in 0.1% w/v solution in distilled water, and allowed to air-dry), thus coating both sides at once, and thereby avoiding the necessity to mark the side of the slide that has been coated.

Cut sections may be stored for several weeks in a normal freezer (–20 °C). Frozen sections should be allowed to warm up to room temperature, and the condensation that forms on the slide allowed to dry, before being stained.

16.3.2 Whole-mounts

The whole-mount should be prepared and then fixed (as described in Section 16.2), or be prepared from fixed tissue. The staining solution should be made up freshly, and it is beneficial to choose a method that incorporates detergent, in order to increase the permeability of the tissue. As a method to increase the permeability it is not recommended to dehydrate the whole-mount through graded alcohols, and rehydrate before staining, as this may inhibit the diaphorase staining reaction.

16.4 Staining protocol

A general staining protocol is given here, and following it, in stepwise order, are comments on variations.

1 Prepare solvent: buffer ± detergent.
2 Add and dissolve tetrazolium salt and other additives if used (e.g. malate, lactate or metal salt).
3 Add and dissolve NADPH or NADP.
4 Apply to tissue (sections or whole-mounts), incubate at 37 °C in the dark for 30–60 min.
5 Wash with buffer.
6 Dehydrate through graded alcohols, clear in a proprietary clearing agent such as Histoclear, or xylene, and mount in synthetic resin (e.g. DPX, Gurr). Alternatively, do not dehydrate, but mount in aqueous mounting medium (Aquamount), or glycerine jelly.

16.4.1 Step 1

Commonly used buffers are PBS and Tris-HCl. Various groups use pH values ranging from 7.3 to 8.0. Because of the non-specific production of formazan, or 'nothing diaphorase' activity being accelerated by alkaline pH (Pearse, 1972), theoretically it would seem appropriate to use pH 7.3 rather than 8.0. However, good results are obtained at pH 8.0 (for example Vincent, Staines & Fibiger, 1983; Chayen & Bitensky, 1991; Sancesario *et al.*, 1993; Southam & Garthwaite, 1993).

It is a good idea to filter the tetrazolium solution, before adding the other components, in order to remove undissolved particles. For small volumes a filter that fits on to the end of a standard syringe is ideal (Minisart NML, Sartorius, Germany). The absolute final concentration will be unknown, but this is not important as it will be close to the nominal concentration, which is not critical. The benefit of a clearer staining pattern outweighs this trivial disadvantage.

Some staining solutions call for the addition of L-malate to the reaction mixture, even though NADPH rather than NADP is being used as the substrate (see Table 16.1). Because of the high concentration of NADPH in the reaction mixture, and the relatively short incubation time needed, the use of L-malate is strictly unnecessary, and the contribution of the regenerated NADPH (see Chapter 12) to the overall production of formazan is probably negligible. In the absence of a systematic study this must remain a questionable practice, but nevertheless it is harmless, and some authors prefer to use it in this way.

Two recipes given in Table 16.1 (Altman, Hoyer & Andersen, 1979; Chayen & Bitensky, 1991) include polyvinyl alcohol (PVA) of a particular grade, in the buffer. Both these protocols were designed for fresh rather than fixed tissues, and the PVA provides a viscous matrix that reduces the loss of soluble components from the tissue section, and it also reduces the rate of solute diffusion. Bearing in mind that in order to reveal NOS by diaphorase activity the tissue must be fixed, there is probably no advantage to be gained by using PVA. However, no systematic study has been performed to show otherwise.

When metal ions are used the salt can be made up into a stock solution at ten or 100 times the required concentration, and diluted in buffer either before or after the tetrazolium salt is added. Tetrazolium salts do not chelate metal ions, only the formazans do.

16.4.2 Steps 1–3

Stock solutions of buffer/detergent/NBT and buffer/NADPH can be made up to twice their working concentration, divided into aliquots and stored frozen (−20 °C). To use, defrost aliquots and mix equal portions immediately prior to use. Discard any excess. Do not freeze–thaw–refreeze. The solutions can be kept frozen for several months. However, some authors maintain that the staining solution must be made up entirely freshly, and prefer to use preweighed vials of NADPH (Sigma Chemical Co.) (Gabbott & Bacon, 1993). These authors also used exceedingly long incubation times (up to 18 h) (Gabbott & Bacon, 1993, 1994a,b), so perhaps ultrafresh reaction solution is necessary.

The reaction mixture containing NBT and NADPH does go ‘off’. If left to stand at room temperature in ambient light a formazan precipitate can be seen to form, but this process is usually slow, taking days before it is easily noticeable. This implies that non-specific staining could be a problem but, in practice, the density of specific staining is much greater (dark blue/very dark blue/almost black) than that of possible non-specific staining (very pale blue), especially when the incubation times are only of the order of 1 h. In order to obviate this problem some authors have preferred to use neotetrazolium (NT) rather than NBT (Chayen & Bitensky, 1991), and using NADP in combination with malate (Vincent *et al.*, 1983; Sancesario *et al.*, 1993) or lactate and lactic dehydrogenase (Nachlas, Wayler & Seligano, 1958) may also help to reduce this problem.

16.4.3 Step 4

To stain the sections, place the slides on a tray, and apply the freshly made staining solution onto the section as a drop that covers the whole section. If the slides are dry the ‘blister’ of solution will remain in place and not run off over the edges of the slide. To ensure that the solution does not run off, the sections can be drawn around with a diamond scribe, with DPX resin, nail varnish or with a PAP pen (Daido Sangyo Co. Ltd., Japan). However, with this single-step staining procedure this precaution is not strictly necessary. The tray with the slides should then be placed in a 37 °C oven, which will also provide darkness.

The reaction proceeds well at room temperature, but will take longer than it does at 37 °C. The incubation time is very variable from tissue to tissue; sections stain more quickly than whole-mounts. The best way to control the development of colour is by eye. Examine the preparations after 30 min and then every 10–15 min. It is very unlikely that they will be over-stained by 30 min.

It is not necessary to stain the tissues in the dark. Usually an oven will provide darkness in any case but, if staining at room temperature, it is not necessary to protect the slides from ambient light. It has also been suggested that the staining should be carried out in an atmosphere of nitrogen (Chayen & Bitensky, 1991), in order to avoid oxidation by atmospheric oxygen. For fixed tissues this is not necessary.

16.4.4 Step 5

This is to stop the reaction by eluting the substrates. In practice, slides stored for longer than a few weeks darken perceptibly. This is due to the substantivity of NBT (see Section 12.4) and its spontaneous conversion to formazan. Since this is a general darkening it usually means that the background develops colour. This can be an advantage as it makes the section easier to find under the microscope. Provided that this general darkening is recognized for what it is, it need not be an undue problem, but it does make interpretation of lightly stained structures more difficult.

16.4.5 Step 6

The dehydration through graded alcohols and clearing (for sections, 2 min in each of: 70%, 80%, 90%, 100%, 100%, 100%, Histoclear) and mounting in a clear resin makes for a permanent slide that can be stored at room temperature. For whole-mounts the periods in each alcohol need to be longer, and minimally 10 min. However, the general darkening noted above will still occur. Furthermore, the reaction product looks more crystalline or granular, and less amorphous, than if the preparation were not dehydrated and were mounted in an aqueous medium instead. The disadvantage of mounting in aqueous medium is that the slides need to be kept refrigerated, and eventually deteriorate through drying-back. In addition the non-specific darkening occurs more rapidly. If permanent slides are not required the aqueous mounting should be preferred.

16.4.6 Controls

The control experiments for NADPH-diaphorase staining are to make sure that the reaction product is not forming spontaneously in the absence of enzyme, and that the formazan production is dependent upon the presence of

NADPH. In this latter case it should be demonstrable that omission of NADPH (or NADP) from the reaction mixture results in a lack of staining. In the former case, it should be demonstrated that the reaction mixture on its own does not produce a formazan precipitate when incubated under the same conditions as those used for staining. Additionally, sections that have been put on a hot-plate at 55 °C for 30 s to 1 min should not stain up because the enzyme will have been denatured.

16.5 NADPH-diaphorase staining for EM

The principles of staining for NADPH-diaphorase activity in order to localize NOS at the EM level are the same as those that apply to the LM level. Problems associated with the localization of NADPH-diaphorase activity at the EM level have been discussed in Chapter 12, and the general warning is that one should be circumspect in one's interpretation of the subcellular distribution of NOS when diaphorase activity is examined.

The only comments to make here are that for EM work the incubation time needs to be extended, in order to ensure that a very heavy reaction product is produced, and that the staining must be carried out before osmification and before embedding. Reaction times of 4–18 h may be employed.

16.6 NOS-immunocytochemical staining for LM

For immunostaining at the LM level the tissue needs to be fixed in much the same way as it is for LM of NADPH-diaphorase activity. The preparation of slide-mounted sections is as for the diaphorase procedure also. For whole-mounts, permeabilization should be considered, in which the preparations are dehydrated through graded alcohols, and then rehydrated (10 min each in: 70%, 80%, 90%, 100%, 90%, 80%, 70%, PBS) and washed with Triton X-100 (0.1–0.3% v/v in PBS).

To stain the sections, place the slides on a tray, and apply the diluted sera or antisera onto the section as a drop that covers the whole section. If the slides are dry the 'blister' of solution will remain in place and not run off over the edge of the slide. To ensure the solution does not run off, the sections can be drawn around with a diamond scribe, with DPX resin, nail varnish or with a PAP pen (Daido Sangyo Co. Ltd., Japan). The washes can be carried out by transferring the slides to a Coplin jar, or a slide rack, and immersing them in the washing solution.

16.7 Staining protocol

There are many successful variations on the three protocols given and discussed in this section. For simplicity it is assumed that the primary antibody has been raised in rabbit, and that the secondary antibody has been raised in goat or donkey against rabbit IgG, although in practice the primary antibody could be a mouse monoclonal and the secondary antibody would be raised against mouse IgM in any non-mouse host species.

16.7.1 Protocol 1: indirect immunofluorescence method

1 Incubate with normal serum from host of secondary antibody, diluted 1:20, for 30–60 min.
2 Brief rinse in PBS.
3 Incubate with primary antibody (of the order of 1:100), 12–18 h at room temperature.
4 Wash (3 × 5 min in PBS, include 0.1% Triton X-100 in final wash).
5 Incubate with secondary antibody (conjugated to fluorescent chromophore), 90 min at room temperature.
6 Wash (3 × 5 min in PBS).
7 Dry off excess fluid from around section, mount in aqueous medium (e.g. Citifluor, City University, London).

16.7.1.1 Step 1

This is often omitted, but experience will tell you whether or not it is beneficial. The aim is block non-specific binding sites of the secondary antibody.

16.7.1.2 Step 3

The optimal dilution of the primary antibody needs to be determined for each type of tissue. Too dilute and the staining pattern will be inadequate, too concentrated and, in addition to wasting a precious resource, the probability of non-specific staining increases, and also if the fluorescence is too intense, flare can occur which provides a diffuse fluorescence around the fluorescing structure, which results in loss of resolution and photogenicity. The primary and secondary antibodies should be diluted in a medium (antibody-diluting medium, or ADM) made up of PBS (0.1 M) that contains bovine serum albumen (0.1% w/v) and sodium azide (0.1% w/v).

The duration of incubation is not critical, but generally the shorter the duration the more concentrated the antibody needs to be. Incubation times as short as 1 h have been used, but most authors use an overnight incubation. Also the temperature used can be variable. Room temperature (approximately 20 °C) is fine, but during warm weather it might be better to incubate the sections in a refrigerator (4–8 °C).

16.7.1.3 Step 4

The detergent reduces the non-specific binding of the secondary antibody, and possibly facilitates elution of unbound primary antibody.

16.7.1.4 Step 5

The dilution of the secondary antibody needs to be optimized by systematic survey, but should be approximately in the range of 1:50–1:250, or four times more concentrated than the primary antibody.

16.7.2 Protocol 2: ABC indirect immunofluorescence method

1 Incubate with normal serum from host of secondary antibody, diluted 1:20, for 30–60 min.
2 Brief rinse in PBS.
3 Incubate with primary antibody, highly diluted (of the order of 1:1000), 12–18 h at room temperature.
4 Wash (3 × 5 min in PBS, include 0.1% Triton X-100 in final wash).
5 Incubate with secondary antibody (conjugated to biotin), 90 min at room temperature, diluted 1:250.
6 Wash (3 × 5 min in PBS).
7 Incubate with third layer of streptavidin conjugated to fluorochrome, 60 min, room temperature, diluted 1:250.
8 Wash (3 × 5 min in PBS).
9 Dry off excess fluid from around section, mount in aqueous medium (e.g. Citifluor, City University, London).

16.7.2.1 Steps 1–4

These are as for Protocol 1.

16.7.2.2 Controls

It is necessary to show that the primary antibody only recognizes that antigen against which it was raised. This is not always entirely possible, so one ends up qualifying the immunostaining with the suffix '-like immunoreactivity', as in 'VIP-like immunoreactivity' etc. However, one can show that the antibody does recognize its specific antigen, and this is by a method known as preadsorption. Many authors erroneously use the term 'preabsorption'. Adsorption and absorption are very different processes: antibodies adsorb, sponges absorb.

For a preadsorption control the primary antisera is diluted to its optimal concentration in ADM containing a high concentration of the antigen. Thus an anti-VIP serum would be diluted in ADM containing VIP at, say, 1 μM. This solution is incubated with gentle agitation for 1 h at room temperature, and then used to stain the tissue in the usual way. If there is no visible staining the antibody has been shown to recognize its antigen, but if staining is still visible it is probable that the antiserum (especially if it is polyclonal) contains an unwanted antibody. A good way to perform the preadsorption control is to use a range of antigen concentrations, from, say, 1 nM up to 10 μM, in steps of half log units (i.e. 1 nM, 3 nM, 10 nM, 30 nM, etc.). If the antibody does recognize only its true antigen there should be a visible concentration-dependent loss of immunoreactivity that is abolished by the higher concentrations.

The specificity of the antibody should also be tested, this time by preadsorbing it with inappropriate antigens. For example, the immunoreactivity of a VIP antiserum should not be affected by preincubation with a high concentration of SP. This test is more important when the antibody has been raised against an antigen that has close relations, e.g. anti-methionine-enkephalin should be tested against leucine-enkephalin, and vice versa.

A third control is to test the non-specific binding of the secondary antibody and the third layer, by simply omitting the incubation with the primary antibody or the incubation with the secondary antibody, respectively. For good immunostaining there should be no visible immunoreactivity when the primary antibody has not been applied, nor should there be any in the ABC method when the secondary antibody has not been applied.

Finally, there is a fourth type of control experiment (positive control) that may be necessary. If a tissue does not show any specific staining, the antibody should be tested on another tissue that is known to contain the antigen. Only if the antibody produces staining in another tissue can one be confident that the lack of staining in the original tissue is due to an absence of antigen, rather than to a loss of affinity of the antibodies in the antiserum.

16.7.3 Protocol 3: peroxidase–antiperoxidase (PAP)

1 On sections, block the endogenous peroxidase activity with hydrogen peroxide (0.3–3% in PBS), rinse.
2 Incubate with primary antiserum, highly diluted (of the order of 1 : 1000), overnight.

3 Wash (3 × 5 min in PBS).
4 Incubate with secondary antibody (unconjugated), 1:50–1:250, 60 min at room temperature.
5 Wash (3 × 5 min in PBS).
6 Incubate with third layer, PAP complex, raised in same species as first layer, 1:50–1:250, 60 min, room temperature.
7 Wash (3 × 5 min in PBS).
8 Incubate with filtered fresh developing solution: PBS containing 3,3′-diaminobenzidine (DAB) (0.5 mg ml^{-1}) and hydrogen peroxide (0.03%), for 1–5 min.
9 Rinse in water.
10 Counterstain with haematoxylin to stain nuclei (optional), dehydrate, clear and mount.

16.7.3.1 Steps 2, 4 and 6

The optimum dilutions need to be established for each antibody and for each type of tissue.

16.7.3.2 Step 8

DAB is thought to be carcinogenic, so take necessary precautions when handling it. Wear a mask and gloves while weighing it out. Immerse utensils such spatulas and weighing boats in a dilute solution of sodium hypochlorite (dilute household bleach) after use, and wash with copious volumes of tap-water.

16.7.3.3 Step 10

Dehydration, clearing and mounting as above (Diaphorase, step 6, Section 16.4.5).

16.7.3.4 Controls

The primary antibody should be tested by preadsorbing it with its appropriate antigen and inappropriate antigens. Non-specific binding and endogenous peroxidase activity should be tested by omitting the secondary antibody and the PAP complex, respectively. Also a positive control might be necessary. (See Section 16.7.2.2 Controls.)

16.8 Selected references

Belai, A., Schmidt, H. H. H. W., Hoyle, C. H. V., Hassall, C. J. S., Saffrey, M. J., Moss, J., Förstermann, U., Murad, F. & Burnstock, G. (1992). Colocalization of nitric oxide synthase and NADPH-diaphorase in the myenteric plexus of the rat gut. *Neuroscience Letters*, **143**, 60–4.

Chayen, J. & Bitensky, L. (1991). *Practical Histochemistry*, second edn., Chichester: John Wiley and Sons.

Pullen, A. H. & Humphreys, P. (1995). Diversity in localization of nitric oxide synthase antigen and NADPH-diaphorase histochemical staining in sacral somatic motor neurones of the cat. *Neuroscience Letters*, **196**, 33–6.

Varndell, I. M. & Polak, J. M. (1986). Electron microscopical immunocytochemistry. In *Immunocytochemistry: Modern Methods and Applications*, second edn., eds. J. M. Polak & S. Van Noorden, pp. 146–66. Bristol: Wright.

Appendix I

Distribution of constitutive nitric oxide synthase as determined by immunocytochemistry

Animal Organ/tissue	Antigen (Type)	References
Bird		
Chicken		
Central nervous system		
brain: various areas	I	Brüning, Funk & Mayer, 1994
Peripheral ganglia		
dorsal root ganglion: cyta (embryo and adult)	I	Ward, Shuttleworth & Kenyon, 1994a
Gastrointestinal tract		
MP and SMP cyta and fibres; muscle coat, fibres	I	Balaskas, Saffrey & Burnstock, 1995
Quail		
Cardiovascular system		
enteric arterioles, adventitial fibres	I	Li, Young & Furness, 1994
Gastrointestinal tract		
MP, cyta and fibres; muscle coat and SMP, fibres	I	Li, Young & Furness, 1994

Bullfrog		
Cardiovascular system		
intracardiac ganglion cells and processes	I	Clark, Kinsberg & Giles, 1994
Gastrointestinal tract		
MP, cyta and fibres; muscle coat, fibres	I	Li *et al.*, 1993
Cat		
Central nervous system		
midbrain and brainstem: various locations including sensory nuclei	I	Maqbool, Batten & McWilliam, 1995
spinal cord: dorsal horn, cyta and fibres	I	Dun *et al.*, 1993
spinal cord: intermediolateral cell column, cyta and processes	I	Dun *et al.*, 1993
spinal cord: lamina X, cyta and fibres	I	Dun *et al.*, 1993
spinal cord: Onuf's nucleus, cyta and fibres	I	Pullen & Humphreys, 1995
spinal cord: ventral horn, cyta and fibres	I	Dun *et al.*, 1993; Pullen & Humphreys, 1995
Peripheral ganglia		
microganglia along glossopharyngeal and carotid sinus nerves	I	Wang *et al.*, 1993b, 1994b
petrosal ganglion: cyta and fibres	I	Wang *et al.*, 1993b, 1994b
Gastrointestinal tract		
lower oesophageal sphincter: circular muscle, fibres	I	Ny *et al.*, 1994
Carotid body, ganglion cells and fibres	I	Wang *et al.*, 1993b, 1994b
Cattle		
Urogenital tract		
Bovine retractor penis muscle, fibres	I	Sheng *et al.*, 1992
Dog		
Peripheral ganglia		
superior cervical ganglion: cyta and processes	I	Hisa *et al.*, 1995

Appendix I (*cont.*)

Animal Organ/tissue	Antigen (Type)	References
Dog		
Blood vessels		
renal arterioles	I	Okamura, Yoshida & Toda, 1995
Gastrointestinal tract		
pyloric sphincter: MP and SMP cyta and fibres; circular muscle, fibres	I	Ward, Xue & Sanders, 1994b
small intestine: MP, cyta and fibres; circular muscle fibres	I	Berezin *et al.*, 1994
ileocolonic sphincter: MP and SMP cyta and fibres; circular muscle, fibres	I	Ward, Xue & Sanders, 1994b
large intestine: interstitial cells	I	Ward *et al.*, 1992c; Xue *et al.*, 1994a
large intestine: interstitial cells	III	Xue *et al.*, 1994a
large intestine: MP and SMP, cyta and fibres; muscle coat, fibres	I	Ward *et al.*, 1992c; Berezin *et al.*, 1994
large intestine: smooth muscles cells and interstitial cells	I	Berezin *et al.*, 1994
Ferret		
Central nervous system		
brain: various regions	I	Matsumoto *et al.*, 1992
Fish		
Teleost		
Central nervous system		
many structures in brain and spinal cord	I	Brüning, Katzbach & Mayer, 1995
Gastrointestinal tract		
MP cyta; circular muscle and SMP, fibres; rectum SMP, cyta	I	Li & Furness, 1993
Eye		
retinal amacrine, ganglion, horizontal and photoreceptor cells	I	Liepe *et al.*, 1994; Ostholm *et al.*, 1994
Vagus nerve	I	Li & Furness, 1993

Guinea-pig		
Central nervous system		
spinal cord: intermediolateral columns, cyta and fibres	I	Anderson *et al.*, 1995; Furness & Anderson, 1994
Peripheral ganglia		
coeliac ganglion: fibres	I	Anderson *et al.*, 1995
inferior mesenteric ganglion: fibres	I	Anderson *et al.*, 1995
nodose ganglion, cyta	I	Fischer *et al.*, 1993
superior cervical ganglion	I	Fischer *et al.*, 1993; Furness & Anderson, 1994
Cardiovascular system		
endocardium and myocardium, fibres	I	Sosunov *et al.*, 1995
intracardiac ganglia in culture	I	Hassall *et al.*, 1992
intracardiac ganglion cells, fibres in myocardium	I	Klimaschewski *et al.*, 1992; Tanaka *et al.*, 1993
coronary arteries, adventitial fibres	I	Klimaschewski *et al.*, 1992; Sosunov *et al.*, 1995
coronary arterioles and capillaries, fibres	I	Sosunov *et al.*, 1995
coronary venules	I	Sosunov *et al.*, 1995
lingual artery, adventitial fibres	I	Kummer & Mayer, 1993
pulmonary arteries and veins, adventitial fibres	I	Klimaschewski *et al.*, 1992; Fischer *et al.*, 1993
renal cortical and medullary vessels, endothelial cells	III	Bachmann, Bosse & Mundel, 1995
renal efferent arterioles, endothelial cells	I	Bachmann, Bosse & Mundel, 1995
submucosal microvessels, endothelial cells	III	O'Brien *et al.*, 1995
tracheal vessels, adventitial fibres	I	Fischer *et al.*, 1993
Gastrointestinal tract		
oesophagus: MP, cyta and fibres; circular muscle and muscularis mucosae, fibres	I	Furness *et al.*, 1994
lower oesophageal sphincter: circular muscle, fibres	I	Furness *et al.*, 1994

Animal Organ/tissue	Antigen (Type)	References
Guinea-pig		
stomach: circular muscle, fibres; muscularis mucosae, fibres	I	Desai *et al.*, 1994; Furness *et al.*, 1994
stomach: MP, cyta and fibres	I	Desai *et al.*, 1994; Schemann, Schaaf & Mader, 1995
small intestine: MP and SMP, cyta and fibres; muscle coat, fibres	I	Costa *et al.*, 1992; Llewellyn Smith *et al.*, 1992; Young *et al.*, 1992; Furness *et al.*, 1994; Li, Young & Furness, 1995
pancreas: ganglia in culture	I	Tay, Moules & Burnstock, 1994
sphincter of Oddi: associated ganglion cells and fibres	I	Wells, Talmage & Mawe, 1995
large intestine: MP and SMP, cyta and fibres; muscle coat, fibres	I	Young *et al.*, 1992; Furness *et al.*, 1994
large intestine: MP in culture, ganglion cells and processes	I	Saffrey *et al.*, 1992
Respiratory tract		
airway smooth muscle and epithelium, fibres	I	Fischer *et al.*, 1993
Urogenital tract		
JGA macula densa and thick ascending limb epithelial cells	I	Bachmann, Bosse & Mundel, 1995
renal nerve fibres	I	Bachmann, Bosse & Mundel, 1995
urinary bladder intramural ganglia in culture, cyta and processes	I	Saffrey *et al.*, 1994
Adrenal gland: medullary ganglion cells and fibres	I	Heym *et al.*, 1995
Human		
Central nervous system		
astrocytic tumours, tumour cells	I,II	Cobbs *et al.*, 1995
spinal cord: superficial dorsal horn, lamina X, intermediolateral column	I	Terenghi *et al.*, 1993
spinal cord: ventral horn, cyta and fibres	I	Terenghi *et al.*, 1993

Peripheral ganglia		
dorsal root ganglia	I	Terenghi *et al.*, 1993
pelvic plexus: cyta and fibres	I	Burnett *et al.*, 1993
Cardiovascular system		
airway vessels, endothelial cells	I	Kobzik *et al.*, 1993
astrocytic tumour vessels, endothelial cells	all	Cobbs *et al.*, 1995
cavernous arteries, adventitial fibres	I	Burnett *et al.*, 1993
circle of Willis, adventitial fibres	I	Nozaki *et al.*, 1993
myometrial vessels, adventitial fibres	I	Burnett *et al.*, 1993; Telfer *et al.*, 1995
myometrial vessels, endothelial cells	III	Telfer *et al.*, 1995
olfactory mucosa, adventitial fibres	I	Kulkarni, Getchell & Getchell, 1994
pancreatic vessels, adventitial fibres	I	Worl *et al.*, 1994b
pancreatic vessels, endothelial cells	III	Worl *et al.*, 1994b
umbilical vessels, endothelial cells	I	Sexton *et al.*, 1995
umbilical vessels, endothelial cells	III	Buttery *et al.*, 1994
ureteral vessels, adventitial fibres	I	Smet *et al.*, 1994
urinary bladder vessels, adventitial fibres	I	Crowe *et al.*, 1995
vaginal wall vessels, adventitial fibres	I	Hoyle *et al.*, 1996
vomeronasal vessel, adventitial fibres	I	Kulkarni, Getchell & Getchell, 1994
Gastrointestinal system		
stomach (foetal and neonatal): MP and SMP, cyta	I	Timmermans *et al.*, 1994
stomach (foetal and neonatal): outer smooth muscle and muscularis mucosae	I	Timmermans *et al.*, 1994
pancreas: ganglion cells; fibres around acini and in islets	I	Worl *et al.*, 1994b
small intestine (foetal and neonatal): circular muscle, fibres	I	Timmermans *et al.*, 1994
small intestine (foetal and neonatal): enteric ganglia and proganglia, cyta	I	Timmermans *et al.*, 1994
ileocaecal junction: MP and SMP, cyta and fibres; circular muscle, fibres	I	Bogers *et al.*, 1994; Faussone Pellegrini *et al.*, 1994

Appendix I (*cont.*)

Animal Organ/tissue	Antigen (Type)	References
Human		
large intestine (foetal and neonatal): MP and SMP, cyta; muscle coat, fibres	I	Timmermans *et al.*, 1994
large intestine: MP and SMP, cyta and fibres; muscle coat, fibres	I	Matini *et al.*, 1995
Respiratory tract		
airways: fibres	I	Kobzik *et al.*, 1993
nasal mucosa: mast cells	I	Bacci *et al.*, 1994
respiratory epithelium	II	Kobzik *et al.*, 1993
vomeronasal and septal glands, fibres	I	Kulkarni, Getchell & Getchell, 1994
Urogenital tract		
bladder neck, fibres	I	Crowe *et al.*, 1995
penis: penile nerve; cavernous and corporeal tissue, fibres	I	Burnett *et al.*, 1993; Leone *et al.*, 1994b
prostate gland: intramural ganglion cells and fibres	I	Burnett *et al.*, 1995a
prostate gland: subepithelial plexus, fibres; epithelial cells	I	Burnett *et al.*, 1995a
ureter: smooth muscle and subepithelial plexus, fibres	I	Smet *et al.*, 1994
urethra: smooth muscle and subepithelial plexus, fibres	I	Leone *et al.*, 1994b
uterus: endometrium, fibres	I	Telfer *et al.*, 1995
uterus: endometrial stroma	III	Telfer *et al.*, 1995
vagina: vaginal wall and papillae, fibres	I	Hoyle *et al.*, 1996
Adrenal medulla: chromaffin cells and fibres	I	Heym, Colombo Benckmann & Mayer, 1994
Eye: ciliary muscle ganglion cells	I	Tamm *et al.*, 1995
Eye: ciliary muscle, trabecular network, Schlemm's canal	III	Nathanson & McKee, 1995
Placenta: syncytiotrophoblast	III	Buttery *et al.*, 1994; Eis *et al.*, 1995
Placenta: amnion fibroblast layer	III	Eis *et al.*, 1995

Species / tissue		Reference
Insect		
Central nervous system		
brain	I	Villar *et al.*, 1994
Monkey		
Central nervous system		
brain: many structures in fore-, mid- and hindbrain	I	Bredt *et al.*, 1991a; Hashikawa *et al.*, 1994; Satoh *et al.*, 1995
visual cortex, cyta and fibres	I	Aoki *et al.*, 1993a
spinal cord: dorsal horn, cyta and fibres	I	Dun *et al.*, 1993
spinal cord: intermediolateral cell column, cyta and processes	I	Dun *et al.*, 1993
spinal cord: lamina X, cyta and fibres	I	Dun *et al.*, 1993
spinal cord: ventral horn, cyta and fibres	I	Dun *et al.*, 1993
Cardiovascular system		
renal arterioles	I	Okamura, Yoshida & Toda, 1995
Mouse		
Central nervous system		
olfactory bulb: plexiform interneurones, periglomerular and granule cells	I	Kishimoto *et al.*, 1993
spinal cord: dorsal horn, cyta and fibres	I	Dun *et al.*, 1993
spinal cord: intermediolateral cell column, cyta and processes	I	Dun *et al.*, 1993
spinal cord: lamina X, cyta and fibres	I	Dun *et al.*, 1993
spinal cord: ventral horn, cyta and fibres	I	Dun *et al.*, 1993
Urogenital tract		
anococcygeus muscle, fibres	I	Brave *et al.*, 1993
vaginal ganglion cells; other regions, fibres	I	Grozdanovic *et al.*, 1994
Pig		
Gastrointestinal tract		
small intestine: SMP, cyta and fibres	I	Krammer *et al.*, 1993
ileocaecal junction: MP and SMP, cyta and fibres; circular muscle, fibres	I	Bogers *et al.*, 1994

Animal Organ/tissue	Antigen (Type)	References
Pig		
large intestine: MP and SMP, cyta and fibres; muscle coat, fibres	I	Barbiers *et al.*, 1994
Urogenital tract		
urethra: muscle coat, fibres	I	Persson *et al.*, 1993, 1995
urinary bladder: detrusor and trigone, fibres	I	Persson *et al.*, 1993, 1995
Rabbit		
Cardiovascular system		
aortic endothelial cells	I	Loesch, Belai & Burnstock, 1993
Eye: retina, many cells types	I	Perez *et al.*, 1995
Rat		
Central nervous system		
cortex, neurones in various regions	I	Bredt, Hwang & Snyder, 1990; Dun *et al.*, 1994b
brainstem structures	I	Bredt, Hwang & Snyder, 1990; Ohta *et al.*, 1993; Dun, Dun & Forstermann, 1994a
fore-, mid- and hindbrain, many structures	I	Rodrigo *et al.*, 1994
hippocampal cells in culture	I	Wendland *et al.*, 1994
hippocampus	I	Valtschanoff *et al.*, 1993a,b; Wendland *et al.*, 1994
hippocampus: CA1 neurones	III	Dinerman *et al.*, 1994; O'Dell *et al.*, 1994
hippocampus (post ischaemia), astrocytes	II	Endoh, Maiese & Wagner, 1994
median eminence microglial cells	II	Van Dam *et al.*, 1995
meningeal macrophages	II	Van Dam *et al.*, 1995
mitochondria	III	Bates *et al.*, 1995

pituitary gonadotrophs and folliculo-stellate cells	I	Hökfelt *et al.*, 1994
spinal cord: dorsal horn superficial laminae, cyta and fibres	I	Dun *et al.*, 1992, 1993; Lee *et al.*, 1993; Herdegen *et al.*, 1994; Saito *et al.*, 1994
spinal cord: intermediolateral cell column, cyta and processes	I	Dun *et al.*, 1992; Anderson *et al.*, 1993; Blottner, Schmidt & Baumgarten, 1993; Dun *et al.*, 1993; Saito *et al.*, 1994; Burnett *et al.*, 1995c; Papka *et al.*, 1995a,b; Vizzard, Erdman & de Groat, 1995
spinal cord: lamina X, cyta and fibres	I	Dun *et al.*, 1992, 1993; Lee *et al.*, 1993; Herdegen *et al.*, 1994; Saito *et al.*, 1994
spinal cord: spinothalamic tract, cyta and fibres	I	Lee *et al.*, 1993
spinal cord: ventral horn, cyta and fibres	I	Dun *et al.*, 1993; Terenghi *et al.*, 1993
spinal cord: around central canal, cyta and fibres	I	Dun *et al.*, 1992; Terenghi *et al.*, 1993
spinal trigeminal nucleus: interneurones	I	Dohrn *et al.*, 1994
Peripheral ganglia		
coeliac/superior mesenteric ganglion: fibres	I	Anderson *et al.*, 1993; Ekblad *et al.*, 1994
dorsal root ganglia: cyta and processes	I	Fiallos Estrada *et al.*, 1993; Nozaki *et al.*, 1993; Vizzard *et al.*, 1994; Dun *et al.*, 1995; Papka *et al.*, 1995b; Vizzard, Erdman & de Groat, 1995
inferior mesenteric ganglion, fibres	I	Anderson *et al.*, 1993
nodose ganglion: cyta and processes	I	Nozaki *et al.*, 1993; Dun *et al.*, 1995
paravertebral ganglia: fibres	I	Anderson *et al.*, 1993
pelvic ganglion: cyta and fibres	I	Anderson *et al.*, 1993; Ding *et al.*, 1993; Domoto & Tsumori, 1994; Vizzard *et al.*, 1994; Ding *et al.*, 1995
pterygopalatine ganglion: cyta and processes	I	Yamamoto *et al.*, 1993b
sphenopalatine ganglion: cyta and processes	I	Nozaki *et al.*, 1993; Ceccatelli *et al.*, 1994
stellate ganglion: fibres	I	Anderson *et al.*, 1993

Appendix I (*cont.*)

Animal Organ/tissue	Antigen (Type)	References
Rat		
submandibular ganglion: cyta and processes	I	Ceccatelli *et al.*, 1994
superior cervical ganglion: cyta and fibres	I	Anderson *et al.*, 1993; Yamamoto *et al.*, 1993b; Dun *et al.*, 1995
trigeminal ganglion: cyta and processes	I	Nozaki *et al.*, 1993; Yamamoto *et al.*, 1993b
Cardiovascular system		
endocardium and myocardium, fibres	I	Sosunov *et al.*, 1995
intracardiac ganglion cells, fibres in myocardium	I	Klimaschewski *et al.*, 1992
airway vessels, endothelial cells	I	Kobzik *et al.*, 1993
aortic endothelial cells	I	Aliev *et al.*, 1995
basilar artery, adventitial fibres and endothelial cells	I,III	Loesch, Belai & Burnstock, 1994; Loesch & Burnstock, 1996
carotid body vessels, fibres	I	Hohler, Mayer & Kummer, 1994
cerebral arteries, adventitial fibres and endothelial cells	I	Nozaki *et al.*, 1993
cerebral arteries, endothelial cells	III	Tomimoto *et al.*, 1994
coronary arteries, adventitial fibres and endothelial cells	I	Klimaschewski *et al.*, 1992; Dikranian, Loesch & Burnstock, 1994; Sosunov *et al.*, 1995
coronary arterioles and capillaries, fibres	I	Sosunov *et al.*, 1995
coronary artery (neonatal), endothelial and smooth muscle cells	I	Loesch & Burnstock, 1995
coronary venules	I	Sosunov *et al.*, 1995
deep dorsal artery, adventitial fibres	I	Alm *et al.*, 1993
deep penile arteries, adventitial fibres	I	Alm *et al.*, 1993
femoral artery, endothelial cells	I	Dikranian, Loesch & Burnstock, 1994

meningeal arteries, endothelial cells	III	Nemade *et al.*, 1995
mesenteric arteries, endothelial cells	I	Ralevic, Dikranian & Burnstock, 1995
nasal mucosa vessels, adventitial fibres	I	Ceccatelli *et al.*, 1994; Hanazawa *et al.*, 1994
nasal mucosa, subepithelial and epithelial fibres	I	Hanazawa *et al.*, 1994
nasal mucosa seromucosal glands, fibres	I	Hanazawa *et al.*, 1994
ocular choroid and limbal vessels, adventitial fibres	I	Yamamoto *et al.*, 1993b
oesophageal vessels, adventitial fibres	I	Worl, Mayer & Neuhuber, 1994a
olfactory mucosa vessels, adventitial fibres	I	Kulkarni, Getchell & Getchell, 1994
pancreatic vessels, adventitial fibres	I	Worl *et al.*, 1994b
pancreatic vessels, endothelial cells	III	Worl *et al.*, 1994b
pulmonary arteries and veins, adventitial fibres	I	Klimaschewski *et al.*, 1992
pulmonary artery (neonatal), endothelial and smooth muscle cells	I	Loesch & Burnstock, 1995
pulmonary vessels, endothelial cells	I	Xue *et al.*, 1994b
renal interlobular arteries, adventitial fibres	I	Liu & Barajas, 1993
renal JGA afferent and efferent arteriole smooth muscle	II	Tojo *et al.*, 1994
reproductive tract vessels, adventitial fibres, endothelial cells	I	Burnett *et al.*, 1995b
salivary gland vessels, adventitial fibres	I	Ceccatelli *et al.*, 1994
Gastrointestinal tract		
salivary glands: fibres in glandular parenchyme	I	Ceccatelli *et al.*, 1994
oesophagus: MP and SMP, cyta and fibres; striated muscle, fibres	I	Ekblad *et al.*, 1994; Worl, Mayer & Neuhuber, 1994a
stomach: MP, cyta and fibres; circular muscle and submucosa, fibres	I	Belai *et al.*, 1992; Forster & Southam, 1993; Ekblad *et al.*, 1994
gastric mucosa epithelial cells	I	Schmidt *et al.*, 1992a
pylorus: MP, cyta and fibres; muscle coat, fibres	I	Ekblad *et al.*, 1994
small intestine: MP, cyta and fibres; muscle coat, fibres; B's glands, fibres	I	Belai *et al.*, 1992; Alm *et al.*, 1993; Ekblad *et al.*, 1994
small intestine, epithelial cells	II	Cook *et al.*, 1994
pancreas: ganglion cells; fibres around acini and in islets	I	Worl *et al.*, 1994b

Appendix I (*cont.*)

Animal Organ/tissue	Antigen (Type)	References
Rat		
pancreas: islet cells	I	Schmidt *et al.*, 1992a
liver: mitochondria	III	Bates *et al.*, 1995
liver: hepatocytes	II	Cook *et al.*, 1994; Sato *et al.*, 1995
large intestine: MP, cyta and fibres; muscle coat and submucosa, fibres	I	Belai *et al.*, 1992; Alm *et al.*, 1993; Ekblad *et al.*, 1994
Respiratory tract		
nasal mucosa: fibres in glandular parenchyme	I	Ceccatelli *et al.*, 1994; Hanazawa *et al.*, 1994; Kulkarni, Getchell & Getchell, 1994
bronchial and bronchiolar epithelial cells	I	Kobzik *et al.*, 1993; Xue *et al.*, 1994b
bronchial and bronchiolar epithelial cells	II	Kobzik *et al.*, 1993; Cook *et al.*, 1994
lung: parenchyme, fibres	I	Kobzik *et al.*, 1993
pulmonary epithelial cells	I	Schmidt *et al.*, 1992a
respiratory epithelium: epithelium	all	Rengasamy, Xue & Johns, 1994
Urogenital tract		
coagulating gland: fibres	I	Burnett *et al.*, 1995b
ejaculatory duct: fibres	I	Burnett *et al.*, 1995b
epididymis: fibres and epithelial cells	I	Burnett *et al.*, 1995b
kidney: hilum and pelvis, fibres	I	Liu & Barajas, 1993
kidney: macula densa	I	Mundel *et al.*, 1992; Schmidt *et al.*, 1992a; Tojo *et al.*, 1994
penis, corpus cavernosum: fibres	I	Alm *et al.*, 1993; Vizzard *et al.*, 1994
urethra, smooth muscle and mucosa: fibres	I	Alm *et al.*, 1993; Vizzard *et al.*, 1994
urinary bladder: fibres	I	Vizzard *et al.*, 1994
urinary bladder, epithelial cells	II	Cook *et al.*, 1994

uterus, fibres	I	Ceccatelli *et al.*, 1994; Papka *et al.*, 1995b
uterus, epithelial cells	I	Schmidt *et al.*, 1992a
vas deferens: fibres	I	Ceccatelli *et al.*, 1994; Burnett *et al.*, 1995b
Adrenal cortex: cyta and fibres	I	Alm *et al.*, 1993; Afework, Ralevic & Burnstock, 1995
Adrenal medulla: cyta and fibres	I	Afework *et al.*, 1992; Alm *et al.*, 1993; Afework, Ralevic & Burnstock, 1995
Carotid bodies: cyta and fibres	I	Hohler, Mayer & Kummer, 1994
Eye: peripheral ocular and anterior uveal fibres	I	Yamamoto *et al.*, 1993a
Eye: retina, amacrine, interplexiform and ganglion cells	I	Yamamoto *et al.*, 1993a; Perez *et al.*, 1995
Lymph nodes	II	Van Dam *et al.*, 1995
Thymus gland: (stimulated) leucocytes	II	Sato *et al.*, 1995
Vagal afferent fibres	I	Dun *et al.*, 1995
Vagal trunk: fibres	I	Forster & Southam, 1993
Salamander		
Eye: retinal amacrine, ganglion, horizontal, photoreceptor and Müller cells	I	Liepe *et al.*, 1994; Kurenni *et al.*, 1995

'Cyta' means nerve cell bodies, 'fibres' means nerve fibres.
B's glands, Brunner's glands.
JGA, juxtaglomerular apparatus.
MP, myenteric plexus.
SMP, submucous plexus.
Order of tissues/organs: central nervous system; peripheral ganglia; cardiovascular system; gastrointestinal tract; respiratory tract; urogenital tract; miscellaneous. Within these divisions tissues are listed in alphabetical order, except for the gastrointestinal tract, in which the tissues are listed in anatomical order from oesophagus to anus.
Although extensive it is not claimed that this table is comprehensive.

Appendix II

Peripheral tissues and organs in which nitric oxide has been identified as a neurotransmitter

Tissue	Function	References
Cat		
Cardiovascular system		
cerebral arteries	↓	Ayajiki, Okamura & Toda, 1994
Gastrointestinal tract		
lower oesophageal sphincter	↓	Kortezova *et al.*, 1994
LOS muscularis mucosae	↓	Dobreva *et al.*, 1994
stomach	↓	Barbier & Lefebvre, 1993
ileum	tonic inhibition	Gustafsson & Delbro, 1993
ileocaecal junction	↓	Mizhorkova *et al.*, 1994
colon	rebound ↑	Venkova & Krier, 1994
Cattle		
Cardiovascular system		
cerebral arteries	↓	Ayajiki, Okamura & Toda, 1993

Urogenital tract		
bovine retractor penis muscle	↓	Martin, Gillespie & Gibson, 1993
Dog		
Cardiovascular system		
cerebral arteries	↓	Toda & Okamura, 1990a,b,c, 1991a; Toda, Ayajiki & Okamura, 1993a; Toda *et al.*, 1993b
mesenteric artery	inhibition of sympathetic excitation	Toda & Okamura, 1990c
ophthalmic arteries	↓, inhibition of sympathetic excitation	Toda, Kitamura & Okamura, 1995a
renal artery	↓, inhibition of sympathetic excitation	Okamura, Yoshida & Toda, 1995
retinal arteries	↓	Toda *et al.*, 1993b; Toda, Kitamura & Okamura, 1994b
saphenous artery	↓	Okamura & Toda, 1994
temporal arteries	↓, inhibition of sympathetic excitation	Toda & Okamura, 1991b, 1993; Okamura, Enokibori & Toda, 1993; Toda, Yoshida & Okamura, 1991b
temporal veins	↓	Toda, Yoshida & Okamura, 1995b
Gastrointestinal tract		
lower oesophageal sphincter	↓	De Man *et al.*, 1991; Boulant *et al.*, 1994
pylorus	↓	Allescher *et al.*, 1992b; Bayguinov & Sanders, 1993; Allescher & Daniel, 1994
duodenum	↓, inhibition of excitatory transmission	Toda, Baba & Okamura, 1990a; Toda, Tanobe & Baba, 1991a; Toda *et al.*, 1992
sphincter of Oddi	↓	Kaufman *et al.*, 1993
ileum	↓	Boeckxstaens *et al.*, 1991b; Bogers *et al.*, 1991
ileocolonic junction	↓	Boeckxstaens *et al.*, 1991b; Bogers *et al.*, 1991; Ward, McKeen & Sanders, 1992c
colon	↓ and rebound ↑	Christinck *et al.*, 1991; Dalziel *et al.*, 1991; Thornbury *et al.*, 1991; Ward *et al.*, 1992a,b; Shuttleworth, Sanders & Keef, 1993a; Shuttleworth *et al.*, 1993b

Appendix II (*cont.*)

Tissue	Function	References
Dog		
Urogenital tract		
urethra	↓	Hashimoto, Kigoshi & Muramatsu, 1993
Ferret		
Gastrointestinal tract		
stomach	↓	Grundy, Gharib Naseri & Hutson, 1993
Guinea-pig		
Cardiovascular system		
pulmonary arteries	↓	Liu *et al.*, 1992a,b
Gastrointestinal tract		
stomach	↓, inhibition of excitatory transmission	Desai, Sessa & Vane, 1991a; Desai *et al.*, 1991b; Lefebvre, Baert & Barbier, 1992a; Desai *et al.*, 1994
sphincter of Oddi	↓	Mourelle *et al.*, 1993; Pauletzki *et al.*, 1993
ileum	inhibition of excitatory transmission	Gustafsson *et al.*, 1990; Wiklund *et al.*, 1993a
ileum	tonic inhibition	Waterman & Costa, 1994; Waterman, Costa & Tonini, 1994
caecum	↓	Shuttleworth, Murphy & Furness, 1991; Wiklund *et al.*, 1993b
colon	↓	Briejer *et al.*, 1992
Respiratory tract		
airway smooth muscle	↓	Li & Rand, 1991; Belvisi *et al.*, 1993
trachea	inhibition of cholinergic ↑	Watson, Maclagan & Barnes, 1993
Human		
Cardiovascular system		

cerebral arteries	↓	Toda, 1993
uterine artery	↓	Toda *et al.*, 1994a
Gastrointestinal tract		
lower oesophageal sphincter	↓	McKirdy *et al.*, 1992; Oliveira *et al.*, 1992; Preiksaitis, Tremblay & Diamant, 1994; Murray *et al.*, 1995
oesophagus	↓	Preiksaitis, Tremblay & Diamant, 1994; Murray *et al.*, 1995
stomach	↓	Lincoln, Hoyle & Burnstock, 1995
pylorus	↓	Oliveira *et al.*, 1992
ileum	↓	Maggi *et al.*, 1991
ileocolonic junction	↓	McKirdy, Marshall & Taylor, 1993
colon	↓	Burleigh, 1992; Boeckxstaens *et al.*, 1993; Keef *et al.*, 1993
internal anal sphincter	↓	Burleigh, 1992; O'Kelly, Brading & Mortensen, 1993
Respiratory tract		
airway smooth muscle	↓	Ellis & Undem, 1992; Belvisi *et al.*, 1992; Bai & Bramley, 1993; Ward *et al.*, 1995
airway smooth muscle	inhibition of cholinergic ↑	Ward *et al.*, 1993
Urogenital tract		
corpus cavernosum	↓	Kim *et al.*, 1991; Pickard, Powell & Zar, 1991; Holmquist, Hedlund & Andersson, 1992; Rajfer *et al.*, 1992; Leone *et al.*, 1994b
urethra	↓	Ehren *et al.*, 1994; Leone *et al.*, 1994b
uterus	tonic inhibition	Buhimschi *et al.*, 1995
Monkey		
Cardiovascular system		
cerebral arteries	↓	Toda & Okamura, 1990a
mesenteric artery	↓, inhibition of sympathetic excitation	Toda & Okamura, 1992
renal artery	↓, inhibition of sympathetic excitation	Okamura, Yoshida & Toda, 1995

Appendix II (*cont.*)

Tissue	Function	References
Monkey		
temporal arteries	↓, inhibition of sympathetic excitation	Toda & Okamura, 1991b,1993
temporal veins		Toda, Yoshida & Okamura, 1995b
Mouse		
Urogenital tract		
anococcygeus muscle	↓	Gibson *et al.*, 1989
Opossum		
Gastrointestinal tract		
oesophagus	↓	Christinck *et al.*, 1991
lower oesophageal sphincter	↓	Tottrup, Knudsen & Gregersen, 1991a; Tottrup, Svane & Forman, 1991b; Paterson, Anderson & Anand, 1992; Conklin *et al.*, 1993
sphincter of Oddi	↓	Allescher *et al.*, 1993
internal anal sphincter	↓	Chakder & Rattan, 1992, 1993a,b; Rattan & Chakder, 1992; Tottrup, Glavind & Svane, 1992; Rattan, Rosenthal & Chakder, 1995
Pig		
Cardiovascular system		
submandibular gland vessels	↓	Modin, Weitzberg & Lundberg, 1994
Respiratory tract		
trachea	↓	Kannan & Johnson, 1992
Urogenital tract		
urethra	↓	Persson & Andersson, 1992; Bridgewater, MacNeil & Brading, 1993; Persson *et al.*, 1993

Rabbit		
Cardiovascular system		
hepatic portal vein	↓	Brizzolara, Crowe & Burnstock, 1993
Gastrointestinal tract		
stomach	inhibition of excitatory transmission	Baccari, Bertini & Calamai, 1993; Baccari, Calamai & Staderini, 1994
sphincter of Oddi	↓	Mourelle *et al.*, 1993
internal anal sphincter	↓	Knudsen, Glavind & Tottrup, 1995
Urogenital tract		
anococcygeus muscle	↓	Graham & Sneddon, 1993; Kaskov, Cellek & Moncada, 1995
corpus cavernosum	↓	Ignarro *et al.*, 1990; Holmquist, Hedlund & Andersson, 1992; Knispel, Goessl & Beckmann, 1992
urethra	↓	Andersson *et al.*, 1991, 1992; Dokita *et al.*, 1991,1994; Zygmunt *et al.*, 1993; Lee, Wein & Levin, 1994
Rat		
Cardiovascular system		
irideal arterioles	inhibition of sympathetic ↑	Hill & Gould, 1995
Gastrointestinal tract		
stomach	↓	Li & Rand, 1990; Boeckxstaens *et al.*, 1991a,1992; D'Amato, Curro & Montuschi, 1992a,b; Lefebvre, De Vriese & Smits, 1992b; McLaren, Li & Rand, 1993; Shimamura *et al.*, 1993; Lefebvre, Hasrat & Gobert, 1994
stomach	inhibition of cholinergic ↑	Lefebvre, De Vriese & Smits, 1992b
pylorus	↓	Soediono & Burnstock, 1994
duodenum	↓	Irie *et al.*, 1991
ileum	↑	Barthó *et al.*, 1992
ileum	↓	Allescher *et al.*, 1992a; Calignano *et al.*, 1992; Kanada *et al.*, 1992

Appendix II (*cont.*)

Tissue	Function	References
Rat		
colon	↓	Hata *et al.*, 1990
Respiratory tract		
airway smooth muscle	inhibition of cholinergic ↑	Sekizawa *et al.*, 1993
Urogenital tract		
anococcygeus muscle	↓	Gillespie, Liu & Martin, 1989; Li & Rand, 1989; Ramagopal & Leighton, 1989
external urethral sphincter	↓	Parlani, Conte & Manzini, 1993
urethra	↓	Persson *et al.*, 1992; Bennett *et al.*, 1995
uterus	↓	Shew *et al.*, 1993
Sheep		
Urogenital tract		
urethra	↓	Garcia Pascual *et al.*, 1991; Thornbury, Hollywood & McHale, 1992; Triguero, Prieto & Garcia Pascual, 1993; Garcia Pascual & Triguero, 1994

↓, Inhibition, relaxation or vasodilatation.
↑, Contraction or vascoconstriction.
Although extensive, this table does not claim to be comprehensive.

References

Abrahamsohn, I. A. & Coffman, R. L. (1995). Cytokine and nitric oxide regulation of the immunosuppression in *Trypanosoma cruzi* infection. *Journal of Immunology*, **155**, 3955–63.

Abrahamsson, T., Brandt, U., Marklund, S. L. & Sjoqvist, P. O. (1992). Vascular bound recombinant extracellular superoxide dismutase type C protects against the detrimental effects of superoxide radicals on endothelium-dependent arterial relaxation. *Circulation Research*, **70**, 264–71.

Abu-Soud, H. M., Feldman, P. L., Clark, P. & Stuehr, D. J. (1994a). Electron transfer in the nitric-oxide synthases: characterization of L-arginine analogs that block heme iron reduction. *Journal of Biological Chemistry*, **269**, 32318–26.

Abu-Soud, H. M. & Stuehr, D. J. (1993). Nitric oxide synthases reveal a role for calmodulin in controlling electron transfer. *Proceedings of the National Academy of Science USA*, **90**, 10769–72.

Abu-Soud, H. M., Yoho, L. L. & Stuehr, D. J. (1994b). Calmodulin controls neuronal nitric-oxide synthase by a dual mechanism. Activation of intra- and interdomain electron transfer. *Journal of Biological Chemistry*, **269**, 32047–50.

Adler, B., Adler, H., Jungi, T. W. & Peterhans, E. (1995). Interferon-α primes macrophages for lipopolysaccharide-induced apoptosis. *Biochemical and Biophysical Research Communications*, **215**, 921–7.

Afework, M., Ralevic, V. & Burnstock, G. (1995). The intra-adrenal distribution of intrinsic and extrinsic nitrergic nerve fibres in the rat. *Neuroscience Letters*, **190**, 109–12.

Afework, M., Tomlinson, A., Belai, A. & Burnstock, G. (1992). Colocalization of nitric oxide synthase and NADPH-diaphorase in rat adrenal gland. *NeuroReport*, **3**, 893–6.

Afework, M., Tomlinson, A. & Burnstock, G. (1994). Distribution and colocalization of nitric oxide synthase and NADPH-diaphorase in adrenal gland of developing, adult and aging Sprague-Dawley rats. *Cell and Tissue Research*, **276**, 133–41.

Aimi, Y., Kimura, H., Kinoshita, T., Minami, Y., Fujimura, M. & Vincent, S. R. (1993). Histochemical localization of nitric oxide synthase in rat enteric nervous system. *Neuroscience*, **53**, 553–60.

Akabane, A., Kato, I., Takasawa, S., Unno, M., Yonekura, H., Yoshimoto, T. &

Okamoto, H. (1995). Nicotinamide inhibits IRF-1 mRNA induction and prevents IL-1-induced nitric oxide synthase expression in pancreatic cells. *Biochemical and Biophysical Research Communications*, **215**, 524–30.

Akaike, T., Yoshida, M., Miyamoto, Y., Sato, K., Kohno, M., Sasamoto, K., Miyazaki, K., Ueda, S. & Maeda, H. (1993). Antagonistic action of imidazolineoxyl N-oxides against endothelium-derived relaxing factor/.NO through a radical reaction. *Biochemistry*, **32**, 827–32.

Aliev, G., Miah, S., Turmaine, M. & Burnstock, G. (1995). An ultrastructural and immunocytochemical study of thoracic aortic endothelium in aged Sprague-Dawley rats. *Journal of Submicroscopic Cytology and Pathology*, **27**, 477–90.

Allescher, H. D. & Daniel, E. E. (1994). Role of NO in pyloric, antral, and duodenal motility and its interaction with other inhibitory mediators. *Digestive Diseases and Sciences*, **39**, 73S–5S.

Allescher, H. D., Lu, S., Daniel, E. E. & Classen, M. (1993). Nitric oxide as putative nonadrenergic noncholinergic inhibitory transmitter in the opossum sphincter of Oddi. *Canadian Journal of Physiology and Pharmacology*, **71**, 525–30.

Allescher, H. D., Sattler, D., Piller, C., Schusdziarra, V. & Classen, M. (1992a). Ascending neural pathways in the rat ileum *in vitro* – effect of capsaicin and involvement of nitric oxide. *European Journal of Pharmacology*, **217**, 153–62.

Allescher, H. D., Tougas, G., Vergara, P., Lu, S. & Daniel, E. E. (1992b). Nitric oxide as a putative nonadrenergic noncholinergic inhibitory transmitter in the canine pylorus *in vivo*. *American Journal of Physiology*, **262**, G695–702.

Alm, P., Larsson, B., Ekblad, E., Sundler, F. & Andersson, K. E. (1993). Immunohistochemical localization of peripheral nitric oxide synthase-containing nerves using antibodies raised against synthesized C- and N-terminal fragments of a cloned enzyme from rat brain. *Acta Physiologica Scandinavica*, **148**, 421–9.

Alonso, J. R., Arevalo, R., Garcia Ojeda, E., Porteros, A., Brinon, J. G. & Aijon, J. (1994). NADPH-diaphorase active and calbindin D-28k-immunoreactive neurons and fibers in the olfactory bulb of the hedgehog (*Erinaceus europaeus*). *Journal of Comparative Neurology*, **351**, 307–27.

Alonso, M. J., Salaices, M., Sanchez Ferrer, C. F. & Marin, J. (1992b). Predominant role for nitric oxide in the relaxation induced by acetylcholine in cat cerebral arteries. *Journal of Pharmacology and Experimental Therapeutics*, **261**, 12–20.

Alonso, M. J., Salaices, M., Sanchez Ferrer, C. F., Ponte, A., Lopez Rico, M. & Marin, J. (1993). Nitric-oxide-related and non-related mechanisms in the acetylcholine-evoked relaxations in cat femoral arteries. *Journal of Vascular Research*, **30**, 339–47.

Alonso, J. R., Sanchez, F., Arevalo, R., Carretero, J., Vazquez, R. & Aijon, J. (1992a). Partial coexistence of NADPH-diaphorase and somatostatin in the rat hypothalamic paraventricular nucleus. *Neuroscience Letters*, **148**, 101–4.

Altman, F. P., Hoyer, P. E. & Andersen, H. (1979). Dehydrogenase histochemistry of lipid-rich tissues: a tetrazolium-metal chelation technique to improve localization. *Histochemical Journal*, **11**, 485–8.

Amano, F. & Noda, T. (1995). Improved detection of nitric oxide radical (NO) production in an activated macrophage culture with a radical scavenger, carboxy PTIO, and Griess reagent. *FEBS Letters*, **368**, 425–8.

Ambache, N., Killick, S. W. & Zar, M. A. (1975). Extraction from ox retractor penis of an inhibitory substance which mimics its atropine-resistant neurogenic relaxation. *British Journal of Pharmacology*, **54**, 409–10.

Amoah-Apraku, B., Chandler, L. J., Harrison, J. K., Tang, S.-S., Ingelfinger, J. R. & Guzman, N. J. (1995). NF-κB and transcriptional control of renal epithelial-inducible nitric oxide synthase. *Kidney International*, **48**, 674–82.

Anderson, C. R., Edwards, S. L., Furness, J. B., Bredt, D. S. & Snyder, S. H. (1993). The distribution of nitric oxide synthase-containing autonomic preganglionic terminals in the rat. *Brain Research*, **614**, 78–85.

Anderson, C. R., Furness, J. B., Woodman, H. L., Edwards, S. L., Crack, P. J. & Smith, A. I. (1995). Characterisation of neurons with nitric oxide synthase immunoreactivity that project to prevertebral ganglia. *Journal of the Autonomic Nervous System*, **52**, 107–16.

Andersson, K. E., Garcia Pascual, A., Forman, A. & Tottrup, A. (1991). Non-adrenergic, non-cholinergic nerve-mediated relaxation of rabbit urethra is caused by nitric oxide. *Acta Physiologica Scandinavica*, **141**, 133–4.

Andersson, K. E., Garcia Pascual, A., Persson, K., Forman, A. & Tottrup, A. (1992). Electrically-induced, nerve-mediated relaxation of rabbit urethra involves nitric oxide. *Journal of Urology*, **147**, 253–9.

Änggård, E. (1994). Nitric oxide: mediator, murderer, and medicine. *Lancet*, **343**, 1199–206.

Angulo, I., Rodríguez, R., García, B., Medina, M., Navarro, J. & Subiza, J. L. (1995). Involvement of nitric oxide in bone marrow-derived natural suppressor activity: its dependence on IFN-γ. *Journal of Immunology*, **155**, 15–26.

Aoki, C., Fenstemaker, S., Lubin, M. & Go, C. G. (1993a). Nitric oxide synthase in the visual cortex of monocular monkeys as revealed by light and electron microscopic immunocytochemistry. *Brain Research*, **620**, 97–113.

Aoki, E., Takeuchi, I. K., Shoji, R. & Semba, R. (1993b). Localization of nitric oxide-related substances in the peripheral nervous tissues. *Brain Research*, **620**, 142–5.

Archer, S. (1993). Measurement of nitric oxide in biological models. *FASEB Journal*, **7**, 349–60.

Archer, S. L. & Cowan, N. J. (1991). Measurement of endothelial cytosolic calcium concentration and nitric oxide production reveals discrete mechanisms of endothelium-dependent pulmonary vasodilatation. *Circulation Research*, **68**, 1569–81.

Arnal, J.-F., Yamin, J., Dockery, S. & Harrison, D. G. (1994). Regulation of endothelial nitric oxide synthase mRNA, protein, and activity during cell growth. *American Journal of Physiology*, **267**, C1381–8.

Assreuy, J., Cunha, F. Q., Liew, F. Y. & Moncada, S. (1993). Feedback inhibition of nitric oxide synthase activity by nitric oxide. *British Journal of Pharmacology*, **108**, 833–7.

Assreuy, J. & Moncada, S. (1992). A perfusion system for the long term study of macrophage activation. *British Journal of Pharmacology*, **107**, 317–21.

Attur, M., Vyas, P., Leszczynska-Piziak, J., Rediske, J., Vora, K., Abramson, S. B. & Amin, A. R. (1995). Human inflammatory cells express mRNA for inducible nitric oxide synthase but lack detectable translated product and nitric oxide synthase activity. *Endothelium*, **3** (Suppl), S57

Attwell, D., Barbour, B. & Szatkowski, M. (1993). Nonvesicular release of neurotransmitter. *Neuron*, **11**, 401–7.

Ayajiki, K., Okamura, T. & Toda, N. (1993). Nitric oxide mediates, and acetylcholine modulates, neurally induced relaxation of bovine cerebral arteries. *Neuroscience*, **54**, 819–25.

Ayajiki, K., Okamura, T. & Toda, N. (1994). Neurogenic relaxations caused by nicotine in isolated cat middle cerebral arteries. *Journal of Pharmacology and Experimental Therapeutics*, **270**, 795–801.

Aylward, G. H. & Findlay, T. J. V. (1971). *Chemical Data Book*, second edn. Chichester: John Wiley & Sons.

Babbedge, R. C., Bland-Ward, P. A., Hart, S. L. & Moore, P. K. (1993a). Inhibition of rat cerebellar nitric oxide synthase by 7-nitro indazole and related substituted indazoles. *British Journal of Pharmacology*, **110**, 225–8.

Babbedge, R. C., Hart, S. L. & Moore, P. K. (1993b). Anti-nociceptive activity of nitric oxide synthase inhibitors in the mouse: dissociation between the effect of L-NAME and L-NMMA. *Journal of Pharmacy and Pharmacology*, **45**, 77–9.

Babbedge, R. C., Wallace, P., Gaffen, Z., Hart, S. L. & Moore, P. K. (1993c). L-N^G-nitro arginine p-nitroanilide (L-NAPNA) is anti-nociceptive in the mouse. *NeuroReport*, **4**, 307–10.

Baccari, M. C., Bertini, M. & Calamai, F. (1993). Effects of L-N^G-nitroarginine on cholinergic transmission in the gastric muscle of the rabbit. *NeuroReport*, **4**, 1102–4.

Baccari, M. C., Calamai, F. & Staderini, G. (1994). Modulation of cholinergic transmission by nitric oxide, VIP and ATP in the gastric muscle. *NeuroReport*, **5**, 905–8.

Bacci, S., Arbi Riccardi, R., Mayer, B., Rumio, C. & Borghi Cirri, M. B. (1994). Localization of nitric oxide synthase immunoreactivity in mast cells of human nasal mucosa. *Histochemistry*, **102**, 89–92.

Bachmann, S., Bosse, H. M. & Mundel, P. (1995). Topography of nitric oxide synthesis by localizing constitutive NO synthases in mammalian kidney. *American Journal of Physiology*, **268**, F885–98.

Bachmann, S. & Mundel, P. (1994). Nitric oxide in the kidney: synthesis, localization and function. *American Journal of Kidney Diseases*, **24**, 112–29.

Bai, T. R. & Bramley, A. M. (1993). Effect of an inhibitor of nitric oxide synthase on neural relaxation of human bronchi. *American Journal of Physiology*, **264**, L425–30.

Baker, R. A., Saccone, G. T., Brookes, S. J. & Toouli, J. (1993). Nitric oxide mediates nonadrenergic, noncholinergic neural relaxation in the Australian possum. *Gastroenterology*, **105**, 1746–53.

Balaskas, C., Saffrey, M. J. & Burnstock, G. (1995). Distribution and colocalization of NADPH-diaphorase activity, nitric oxide synthase immunoreactivity, and VIP immunoreactivity in the newly hatched chicken gut. *Anatomical Record*, **243**, 10–18.

Bani, D., Masini, E., Bello, M. G., Bigazzi, M. & Sacchi, T. B. (1995). Relaxin activates the L-arginine-nitric oxide pathway in human breast cancer cells. *Cancer Research*, **55**, 5272–5.

Bannerman, D. M., Chapman, P. F., Kelly, P. A., Butcher, S. P. & Morris, R. G. (1994). Inhibition of nitric oxide synthase does not impair spatial learning. *Journal of Neuroscience*, **14**, 7404–14.

Barbier, A. J. & Lefebvre, R. A. (1992). Effect of LY 83583 on relaxation induced by non-adrenergic non-cholinergic nerve stimulation and exogenous nitric oxide in the rat gastric fundus. *European Journal of Pharmacology*, **219**, 331–4.

Barbier, A. J. & Lefebvre, R. A. (1993). Involvement of the L-arginine: nitric oxide pathway in nonadrenergic noncholinergic relaxation of the cat gastric fundus. *Journal of Pharmacology and Experimental Therapeutics*, **266**, 172–8.

Barbier, A. J. & Lefebvre, R. A. (1995). Relaxant influence of phosphodiesterase inhibitors in the cat gastric fundus. *European Journal of Pharmacology*, **276**, 41–7.

Barbiers, M., Timmermans, J. P., Scheuermann, D. W., Adriaensen, D., Mayer, B. & De Groodt Lasseel, M. H. (1994). Nitric oxide synthase-containing neurons in the pig large intestine: topography, morphology, and viscerofugal projections. *Microscopy Research and Technique*, **29**, 72–8.

Barnes, P. J. & Belvisi, M. G. (1993). Nitric oxide and lung disease. *Thorax*, **48**, 1034–43.

Barnes, P. J. & Liew, F. Y. (1995). Nitric oxide and asthmatic inflammation. *Immunology Today*, **16**, 128–30.

Barthó, L., Kóczán, G., Pethö, G. & Maggi, C. A. (1992). Blockade of nitric oxide synthase inhibits nerve-mediated contraction in the rat small intestine. *Neuroscience Letters*, **145**, 43–6.

Bates, T. E., Loesch, A., Burnstock, G. & Clark, J. B. (1995). Immunocytochemical evidence for a mitochondrially located nitric oxide synthase in brain and liver. *Biochemical and Biophysical Research Communications*, **213**, 896–900.

Bauer, J. A., Booth, B. P. & Fung, H. L. (1995). Nitric oxide donors: biochemical pharmacology and therapeutics. *Advances in Pharmacology*, **34**, 361–81.

Baydoun, A. R., Knowles, R. G., Hodson, H. F., Moncada, S. & Mann, G. E. (1995). Symmetric and asymmetric dimethylarginine inhibits arginine transport and nitric oxide synthesis in lipopolysaccharide-activated J774 cells. *Endothelium*, **3**(Suppl), S35.

Bayguinov, O. & Sanders, K. M. (1993). Role of nitric oxide as an inhibitory neurotransmitter in the canine pyloric sphincter. *American Journal of Physiology*, **264**, G975–83.

Beckman, J. S., Chen, J., Crow, J. P. & Ye, Y. Z. (1994a). Reactions of nitric oxide, superoxide and peroxynitrite with superoxide dismutase in neurodegeneration. *Progress in Brain Research*, **103**, 371–80.

Beckman, J. S., Chen, J., Ischiropoulos, H. & Crow, J. P. (1994b). Oxidative chemistry of peroxynitrite. *Methods in Enzymology*, **233**, 229–40.

Bekkers, J. M. & Stevens, C. F. (1990). Presynaptic mechanisms for long-term potentiation in the hippocampus. *Nature*, **346**, 724–9.

Belai, A., Schmidt, H. H. H. W., Hoyle, C. H. V., Hassall, C. J., Saffrey, M. J., Moss, J., Förstermann, U., Murad, F. & Burnstock, G. (1992). Colocalization of nitric oxide synthase and NADPH-diaphorase in the myenteric plexus of the rat gut. *Neuroscience Letters*, **143**, 60–4.

Belvisi, M. G., Miura, M., Stretton, D. & Barnes, P. J. (1993). Endogenous vasoactive intestinal peptide and nitric oxide modulate cholinergic neurotransmission in guinea-pig trachea. *European Journal of Pharmacology*, **231**, 97–102.

Belvisi, M. G., Stretton, C. D., Miura, M., Verleden, G. M., Tadjkarimi, S., Yacoub, M. H. & Barnes, P. J. (1992). Inhibitory NANC nerves in human tracheal smooth muscle: a quest for the neurotransmitter. *Journal of Applied Physiology*, **73**, 2505–10.

Bennett, B. C., Kruse, M. N., Roppolo, J. R., Flood, H. D., Fraser, M. & de Groat, W. C. (1995). Neural control of urethral outlet activity in vivo: role of nitric oxide. *Journal of Urology*, **153**, 2004–9.

Berezin, I., Snyder, S. H., Bredt, D. S. & Daniel, E. E. (1994). Ultrastructural localization of nitric oxide synthase in canine small intestine and colon. *American Journal of Physiology*, **266**, C981–9.

Berkowitz, B. A. & Ohlstein, E. H. (1992). Evidence that endothelium-derived relaxing factor released from endothelial cells is not nitric oxide. In *Endothelial Regulation of Vascular Tone*, ed. U. S. Ryan & G. M. Rubanyi, pp. 51–6. New York: Marcel Dekker Inc.

Berlin, R. (1987). Historical aspects of nitrate therapy. *Drugs*, **33** (Suppl 4), 1–4.

Bevan, J. A. & Osher, J. V. (1972). A direct method for recording tension changes in the wall of small blood vessels *in vitro*. *Agents and Actions*, **2**, 257–60.

Bjorkman, R. (1995). Central antinociceptive effects of non-steroidal anti-inflammatory drugs and paracetamol. Experimental studies in the rat. *Acta Anaesthesiologica Scandinavica Supplement*, **103**, 1–44.

Bjorling, D. E., Saban, R., Tengowski, M. W., Gruel, S. M. & Rao, V. K. (1992). Removal of venous endothelium with air. *Journal of Pharmacological and Toxicological Methods*, **28**, 149–57.

Blatter, L. A., Taha, Z., Mesaros, S., Shacklock, P. S., Wier, W. G. & Malinski, T. (1995). Simultaneous measurements of Ca^{2+} and nitric oxide in bradykinin-stimulated vascular endothelial cells. *Circulation Research*, **76**, 922–4.

Bliss, T. V. P. & Lømo, T. (1973). Long-lasting potentiation of synaptic transmission in the dentate area of the anaesthetized rabbit following stimulation of the perforant path. *Journal of Physiology (London)*, **232**, 331–56.

Blottner, D., Schmidt, H. H. H. W. & Baumgarten, H. G. (1993). Nitroxergic autonomic neurones in rat spinal cord. *NeuroReport*, **4**, 923–6.

Boeckxstaens, G. E., De Man, J. G., De Winter, B. Y., Herman, A. G. & Pelckmans, P. A. (1994). Pharmacological similarity between nitric oxide and the nitrergic neurotransmitter in the canine ileocolonic junction. *European Journal of Pharmacology*, **264**, 85–9.

Boeckxstaens, G. E., De Man, J. G., De Winter, B. Y., Moreels, T. G., Herman, A. G. & Pelckmans, P. A. (1995). Bioassay and pharmacological characterization of the nitrergic neurotransmitter. *Archives Internationales de Pharmacodynamie et Therapie*, **329**, 11–26.

Boeckxstaens, G. E., Pelckmans, P. A., Bogers, J. J., Bult, H., De Man, J. G., Oosterbosch, L., Herman, A. G. & Van Maercke, Y. M. (1991a). Release of nitric oxide upon stimulation of nonadrenergic noncholinergic nerves in the rat gastric fundus. *Journal of Pharmacology and Experimental Therapeutics*, **256**, 441–7.

Boeckxstaens, G. E., Pelckmans, P. A., Bult, H., De Man, J. G., Herman, A. G. & Van Maercke, Y. M. (1991b). Evidence for nitric oxide as mediator of non-adrenergic non-cholinergic relaxations induced by ATP and GABA in the canine gut. *British Journal of Pharmacology*, **102**, 434–8.

Boeckxstaens, G. E., Pelckmans, P. A., De Man, J. G., Bult, H., Herman, A. G. & Van Maercke, Y. M. (1992). Evidence for a differential release of nitric oxide and vasoactive intestinal polypeptide by nonadrenergic noncholinergic nerves in the rat gastric fundus. *Archives Internationales de Pharmacodynamie et Therapie*, **318**, 107–15.

Boeckxstaens, G. E., Pelckmans, P. A., Herman, A. G. & Van Maercke, Y. M. (1993). Involvement of nitric oxide in the inhibitory innervation of the human isolated colon. *Gastroenterology*, **104**, 690–7.

Bogdan, C., Werner, E., Stenger, S., Wachter, H., Röllinghoff, M. & Werner-Felmayer, G. (1995). 2,4-Diamino-6-hydroxypyrimidine, an inhibitor of tetrahydrobiopterin synthesis, downregulates the expression of iNOS protein and mRNA in primary murine macrophages. *FEBS Letters*, **363**, 69–74.

Bogers, J. J., Pelckmans, P. A., Boeckxstaens, G. E., De Man, J. G., Herman, A. G. & Van Maercke, Y. M. (1991). The role of nitric oxide in serotonin-induced relaxations in the canine terminal ileum and ileocolonic junction. *Naunyn Schmiedeberg's Archives of Pharmacology*, **344**, 716–9.

Bogers, J. J., Timmermans, J. P., Scheuermann, D. W., Pelckmans, P. A., Mayer, B. & van Marck, E. A. (1994). Localization of nitric oxide synthase in enteric neurons of the porcine and human ileocaecal junction. *Anatomischer Anzeiger*, **176**, 131–5.

Bogle, R. G., Baydoun, A. R., Pearson, J. D. & Mann, G. E. (1996). Regulation of L-arginine transport and nitric oxide release in superfused porcine aortic endothelial cells. *Journal of Physiology (London)*, **490**, 229–41.

Bogle, R. G., Coade, S. B., Moncada, S., Pearson, J. D. & Mann, G. E. (1991). Bradykinin and ATP stimulate L-arginine uptake and nitric oxide release in vascular endothelial cells. *Biochemical and Biophysical Research Communications*, **180**, 926–32.

Bogle, R. G., Moncada, S., Pearson, J. D. & Mann, G. E. (1992). Identification of inhibitors of nitric oxide synthase that do not interact with the endothelial cell L-arginine transporter. *British Journal of Pharmacology*, **105**, 768–70.

Bohme, G. A., Bon, C., Lemaire, M., Reibaud, M., Piot, O., Stutzmann, J. M., Doble, A. & Blanchard, J. C. (1993). Altered synaptic plasticity and memory formation in nitric oxide synthase inhibitor-treated rats. *Proceedings of the National Academy of Sciences USA*, **90**, 9191–4.

Bohme, G. A., Bon, C., Stutzmann, J. M., Doble, A. & Blanchard, J. C. (1991). Possible involvement of nitric oxide in long-term potentiation. *European Journal of Pharmacology*, **199**, 379–81.

Bonhoeffer, T., Staiger, V. & Aersten, A. (1989). Synaptic plasticity in the rat hippocampal slice cultures: local 'Hebbian' conjunction or pre- and postsynaptic stimulation leads to distributed synaptic enhancement. *Proceedings of the National Academy of Sciences USA*, 86, 8113–7.

Bonnardeaux, A., Nadaud, S., Charru, A., Jeunemaitre, X., Corvol, P. & Soubrier, F. (1995). Lack of evidence for linkage of the endothelial cell nitric oxide synthase gene to essential hypertension. *Circulation*, **91**, 96–102.

Boughton-Smith, N. K., Evans, S. M., Hawkey, C. J., Cole, A. T., Balsitis, M., Whittle, B. J. R. & Moncada, S. (1993). Nitric oxide synthase activity in ulcerative colitis and Crohn's disease. *Lancet*, **342**, 338–40.

Boulant, J., Fioramonti, J., Dapoigny, M., Bommelaer, G. & Bueno, L. (1994). Cholecystokinin and nitric oxide in transient lower esophageal sphincter relaxation to gastric distention in dogs. *Gastroenterology*, **107**, 1059–66.

Boulton, C. L., Irving, A. J., Southam, E., Potier, B., Garthwaite, J. & Collingridge, G. L. (1994). The nitric oxide – cyclic GMP pathway and synaptic depression in rat hippocampal slices. *European Journal of Neuroscience*, **6**, 1528–35.

Boutard, V., Havouis, R., Fouqueray, B., Philippe, C., Moulinoux, J.-P. & Baud, L. (1995). Transforming growth factor-β stimulates arginase activity in macrophages. *Journal of Immunology*, **155**, 2077–84.

Bowman, A. & Drummond, A. H. (1984). Cyclic GMP mediates neurogenic relaxation in the bovine retractor penis muscle. *British Journal of Pharmacology*, **81**, 665–74.

Bowman, A. & Gillespie, J. S. (1982). Block of some non-adrenergic inhibitory responses of smooth muscle by a substance from haemolysed erythrocytes. *Journal of Physiology (London)*, **328**, 11–25.

Bowman, A., Gillespie, J. S. & Pollock, D. (1982). Oxyhaemoglobin blocks non-adrenergic non-cholinergic inhibition in the bovine retractor penis muscle. *European Journal of Pharmacology*, **85**, 221–4.

Brain, S. D., Hughes, S. R., Cambridge, H. & O'Driscoll, G. (1993). The contribution of calcitonin gene-related peptide (CGRP) to neurogenic vasodilator responses. *Agents and Actions*, **38**, C19–21.

Brave, S. R., Tucker, J. F., Gibson, A., Bishop, A. E., Riveros Moreno, V., Moncada, S. & Polak, J. M. (1993). Localisation of nitric oxide synthase within non-adrenergic, non-cholinergic nerves in the mouse anococcygeus. *Neuroscience Letters*, **161**, 93–6.

Bredt, D. S., Ferris, C. D. & Snyder, S. H. (1992). Nitric oxide synthase regulatory sites: phosphorylation by cyclic AMP dependent protein kinases, protein kinase C, calcium/calmodulin protein kinase; identification of flavin and calmodulin binding sites. *Journal of Biological Chemistry*, **267**, 10976–81.

Bredt, D. S., Glatt, C. E., Hwang, P. M., Fotuhi, M., Dawson, T. M. & Snyder, S. H. (1991a). Nitric oxide synthase protein and mRNA are discretely localized in neuronal populations of the mammalian CNS together with NADPH diaphorase. *Neuron*, **7**, 615–24.

Bredt, D. S., Hwang, P. M., Glatt, C., Lowenstein, C., Reed, R. R. & Snyder, S. H. (1991b). Cloned and expressed nitric oxide synthase structurally resembles cytochrome P-450 reductase. *Nature*, **351**, 714–8.

Bredt, D. S., Hwang, P. M. & Snyder, S. H. (1990). Localization of nitric oxide synthase indicating a neural role for nitric oxide. *Nature*, **347**, 768–70.

Bredt, D. S. & Snyder, S. H. (1989). Nitric oxide mediates glutamate-linked enhancement of cGMP levels in the cerebellum. *Proceedings of the National Academy of Sciences USA*, **86**, 9030–3.

Bredt, D. S. & Snyder, S. H. (1990). Isolation of nitric oxide synthetase, a calmodulin-requiring enzyme. *Proceedings of the National Academy of Science USA*, **87**, 682–5.

Bredt, D. S. & Snyder, S. H. (1992). Nitric oxide, a novel neuronal messenger. *Neuron*, **8**, 3–11.

Brenman, J. E., Chao, D. S., Xia, H., Aldape, K. & Bredt, D. S. (1995). Nitric oxide synthase complexed with dystrophin and absent from skeletal muscle sarcolemma in Duchenne muscular dystrophy. *Cell*, **82**, 743–52.

Bridgewater, M., MacNeil, H. F. & Brading, A. F. (1993). Regulation of tone in pig urethral smooth muscle. *Journal of Urology*, **150**, 223–8.

Briejer, M. R., Akkermans, L. M., Meulemans, A. L., Lefebvre, R. A. & Schuurkes, J. A. (1992). Nitric oxide is involved in 5-HT-induced relaxations of the guinea-pig colon ascendens *in vitro*. *British Journal of Pharmacology*, **107**, 756–61.

Brien, J. F., McLaughlin, B. E., Nakatsu, K. & Marks, G. S. (1991). Quantitation of nitric oxide formation from nitrovasodilator drugs by chemiluminescence analysis of headspace gas. *Journal of Pharmacological Methods*, **25**, 19–27.

Brizzolara, A. L., Crowe, R. & Burnstock, G. (1993). Evidence for the involvement of both ATP and nitric oxide in non-adrenergic, non-cholinergic inhibitory neurotransmission in the rabbit portal vein. *British Journal of Pharmacology*, **109**, 606–8.

Brorson, J. R., Manzolillo, P. A. & Miller, R. J. (1994). Ca^{2+} entry via AMPA/KA receptors and excitotoxicity in cultured cerebellar Purkinje cells. *Journal of Neuroscience*, **14**, 187–97.

Brorson, J. R., Marcuccilli, C. J. & Miller, R. J. (1995). Delayed antagonism of calpain reduces excitotoxicity in cultured neurons. *Stroke*, **26**, 1259–66.

Brosnan, C. F., Battistini, L., Raine, C. S., Dickson, D. W., Casadevall, A. & Lee, S. C. (1994). Reactive nitrogen intermediates in human neuropathology: an overview. *Developmental Neuroscience*, **16**, 152–61.

Brüne, B., Dimmeler, S., Molina y Vedia, L. & Lapetina, E. G. (1994). Nitric oxide: a signal for ADP-ribosylation of proteins. *Life Sciences*, **54**, 61–70.

Brüne, B. & Lapetina, E. G. (1989). Activiation of a cytosolic ADP-ribosyltransferase by nitric oxide generating agents. *Journal of Biological Chemistry*, **264**, 8455–8.

Brüne, B. & Lapetina, E. G. (1991). Phosphorylation of nitric oxide synthase by protein kinase A. *Biochemical and Biophysical Research Communications*, **181**, 921–6.

Brüne, B., Mohr, S. & Messmer, U. K. (1996). Protein thiol modification and apoptotic cell death as cGMP-independent nitric oxide (NO) signaling pathways. *Reviews in Physiology, Biochemistry and Pharmacology*, **127**, 1–30.

Brüning, G. (1993). NADPH-diaphorase histochemistry in the postnatal mouse cerebellum suggests specific developmental functions for nitric oxide. *Journal of Neuroscience Research*, **36**, 580–7.

Brüning, G., Funk, U. & Mayer, B. (1994). Immunocytochemical localization of nitric oxide synthase in the brain of the chicken. *NeuroReport*, **5**, 2425–8.

Brüning, G., Katzbach, R. & Mayer, B. (1995). Histochemical and immunocytochemical localization of nitric oxide synthase in the central nervous system of the goldfish, carassius auratus. *Journal of Comparative Neurology*, **358**, 353–82.

Bucala, R., Tracey, K. J. & Cerami, A. (1991). Advanced glycosylation products quench nitric oxide and mediate defective endothelium-dependent vasodilatation in experimental diabetes. *Journal of Clinical Investigation*, **87**, 432–8.

Buga, G. M., Gold, M. E., Fukuto, J. M. & Ignarro, L. J. (1991). Shear stress-induced release of nitric oxide from endothelial cells grown on beads. *Hypertension*, **17**, 187–93.

Buga, G. M., Griscavage, J. M., Rogers, N. E. & Ignarro, L. J. (1993). Negative feedback regulation of endothelial cell function by nitric oxide. *Circulation Research*, **73**, 808–12.

Buhimschi, I., Yallampalli, C., Dong, Y. L. & Garfield, R. E. (1995). Involvement of a nitric oxide-cyclic guanosine monophosphate pathway in control of human uterine contractility during pregnancy. *American Journal of Obstetrics and Gynecology*, **172**, 1577–84.

Bukrinsky, M. I., Nottet, H. S. L. M., Schmidtmayerova, H., Dubrovsky, L., Flanagan, C. R., Mullins, M. E., Lipton, S. A. & Gendelman, H. E. (1995). Regulation of nitric oxide synthase activity in human immunodeficiency virus type 1 (HIV-1)-infected monocytes: implications for HIV-associated neurological disease. *Journal of Experimental Medicine*, **181**, 735–45.

Bult, H., Boeckxstaens, G. E., Pelckmans, P. A., Jordaens, F. H., Van Maercke, Y. M. & Herman, A. G. (1990). Nitric oxide as an inhibitory non-adrenergic non-cholinergic neurotransmitter. *Nature*, **345**, 346–7.

Burleigh, D. E. (1992). N^G-nitro-L-arginine reduces nonadrenergic, noncholinergic relaxations of human gut. *Gastroenterology*, **102**, 679–83.

Burnett, A. L., Lowenstein, C. J., Bredt, D. S., Chang, T. S. & Snyder, S. H. (1992). Nitric oxide: a physiologic mediator of penile erection. *Science,* **257**, 401–3.

Burnett, A. L., Maguire, M. P., Chamness, S. L., Ricker, D. D., Takeda, M., Lepor, H. & Chang, T. S. (1995a). Characterization and localization of nitric oxide synthase in the human prostate. *Urology*, **45**, 435–9.

Burnett, A. L., Ricker, D. D., Chamness, S. L., Maguire, M. P., Crone, J. K., Bredt, D. S., Snyder, S. H. & Chang, T. S. (1995b). Localization of nitric oxide synthase in the reproductive organs of the male rat. *Biology of Reproduction*, **52**, 1–7.

Burnett, A. L., Saito, S., Maguire, M. P., Yamaguchi, H., Chang, T. S. & Hanley, D. F. (1995c). Localization of nitric oxide synthase in spinal nuclei innervating pelvic ganglia. *Journal of Urology*, **153**, 212–7.

Burnett, A. L., Tillman, S. L., Chang, T. S., Epstein, J. I., Lowenstein, C. J., Bredt, D. S., Snyder, S. H. & Walsh, P. C. (1993). Immunohistochemical localization of nitric oxide synthase in the autonomic innervation of the human penis. *Journal of Urology*, **150**, 73–6.

Burnstock, G. (1981). Neurotransmitters and trophic factors in the autonomic nervous system. *Journal of Physiology (London)*, **313**, 1–35.

Burnstock, G. (1996). A unifying purinergic hypothesis for the initiation of pain. *Lancet*, **347**, 1064–5.

Burstone, M. S. (1962). *Enzyme Histochemistry and its Application in the Study of Neoplasms.* New York: Academic Press.

Busconi, L. & Michel, T. (1993). Endothelial nitric oxide synthase. N-terminal myristoylation determines subcellular localization. *Journal of Biological Chemistry*, **268**, 8410–3.

Busconi, L. & Michel, T. (1995). Recombinant endothelial nitric oxide synthase: post-translational modifications in a baculovirus expression system. *Molecular Pharmacology*, **47**, 655–9.

Bush, P. A., Aronson, W. J., Buga, G. M., Rajfer, J. & Ignarro, L. J. (1992a). Nitric oxide is a potent relaxant of human and rabbit corpus cavernosum. *Journal of Urology*, **147**, 1650–5.

Bush, P. A., Gonzalez, N. E. & Ignarro, L. J. (1992b). Biosynthesis of nitric oxide and citrulline from L-arginine by constitutive nitric oxide synthase present in rabbit corpus cavernosum. *Biochemical and Biophysical Research Communications*, **186**, 308–14.

Busse, R., Fleming, I. & Hecker, M. (1993). Signal transduction in endothelium-dependent vasodilatation. *European Heart Journal*, **14** (Suppl I), 2–9.

Busse, R., Hecker, M. & Fleming, I. (1994). Control of nitric oxide and prostacyclin synthesis in endothelial cells. *Arzneimittel-Forschung*, **44** (Suppl 3A), 392–6.

Buttery, L. D. K., McCarthy, A., Springall, D. R., Sullivan, M. H., Elder, M. G., Michel, T. & Polak, J. M. (1994). Endothelial nitric oxide synthase in the human placenta: regional distribution and proposed regulatory role at the feto-maternal interface. *Placenta*, **15**, 257–65.

Buttery, L. D. K., Springall, D. R., Evans, T. J., Parums, D. V., Standfield, N. & Polak, J. M. (1995). Inducible nitric oxide synthase (iNOS) is present in atherosclerotic vessels and relates to the severity of the lesion. *Endothelium*, **3** (Suppl), S10.

Calapai, G., Squadrito, F., Altavilla, D., Zingarelli, B., Campo, G. M., Cilia, M. & Caputi, A. P. (1992). Evidence that nitric oxide modulates drinking behaviour. *Neuropharmacology*, **31**, 761–4.

Caldwell, M., O'Neill, M., Earley, B. & Leonard, B. (1994). N^G-Nitro-L-arginine protects against ischaemia-induced increases in nitric oxide and hippocampal neurodegeneration in the gerbil. *European Journal of Pharmacology*, **260**, 191–200.

Calignano, A., Whittle, B. J., Di Rosa, M. & Moncada, S. (1992). Involvement of endogenous nitric oxide in the regulation of rat intestinal motility *in vivo*. *European Journal of Pharmacology*, **229**, 273–6.

Calka, J., Wolf, G. & Brosz, M. (1994). Ultrastructural demonstration of NADPH-diaphorase histochemical activity in the supraoptic nucleus of normal and dehydrated rats. *Brain Research Bulletin*, **34**, 301–8.

Calver, A., Collier, J., Moncada, S. & Vallance, P. (1992a). Effect of local intra-arterial N^G-monomethyl-L-arginine in patients with hypertension: the nitric oxide dilator mechanism appears impaired. *Journal of Hypertension*, **10**, 1025–31.

Calver, A., Collier, J. & Vallance, P. (1992b). Inhibition and stimulation of nitric oxide synthesis in the human forearm arterial bed of patients with insulin-dependent diabetes. *Journal of Clinical Investigation*, **90**, 2548–54.

Calver, A., Collier, J. & Vallance, P. (1994). Forearm blood flow responses to a nitric oxide synthase inhibitor in patients with treated essential hypertension. *Cardiovascular Research*, **28**, 1720–5.

Cameron, N. E. & Cotter, M. A. (1994). Impaired contraction and relaxation in aorta from streptozotocin-diabetic rats: role of polyol pathway. *Diabetologia*, **35**, 1011–19.

Capasso, F., Mascolo, N., Izzo, A. A. & Gaginella, T. S. (1994). Dissociation of castor oil-induced diarrhoea and intestinal mucosal injury in rat: effect of N^G-nitro-L-arginine methyl ester. *British Journal of Pharmacology*, **113**, 1127–30.

Carlberg, M. (1994). Assay of neuronal nitric oxide synthase by HPLC determination of citrulline. *Journal of Neuroscience Methods*, **52**, 165–7.

Carreau, A., Duval, D., Poignet, H., Scatton, B., Vige, X. & Nowicki, J. P. (1994). Neuroprotective efficacy of N^{ω}-nitro-L-arginine after focal cerebral ischemia in the mouse and inhibition of cortical nitric oxide synthase. *European Journal of Pharmacology*, **256**, 241–9.

Casino, P. R., Kilcoyne, C. M., Quyyumi, A. A., Hoeg, J. M. & Panza, J. A. (1994). Investigation of decreased availability of nitric oxide precursor as the mechanism responsible for impaired endothelium-dependent vasodilation in hypercholesterolemic patients. *Journal of the American College of Cardiology*, **23**, 844–50.

Ceccatelli, S., Lundberg, J. M., Zhang, X., Aman, K. & Hokfelt, T. (1994). Immunohistochemical demonstration of nitric oxide synthase in the peripheral autonomic nervous system. *Brain Research*, **656**, 381–95.

Cellek, S., Kasakov, L. & Moncada, S. (1996). Inhibition of nitrergic relaxations by a selective inhibitor of the soluble guanylate cyclase. *British Journal of Pharmacology*, **118**, 137–40.

Chakder, S. & Rattan, S. (1992). Neurally mediated relaxation of opossum internal anal sphincter: influence of superoxide anion generator and the scavenger. *Journal of Pharmacology and Experimental Therapeutics*, **260**, 1113–8.

Chakder, S. & Rattan, S. (1993a). Release of nitric oxide by activation of nonadrenergic noncholinergic neurons of internal anal sphincter. *American Journal of Physiology*, **264**, G7–12.

Chakder, S. & Rattan, S. (1993b). Involvement of cAMP and cGMP in relaxation of internal anal sphincter by neural stimulation, VIP, and NO. *American Journal of Physiology*, **264**, G702–7.

Chakder, S., Rosenthal, G. J. & Rattan, S. (1995). *In vivo* and *in vitro* influence of human recombinant hemoglobin on esophageal function. *American Journal of Physiology*, **268**, G443–50.

Charles, I. G., Chubb, A., Gill, R., Clare, J., Lowe, P. N., Holmes, L. S., Page, M., Keeling, J. G., Moncada, S. & Riveros-Moreno, V. (1993). Cloning and expression of a rat neuronal nitric oxide synthase coding sequence in a baculovirus/insect cell system. *Biochemical and Biophysical Research Communications*, **196**, 1481–9.

Chartrain, N. A., Gellers, D. A., Koty, P. P., Sitrin, N. F., Nusslers, A. K., Hoffman, E. P., Billiar, T. R., Hutchinson, N. I. & Mudgett, J. S. (1994). Molecular cloning, structure, and chromosomal localization of the human inducible nitric oxide synthase gene. *Journal of Biological Chemistry*, **269**, 6765–72.

Chayen, J. & Bitensky, L. (1991). *Practical Histochemistry*, second edn., Chichester: John Wiley and Sons.

Chen, C. & Schofield, G. G. (1995). Nitric oxide donors enhanced Ca^{2+} currents and blocked noradrenaline-induced Ca^{2+} current inhibition in rat sympathetic neurons. *Journal of Physiology (London)*, **482**, 521–31.

Chen, F., Sun, S.-C., Kuh, D. C., Gaydos, L. J. & Demers, L. M. (1995). Essential role of NF-κB activation in silica-induced inflammatory mediator production in macrophages. *Biochemical and Biophysical Research Communications*, **214**, 985–92.

Chen, F. Y. & Lee, T. J. (1993). Role of nitric oxide in neurogenic vasodilation of porcine cerebral artery. *Journal of Pharmacology and Experimental Therapeutics*, **265**, 339–45.

Chen, R. Y. & Guth, P. H. (1995). Interaction of endogenous nitric oxide and CGRP in sensory neuron-induced gastric vasodilation. *American Journal of Physiology*, **268**, G791–6.

Chen, S. & Aston Jones, G. (1994). Cerebellar injury induces NADPH diaphorase in Purkinje and inferior olivary neurons in the rat. *Experimental Neurology*, **126**, 270–6.

Chiang, L. W., Schweizer, F. E., Tsien, R. W. & Schulman, H. (1994). Nitric oxide synthase expression in single hippocampal neurons. *Brain Research and Molecular Brain Research*, **27**, 183–8.

Chiesi, M. & Schwaller, R. (1995). Inhibition of constitutive endothelial NO-synthase activity by tannin and quercetin. *Biochemical Pharmacology*, **49**, 495–501.

Cho, H. J., Martin, E., Xie, Q.-W., Sassa, S. & Nathan, C. (1995). Inducible nitric oxide synthase: identification of amino acid residues essential for dimerization and binding of tetrahydrobiopterin. *Proceedings of the National Academy of Science USA*, **92**, 11514–8.

Choi, D. W. (1993). Nitric oxide: foe or friend to the injured brain? *Proceedings of the National Academy of Science USA*, **90**, 9741–3.

Christensen, J. & Fang, S. (1994). Colocalization of NADPH-diaphorase activity and certain neuropeptides in the esophagus of opossum (*Didelphis virginiana*). *Cell and Tissue Research*, **278**, 557–62.

Christinck, F., Jury, J., Cayabyab, F. & Daniel, E. E. (1991). Nitric oxide may be the final mediator of nonadrenergic, noncholinergic inhibitory junction potentials in the gut. *Canadian Journal of Physiology and Pharmacology*, **69**, 1448–58.

Chu, S. C., Wu, H.-P., Banks, T. C., Eissa, N. T. & Moss, J. (1995). Structural diversity in the 5′-untranslated region of cytokine-stimulated human inducible nitric oxide synthase mRNA. *Journal of Biological Chemistry*, **270**, 10625–30.

Chung, S.-J. & Fung, H.-L. (1990). Identification of the subcellular site for nitroglycerin metabolism to nitric oxide in bovine coronary smooth muscle cells. *Journal of Pharmacology and Experimental Therapeutics*, **253**, 614–9.

Clark, R. B., Kinsberg, E. R. & Giles, W. R. (1994). Histochemical localization of nitric oxide synthase in the bullfrog intracardiac ganglion. *Neuroscience Letters*, **182**, 255–8.

Closs, E. J., Basha, F. Z., Habermeier, A. & Förstermann, U. (1995). Transport of NO synthase inhibitors and inhibition of L-arginine transport. *Endothelium*, **3** (Suppl), S31.

Clowes, A. W., Reidy, M. A. & Clowes, M. M. (1983). Kinetics of cellular proliferation after arterial injury. I. Smooth muscle growth in the absence of endothelium. *Laboratory Investigations*, **49**, 327–33.

Clowry, G. J. (1993). Axotomy induces NADPH diaphorase activity in neonatal but not adult motoneurones. *NeuroReport*, **5**, 361–4.

Cobbs, C. S., Brenman, J. E., Aldape, K. D., Bredt, D. S. & Israel, M. A. (1995). Expression of nitric oxide synthase in human central nervous system tumors. *Cancer Research*, **55**, 727–30.

Cockcroft, J. R., Chowienczyk, P. J., Benjamin, N. & Ritter, J. M. (1994). Preserved endothelium-dependent vasodilatation in patients with essential hypertension. *New England Journal of Medicine*, **330**, 1036–40.

Coggi, G., Dell'Orto, P. & Viale, G. (1986). Avidin-biotin methods. In *Immunocytochemistry: Modern Methods and Applications*, second edn., eds. J. M. Polak & S. Van Noorden, pp. 54–70. Bristol: Wright.

Colasanti, M., Persichini, T., Menegazzi, M., Mariotto, S., Giordano, E., Caldarera, C. M., Sogos, V., Lauro, G. M. & Suzuki, H. (1995). Induction of nitric oxide synthase mRNA expression: suppression by exogenous nitric oxide. *Journal of Biological*

Chemistry, **270**, 26731–3.

Conklin, J. L., Du, C., Murray, J. A. & Bates, J. N. (1993). Characterization and mediation of inhibitory junction potentials from opossum lower esophageal sphincter. *Gastroenterology*, **104**, 1439–44.

Conklin, J. L., Murray, J., Ledlow, A., Clark, E., Hayek, B., Picken, H. & Rosenthal, G. (1995). Effects of recombinant human hemoglobin on motor functions of the opossum esophagus. *Journal of Pharmacology and Experimental Therapeutics*, **273**, 762–7.

Connor, J. R., Manning, P. T., Settle, S. L., Moore, W. M., Jerome, G. M., Webber, R. K., Tjoeng, S. T. & Currie, M. G. (1995). Suppression of adjuvant-induced arthritis by selective inhibition of inducible nitric oxide synthase. *European Journal of Pharmacology*, **273**, 15–24.

Conrad, K. P., Joffe, G. M., Kruszyna, H., Kruszyna, R., Rochelle, L. G., Smith, R. P., Chavez, J. E. & Mosher, M. D. (1993). Identification of increased nitric biosynthesis during pregnancy in rats. *FASEB Journal*, **7**, 566–71.

Cook, H. T., Bune, A. J., Jansen, A. S., Taylor, G. M., Loi, R. K. & Cattell, V. (1994). Cellular localization of inducible nitric oxide synthase in experimental endotoxic shock in the rat. *Clinical Science*, **87**, 179–86.

Cooke, J. P., Singer, A. H., Tsao, P., Zera, P., Rowan, R. A. & Billingham, M. E. (1992). Antiatherogenic effects of L-arginine in the hypercholesterolemic rabbit. *Journal of Clinical Investigation*, **90**, 1168–72.

Cooke, J. P., Tsao, P., Singer, A., Wang, B., Kosek, J. & Drexler, H. (1993). Antiatherogenic effect of nuts: is the answer NO? *Archives of Internal Medicine*, **153**, 896–9.

Coons, A. H., Leduc, E. H. & Connolly, J. M. (1955). Studies on antibody production. I. A method for the histochemical demonstation of specific antibody and its application to a study of the hyperimmune rabbit. *Journal of Experimental Medicine*, **102**, 46–60.

Corbett, J. A., Lancaster, J. R. J., Sweetland, M. A. & McDaniel, M. L. (1991). Interleukin-1β-induced formation of EPR-detectable iron-nitrosyl complexes in islets of Langerhans. *Journal of Biological Chemistry*, **266**, 21351–4.

Corbett, J. A. & McDaniel, M. L. (1992). Does nitric oxide mediate autoimmune destruction of beta-cells? Possible therapeutic interventions in IDDM. *Diabetes*, **41**, 897–903.

Corbett, J. A., Tilton, R. G., Chang, K., Hasan, K. S., Ido, Y., Wang, J. L., Sweetland, M. A., Lancaster, J. R. J., Williamson, J. R. & McDaniel, M. L. (1992a). Aminoguanidine, a novel inhibitor of nitric oxide formation, prevents diabetic vascular dysfunction. *Diabetes*, **41**, 552–6.

Corbett, J. A., Wang, J. L., Hughes, J. H., Wolf, B. A., Sweetland, M. A., Lancaster, J. R. J. & McDaniel, M. L. (1992b). Nitric oxide and cyclic GMP formation induced by interleukin 1β in islets of Langerhans. *Biochemical Journal*, **287**, 229–35.

Corbett, J. A., Wang, J. L., Sweetland, M. A., Lancaster, J. R. J. & McDaniel, M. L.

(1992c). Interleukin 1β induces the formation of nitric oxide by β-cells purified from rodent islets of Langerhans. *Journal of Clinical Investigation*, **90**, 2384–91.

Cosentino, F. & Katusic, Z. S. (1995). Tetrahydrobiopterin and dysfunction of endothelial nitric oxide synthase in coronary arteries. *Circulation*, **91**, 139–44.

Costa, M., Furness, J. B., Pompolo, S., Brookes, S. J. H., Bornstein, J. C., Bredt, D. S. & Snyder, S. H. (1992). Projections and chemical coding of neurons with immunoreactivity for nitric oxide synthase in the guinea-pig small intestine. *Neuroscience Letters*, **148**, 121–5.

Cowley, A. W. J., Mattson, D. L., Lu, S. & Roman, R. J. (1995). The renal medulla and hypertension. *Hypertension*, **25**, 663–73.

Crepel, F. & Jaillard, D. (1990). Protein kinases, nitric oxide and long-term depression of synapses in the cerebellum. *NeuroReport*, **1**, 133–6.

Cross, A. H., Misko, T. P., Lin, R. F., Hickey, W. F., Trotter, J. L. & Tilton, R. G. (1994). Aminoguanidine, an inhibitor of inducible nitric oxide synthase, ameliorates experimental autoimmune encephalomyelitis in SJL mice. *Journal of Clinical Investigation*, **93**, 2684–90.

Crowe, R., Noble, J., Robson, T., Soediono, P., Milroy, E. J. & Burnstock, G. (1995). An increase of neuropeptide Y but not nitric oxide synthase-immunoreactive nerves in the bladder neck from male patients with bladder neck dyssynergia. *Journal of Urology*, **154**, 1231–6.

Currò, D., Volpe, A. R. & Preziosi, P. (1996). Nitric oxide synthase activity and non-adrenergic non-cholinergic relaxation in the rat gastric fundus. *British Journal of Pharmacology*, **117**, 717–23.

D'Amato, M., Curro, D. & Montuschi, P. (1992a). Effects of nitric oxide synthase inhibitors on the relaxation induced by non-adrenergic non-cholinergic nerve-stimulation in the rat gastric fundus. *Pharmacological Research*, **25** (Suppl 1), 1–2.

D'Amato, M., Curro, D. & Montuschi, P. (1992b). Evidence for dual components in the non-adrenergic non-cholinergic relaxation in the rat gastric fundus: role of endogenous nitric oxide and vasoactive intestinal polypeptide. *Journal of the Autonomic Nervous System*, **37**, 175–86.

Dail, W. G., Galloway, B. & Bordegaray, J. (1993). NADPH diaphorase innervation of the rat anococcygeus and retractor penis muscles. *Neuroscience Letters*, **160**, 17–20.

Dalziel, H. H., Thornbury, K. D., Ward, S. M. & Sanders, K. M. (1991). Involvement of nitric oxide synthetic pathway in inhibitory junction potentials in canine proximal colon. *American Journal of Physiology*, **260**, G789–92.

Daniel, H., Hemart, N., Jaillard, D. & Crepel, F. (1993). Long-term depression requires nitric oxide and guanosine 3′ : 5′ cyclic monophosphate production in rat cerebellar Purkinje cells. *European Journal of Neuroscience*, **5**, 1079–82.

Davies, M. G., Kim, J. H., Dalen, H., Makhoul, R. G., Svendsen, E. & Hagen, P.-O. (1994). Reduction of experimental vein graft intimal hyperplasia and preservation of nitric oxide-mediated relaxation by the nitric oxide precursor L-arginine. *Surgery*, **116**, 557–68.

Davies, P. F. (1993). Endothelium as a signal transduction interface for flow forces: cell surface dynamics. *Thrombosis and Haemostasis*, **70**, 124–8.

Dawson, D. A. (1994). Nitric oxide and focal cerebral ischemia: multiplicity of actions and diverse outcome. *Cerebrovascular and Brain Metabolism Reviews*, **6**, 299–324.

Dawson, T. M., Bredt, D. S., Fotuhi, M., Hwang, P. M. & Snyder, S. H. (1991). Nitric oxide synthase and neuronal NADPH diaphorase are identical in brain and peripheral tissues. *Proceedings of the National Academy of Sciences USA*, **88**, 7797–801.

Dawson, T. M. & Snyder, S. H. (1994). Gases as biological messengers: nitric oxide and carbon monoxide in the brain. *Journal of Neuroscience*, **14**, 5147–59.

Day, B. J., Jia, L., Arnelle, D. R., Crapo, J. D. & Stamler, J. S. (1995). Transduction of NOS activity from endothelium to vascular smooth muscle: involvement of *S*-nitrosothiols. *Endothelium*, **3** (Suppl), S72.

De Belder, A. J., MacAllister, R., Radomski, M. W., Moncada, S. & Vallance, P. J. T. (1994). Effects of *S*-nitroso-glutathione in the human forearm circulation: evidence for selective inhibition of platelet activation. *Cardiovascular Research*, **28**, 691–4.

De Graaf, J. C., Banga, J. D., Moncada, S., Palmer, R. M. J., De Groot, P. G. & Sixma, J. J. (1992). Nitric oxide functions as an inhibitor of platelet adhesion under flow conditions. *Circulation*, **85**, 2284–90.

De Groote, M. A., Granger, D., Xu, Y., Campbell, G., Prince, R. & Fang, F. C. (1995). Genetic and redox determinants of nitric oxide cytotoxicity in a *Salmonella typhimurium* model. *Proceedings of the National Academy of Science USA*, **92**, 6399–403.

De Kimpe, S. J., Kengatharan, M., Thiemermann, C. & Vane, J. R. (1995). Effects of aminoguanide on multiple organ failure elicited by lipoteichoic acid and peptidoglycan in anaesthetized rats. *Endothelium*, **3** (Suppl), S116.

De Man, J. G., Boeckxstaens, G. E., De Winter, B. Y., Moreels, T. G., Misset, M. E., Herman, A. G. & Pelckmans, P. A. (1995). Comparison of the pharmacological profile of *S*-nitrosothiols, nitric oxide and the nitrergic neurotransmitter in the canine ileocolonic junction. *British Journal of Pharmacology*, **114**, 1179–84.

De Man, J. G., Pelckmans, P. A., Boeckxstaens, G. E., Bult, H., Oosterbosch, L., Herman, A. G. & Van Maercke, Y. M. (1991). The role of nitric oxide in inhibitory non-adrenergic non-cholinergic neurotransmission in the canine lower oesophageal sphincter. *British Journal of Pharmacology*, **103**, 1092–6.

De Vente, J. & Steinbusch, H. W. (1992). On the stimulation of soluble and particulate guanylate cyclase in the rat brain and the involvement of nitric oxide as studied by cGMP immunocytochemistry. *Acta Histochemica*, **92**, 13–38.

Deguchi, T. & Yoshioka, M. (1982). L-Arginine identified as an endogenous activator for soluble guanylate cyclase from neuroblastoma cells. *Journal of Biological Chemistry*, **257**, 10147–51.

Demerle Pallardy, C., Lonchampt, M. O., Chabrier, P. E. & Braquet, P. (1991). Absence of implication of L-arginine/nitric oxide pathway on neuronal cell injury induced by L-glutamate or hypoxia. *Biochemical and Biophysical Research Communications*, **181**, 456–64.

Denis, M. (1994). Human monocytes/macrophages: NO or no NO? *Journal of Leukocyte Biology*, **55**, 682–4.

Denlinger, L. C., Fisette, P. L., Garis, K. A., Kwon, G., Vazquez-Torres, A., Simon, A. D., Nguyen, B., Proctor, R. A., Bertics, P. J. & Corbett, J. A. (1996). Regulation of inducible nitric oxide synthase expression by macrophage purinoceptors and calcium. *Journal of Biological Chemistry*, **271**, 337–42.

Desai, K. M., Sessa, W. C. & Vane, J. R. (1991a). Involvement of nitric oxide in the reflex relaxation of the stomach to accommodate food or fluid. *Nature*, **351**, 477–9.

Desai, K. M., Warner, T. D., Bishop, A. E., Polak, J. M. & Vane, J. R. (1994). Nitric oxide, and not vasoactive intestinal peptide, as the main neurotransmitter of vagally induced relaxation of the guinea pig stomach. *British Journal of Pharmacology*, **113**, 1197–202.

Desai, K. M., Zembowicz, A., Sessa, W. & Vane, J. R. (1991b). Nitroxergic nerves mediate vagally induced relaxation in the isolated stomach of the guinea pig. *Proceedings of the National Academy of Sciences USA*, **88**, 11490–4.

Dey, R. D., Mayer, B. & Said, S. I. (1993). Colocalization of vasoactive intestinal peptide and nitric oxide synthase in neurons of the ferret trachea. *Neuroscience*, **54**, 839–43.

Diamond, J. & Blisard, K. S. (1976). Effects of stimulant and relaxant drugs on tension and cyclic nucleotide levels in canine femoral artery. *Molecular Pharmacology*, **12**, 688–92.

Dikranian, K., Loesch, A. & Burnstock, G. (1994). Localisation of nitric oxide synthase and its colocalisation with vasoactive peptides in coronary and femoral arteries. An electron microscope study. *Journal of Anatomy*, **184**, 583–90.

Dinerman, J. L., Dawson, T. M., Schell, M. J., Snowman, A. & Snyder, S. H. (1994). Endothelial nitric oxide synthase localized to hippocampal pyramidal cells: implications for synaptic plasticity. *Proceedings of the National Academy of Sciences USA*, **91**, 4214–8.

Dinerman, J. L., Lowenstein, C. J. & Snyder, S. H. (1993). Molecular mechanisms of nitric oxide regulation: potential relevance to cardiovascular disease. *Circulation Research*, **73**, 217–22.

Ding, Y. Q., Takada, M., Kaneko, T. & Mizuno, N. (1995). Colocalization of vasoactive intestinal polypeptide and nitric oxide in penis-innervating neurons in the major pelvic ganglion of the rat. *Neuroscience Research*, **22**, 129–31.

Ding, Y. Q., Wang, Y. Q., Qin, B. Z. & Li, J. S. (1993). The major pelvic ganglion is the main source of nitric oxide synthase-containing nerve fibers in penile erectile tissue of the rat. *Neuroscience Letters*, **164**, 187–9.

Dobreva, G., Mizhorkova, Z., Kortezova, N. & Papasova, M. (1994). Some characteristics of the muscularis mucosae of the cat lower esophageal sphincter. *General*

Pharmacology, **25**, 639–43.

Dohrn, C. S., Mullett, M. A., Price, R. H. & Beitz, A. J. (1994). Distribution of nitric oxide synthase-immunoreactive interneurons in the spinal trigeminal nucleus. *Journal of Comparative Neurology*, **346**, 449–60.

Dokita, S., Morgan, W. R., Wheeler, M. A., Yoshida, M., Latifpour, J. & Weiss, R. M. (1991). N^G-nitro-L-arginine inhibits non-adrenergic, non-cholinergic relaxation in rabbit urethral smooth muscle. *Life Sciences*, **48**, 2429–36.

Dokita, S., Smith, S. D., Nishimoto, T., Wheeler, M. A. & Weiss, R. M. (1994). Involvement of nitric oxide and cyclic GMP in rabbit urethral relaxation. *European Journal of Pharmacology*, **266**, 269–75.

Domoto, T. & Tsumori, T. (1994). Co-localization of nitric oxide synthase and vasoactive intestinal peptide immunoreactivity in neurons of the major pelvic ganglion projecting to the rat rectum and penis. *Cell and Tissue Research*, **278**, 273–8.

Dong, Y., Chao, A. C., Kouyama, K., Hsu, Y., Bocian, R. C., Moss, R. B. & Gardner, P. (1995). Activation of CFTR chloride current by nitric oxide in human T lymphocytes. *The EMBO Journal*, **14**, 2700–7.

Draijer, R., Atsma, D. E., van der Laarse, A. & van Hinsbergh, V. W. M. (1995). cGMP and nitric oxide modulate thrombin-induced endothelial permeability: regulation via different pathways in human aortic and umbilical vein endothelial cells. *Circulation Research*, **76**, 199–208.

Dudek, R. R., Conforto, A., Pinto, V., Wildhirt, S. & Suzuki, H. (1995). Inhibition of endothelial nitric oxide synthase by cytochrome P-450 reductase inhibitors. *Proceedings of the Society for Experimental Biology and Medicine*, **209**, 60–4.

Dun, N. J., Dun, S. L., Chiba, T. & Förstermann, U. (1995). Nitric oxide synthase-immunoreactive vagal afferent fibers in rat superior cervical ganglia. *Neuroscience*, **65**, 231–9.

Dun, N. J., Dun, S. L. & Förstermann, U. (1994a). Nitric oxide synthase immunoreactivity in rat pontine medullary neurons. *Neuroscience*, **59**, 429–45.

Dun, N. J., Dun, S. L., Förstermann, U. & Tseng, L. F. (1992). Nitric oxide synthase immunoreactivity in rat spinal cord. *Neuroscience Letters*, **147**, 217–20.

Dun, N. J., Dun, S. L., Wu, S. Y., Förstermann, U., Schmidt, H. H. H. W. & Tseng, L. F. (1993). Nitric oxide synthase immunoreactivity in the rat, mouse, cat and squirrel monkey spinal cord. *Neuroscience*, **54**, 845–57.

Dun, N. J., Huang, R., Dun, S. L. & Förstermann, U. (1994b). Infrequent co-localization of nitric oxide synthase and calcium binding proteins immunoreactivity in rat neocortical neurons. *Brain Research*, **666**, 289–94.

Dykhuizen, R. S., Mowat, N. A. G., Smith, C. C., Douglas, J. G., Smith, L. & Benjamin, N. (1995). Nitric oxide production in infective gastro-enteritis and inflammatory bowel disease. *Endothelium*, **3** (Suppl), S105.

East, S. J. & Garthwaite, J. (1990). Nanomolar N^G-nitroarginine inhibits NMDA-induced cyclic GMP formation in rat cerebellum. *European Journal of Pharmacology*, **184**, 311–3.

East, S. J. & Garthwaite, J. (1991). NMDA receptor activation in rat hippocampus induces cyclic GMP formation through the L-arginine-nitric oxide pathway. *Neuroscience Letters*, **123**, 17–9.

Eccles, J. C. (1964). *The Physiology of Synapses*. New York: Academic Press.

Egberongbe, Y. I., Gentleman, S. M., Falkai, P., Bogerts, B., Polak, J. M. & Roberts, G. W. (1994). The distribution of nitric oxide synthase immunoreactivity in the human brain. *Neuroscience*, **59**, 561–78.

Ehren, I., Iversen, H., Jansson, O., Adolfsson, J. & Wiklund, N. P. (1994). Localization of nitric oxide synthase activity in the human lower urinary tract and its correlation with neuroeffector responses. *Urology*. **44**, 683–7.

Eigler, A., Moeller, J. & Endres, S. (1995). Exogenous and endogenous nitric oxide attenuates tumor necrosis factor synthesis in murine macrophage cell line RAW 264.7. *Journal of Immunology*, **154**, 4048–54.

Eis, A. L., Brockman, D. E., Pollock, J. S. & Myatt, L. (1995). Immunohistochemical localization of endothelial nitric oxide synthase in human villous and extravillous trophoblast populations and expression during syncytiotrophoblast formation *in vitro*. *Placenta*, **16**, 113–26.

Ekblad, E., Mulder, H., Uddman, R. & Sundler, F. (1994). NOS-containing neurons in the rat gut and coeliac ganglia. *Neuropharmacology*, **33**, 1323–31.

Ellis, J. L. & Conanan, N. D. (1995). Modulation of relaxant responses evoked by a nitric oxide donor and by nonadrenergic, noncholinergic stimulation by isozyme-selective phosphodiesterase inhibitors in guinea pig trachea. *Journal of Pharmacology and Experimental Therapeutics*, **272**, 997–1004.

Ellis, J. L. & Undem, B. J. (1992). Inhibition by L-N^G-nitro-L-arginine of nonadrenergic-noncholinergic-mediated relaxations of human isolated central and peripheral airway. *American Reviews of Respiratory Diseases*, **146**, 1543–7.

Endoh, M., Maiese, K. & Wagner, J. (1994). Expression of the inducible form of nitric oxide synthase by reactive astrocytes after transient global ischemia. *Brain Research*, **651**, 92–100.

Evans, T.J., Spink, J. & Cohen, J. (1995). Inducible nitric oxide synthase (iNOS) gene promoter: binding of interferon regulator factors (IRF) 1 and 2. *Endothelium*, **3** (Suppl), S50.

Fang, S., Christensen, J., Conklin, J. L., Murray, J. A. & Clark, G. (1994). Roles of Triton X-100 in NADPH-diaphorase histochemistry. *Journal of Histochemistry and Cytochemistry*, **42**, 1519–24.

Farber, E. & Louviere, C. D. (1956). Histochemical localization of specific oxidative enzymes. IV. Soluble oxidation-reduction dyes as aids in the histochemical localization of oxidative enzymes with tetrazolium salts. *Journal of Histochemistry and Cytochemistry*, **4**, 347–56.

Farber, E., Sternberg, W. H. & Dunlap, C. E. (1956). Histochemical localization of specific oxidative enzymes. I. Tetrazolium stains for diphosphopyridine nucleotide diaphorase and triphosphopyridine nucleotide diaphorase. *Journal of Histochemistry and Cytochemistry*, **4**, 254–65.

Faussone Pellegrini, M. S., Bacci, S., Pantalone, D., Cortesini, C. & Mayer, B. (1994).

Nitric oxide synthase immunoreactivity in the human ileocecal region. *Neuroscience Letters*, **170**, 261–5.

Feelisch, M. (1991). The biochemical pathways of nitric oxide formation from nitrovasodilators: appropriate choice of exogenous NO donors and aspects of preparation and handling of aqueous NO solutions. *Journal of Cardiovascular Pharmacology*, **17** (Suppl 3), S25–33.

Feelisch, M. (1993). Biotransformation to nitric oxide of organic nitrates in comparison to other vasodilators. *European Heart Journal*, **14** (Suppl I), 123–32.

Feelisch, M., te Poel, M., Zamora, R., Deussen, A. & Moncada, S. (1994). Understanding the controversy over the identity of EDRF. *Nature*, **368**, 62–5.

Fehsel, K., Kröncke, K.-D., Meyer, K. L., Huber, H., Wahn, V. & Kolb-Bachofen, V. (1995). Nitric oxide induces apoptosis in mouse thymocytes. *Journal of Immunology*, **155**, 2858–65.

Fernandes, L. B., Ellis, J. L. & Undem, B. J. (1994). Potentiation of nonadrenergic noncholinergic relaxation of human isolated bronchus by selective inhibitors of phosphodiesterase isozymes. *American Journal of Respiratory and Critical Care Medicine*, **150**, 1384–90.

Fiallos Estrada, C. E., Kummer, W., Mayer, B., Bravo, R., Zimmermann, M. & Herdegen, T. (1993). Long-lasting increase of nitric oxide synthase immunoreactivity, NADPH-diaphorase reaction and c-JUN co-expression in rat dorsal root ganglion neurons following sciatic nerve transection. *Neuroscience Letters*, **150**, 169–73.

Filep, J. G., Földes-Filep, É. & Sirois, P. (1993). Nitric oxide modulates vascular permeability in the rat coronary circulation. *British Journal of Pharmacology*, **108**, 323–6.

Finberg, J. P. M., Levy, S. & Vardi, Y. (1993). Inhibition of nerve stimulation-induced vasodilatation in corpora cavernosa of the pithed rat by blockade of nitric oxide synthase. *British Journal of Pharmacology*, **108**, 1038–42.

Fischer, A., Mundel, P., Mayer, B., Preissler, U., Philippin, B. & Kummer, W. (1993). Nitric oxide synthase in guinea pig lower airway innervation. *Neuroscience Letters*, **149**, 157–60.

Ford, P. C., Wink, D. A. & Stanbury, D. M. (1993). Autoxidation kinetics of aqueous nitric oxide. *FEBS Letters*, **326**, 1–3.

Forster, E. R. & Southam, E. (1993). The intrinsic and vagal extrinsic innervation of the rat stomach contains nitric oxide synthase. *NeuroReport*, **4**, 275–8.

Förstermann, U. (1994). Biochemistry and molecular biology of nitric oxide synthases. *Arzneimittel-Forschung*, **44** (Suppl 3a), 402–7.

Förstermann, U., Gorsky, L. D., Pollock, J. S., Schmidt, H. H. H. W., Heller, M. & Murad, F. (1990). Regional distribution of EDRF/NO-synthesizing enzyme(s) in rat brain. *Biochemical and Biophysical Research Communications*, **168**, 727–32.

Förstermann, U., Pollock, J. S., Schmidt, H. H. H. W., Heller, M. & Murad, F. (1991a). Calmodulin-dependent endothelium-derived relaxing factor/nitric oxide synthase activity is present in the particulate and cytosolic fractions of bovine aortic endothelial cells. *Proceedings of the National Academy of Science USA*, **88**, 1788–92.

Förstermann, U., Pollock, J. S., Tracey, W. R. & Nakane, M. (1994). Isoforms of nitric-oxide synthase: purification and regulation. *Methods in Enzymology*, **233**, 258–64.

Förstermann, U., Schmidt, H. H. H. W., Pollock, J. S., Sheng, H., Mitchell, J. A., Warner, T. D., Nakane, M. & Murad, F. (1991b). Isoforms of nitric oxide synthase: characterization and purification from different cell types. *Biochemical Pharmacology*, **42**, 1849–57.

Frey, C., Narayanan, K., McMillan, K., Spack, L., Gross, S. S., Masters, B. S. & Griffith, O. W. (1994). L-Thiocitrulline: a stereospecific, heme-binding inhibitor of nitric-oxide synthases. *Journal of Biological Chemistry*, **269**, 26083–91.

Fujimori, H. & Pan Hou, H. (1991). Effect of nitric oxide on L-[^{3}H]glutamate binding to rat brain synaptic membranes. *Brain Research*, **554**, 355–7.

Fukuto, J. M. & Chaudhuri, G. (1995). Inhibition of constitutive and inducible nitric oxide synthase: potential selective inhibition. *Annual Reviews of Pharmacology and Toxicology*, **35**, 165–94.

Furchgott, R. F. (1981). The requirement for endothelial cells in the relaxation of arteries by acetylcholine and some other vasodilators. *Trends in Pharmacological Sciences*, **10**, 173–6.

Furchgott, R. F. (1983). Role of endothelium in responses of vascular smooth muscle. *Circulation Research*, **53**, 557–73.

Furchgott, R. F., Jothianandan, D. & Ansari, N. (1995). Differentiation between EDRF and NO in organ chamber experiments on rings of rabbit aorta. *Endothelium*, **3** (Suppl), S73.

Furchgott, R. F. & Vanhoutte, P. M. (1989). Endothelium-derived relaxing and contracting factors. *FASEB Journal*, **3**, 2007–18.

Furchgott, R. F. & Zawadzki, J. V. (1980). The obligatory role of endothelial cells in the relaxation of arterial smooth muscle by acetylcholine. *Nature*, **288**, 373–6.

Furfine, E. S., Harmon, M. F., Paith, J. E., Knowles, R. G., Salter, M., Kiff, R. J., Duffy, C., Hazelwood, R., Oplinger, J. A. & Garvey, E. P. (1994). Potent and selective inhibition of human nitric oxide synthases: selective inhibition of neuronal nitric oxide synthase by *S*-ethyl-L-thiocitrulline. *Journal of Biological Chemistry*, **269**, 26677–83.

Furman, B. L. & Sneddon, P. (1993). Endothelium-dependent vasodilator responses of the isolated mesenteric bed are preserved in long-term streptozotocin diabetic rats. *European Journal of Pharmacology*, **232**, 29–34.

Furness, J. B. & Anderson, C. R. (1994). Origins of nerve terminals containing nitric oxide synthase in the guinea-pig coeliac ganglion. *Journal of the Autonomic Nervous System*, **46**, 47–54.

Furness, J. B., Li, Z. S., Young, H. M. & Förstermann, U. (1994). Nitric oxide synthase in the enteric nervous system of the guinea-pig: a quantitative description. *Cell and Tissue Research*, **277**, 139–49.

Fye, W. B. (1986). T. Lauder Brunton and amyl nitrite: a Victorian vasodilator. *Circulation*, **74**, 222–9.

Gabbott, P. L. & Bacon, S. J. (1993). Histochemical localization of NADPH-dependent diaphorase (nitric oxide synthase) activity in vascular endothelial cells in the rat brain. *Neuroscience*, **57**, 79–95.

Gabbott, P. L. & Bacon, S. J. (1994a). Two types of interneuron in the dorsal lateral geniculate nucleus of the rat: a combined NADPH diaphorase histochemical and GABA immunocytochemical study. *Journal of Comparative Neurology*, **350**, 281–301.

Gabbott, P. L. & Bacon, S. J. (1994b). An oriented framework of neuronal processes in the ventral lateral geniculate nucleus of the rat demonstrated by NADPH diaphorase histochemistry and GABA immunocytochemistry. *Neuroscience*, **60**, 417–40.

Gabriel, R. (1991). A method for the demonstration of NADPH-diaphorase activity in anuran species using unfixed retinal wholemounts. *Archives of Histology and Cytology*, **54**, 207–11.

Garcia Pascual, A., Costa, G., Garcia Sacristan, A. & Andersson, K. E. (1991). Relaxation of sheep urethral muscle induced by electrical stimulation of nerves: involvement of nitric oxide. *Acta Physiologica Scandinavica*, **141**, 531–9.

Garcia Pascual, A. & Triguero, D. (1994). Relaxation mechanisms induced by stimulation of nerves and by nitric oxide in sheep urethral muscle. *Journal of Physiology (London)*, **476**, 333–47.

Garg, U. C. & Hassid, A. (1989). Nitric oxide-generating vasodilators and 8-bromo-cyclic guanosine monophosphate inhibit mitogenesis and proliferation of cultured rat vascular smooth muscle cells. *Journal of Clinical Investigation*, **83**, 1774–7.

Garthwaite, J. (1991). Glutamate, nitric oxide and cell–cell signalling in the nervous system. *Trends in Neuroscience*, **14**, 60–7.

Garthwaite, J. (1993). Nitric oxide signalling in the nervous system. *Seminars in the Neurosciences*, **5**, 171–80.

Garthwaite, J. & Boulton, C. L. (1995). Nitric oxide signaling in the central nervous system. *Annual Reviews of Physiology*, **57**, 683–706.

Garthwaite, J., Charles, S. L. & Chess-Williams, R. (1988). Endothelium-derived relaxing factor release on activation of NMDA receptors suggests role as intercellular messenger in the brain. *Nature*, **336**, 385–8.

Garthwaite, J., Garthwaite, G., Palmer, R. M. & Moncada, S. (1989). NMDA receptor activation induces nitric oxide synthesis from arginine in rat brain slices. *European Journal of Pharmacology*, **172**, 413–6.

Garthwaite, J., Southam, E., Boulton, C. L., Nielsen, E. B., Schmidt, K. & Mayer, B. (1995). Potent and selective inhibition of nitric oxide-sensitive guanylyl cyclase by 1*H*-[1,2,4]oxadiazolo[4,3-*a*]quinoxalin-1-one. *Molecular Pharmacology*, **48**, 184–8.

Garvey, E. P., Oplinger, J. A., Tanoury, G. J., Sherman, P. A., Fowler, M., Marshall, S., Harmon, M. F., Paith, J. E. & Furfine, E. S. (1994). Potent and selective inhibition of human nitric oxide synthases: inhibition by non-amino acid isothioureas. *Journal of Biological Chemistry*, **269**, 26669–76.

Gaw, A. J., Aberdeen, J., Humphrey, P. P., Wadsworth, R. M. & Burnstock, G. (1991). Relaxation of sheep cerebral arteries by vasoactive intestinal polypeptide and neurogenic stimulation: inhibition by L-N^G-monomethyl arginine in endothelium-denuded vessels. *British Journal of Pharmacology*, **102**, 567–72.

Geller, D. A., de Vera, M. E., Russell, D. A., Shapiro, R. A., Nussler, A. K., Simmons, R. L. & Billiar, T. R. (1995a). A central role for IL-1 in the *in vitro* and *in vivo* regulation of hepatic inducible nitric oxide synthase: IL-1 induces hepatic nitric oxide synthesis. *Journal of Immunology*, **155**, 4890–8.

Geller, D. A., de Vera, M. E., Shapiro, R. A., Nussler, A. K., Mudgett, J. S. & Billiar, T. R. (1995b). Molecular analysis of the human inducible nitric oxide synthase gene promoter. *Endothelium*, **3** (Suppl), S3.

Geng, Y. & Lotz, M. (1995). Increased intracellular Ca^{2+} selectively suppresses IL-1-induced NO production by reducing iNOS mRNA stability. *Journal of Cell Biology*, **129**, 1651–7.

Gibson, A., Babbedge, R., Brave, S. R., Hart, S. L., Hobbs, A. J., Tucker, J. F., Wallace, P. & Moore, P. K. (1992). An investigation of some *S*-nitrosothiols, and of hydroxy-arginine, on the mouse anococcygeus. *British Journal of Pharmacology*, **107**, 715–21.

Gibson, A., Brave, S. R., McFadzean, I., Tucker, J. F. & Wayman, C. (1995). The nitrergic transmitter of the anococcygeus – NO or not? *Archives Internationales de Pharmacodynamie et Therapie*, **329**, 39–51.

Gibson, A., Mirzazadeh, S., Al-Swayeh, O. A., Chong, N. W. S. & Moore, P. K. (1989). L-N^G-Nitroarginine is a potent, L-arginine reversible, inhibitor of NANC relaxations in the mouse anococcygeus. *British Journal of Pharmacology*, **98**, 904P.

Gillespie, J. S., Liu, X. R. & Martin, W. (1989). The effects of L-arginine and N^G-monomethyl L-arginine on the response of the rat anococcygeus muscle to NANC nerve stimulation. *British Journal of Pharmacology*, **98**, 1080–2.

Gillespie, J. S. & Martin, W. (1980). A smooth muscle inhibitory material from the bovine retractor penis and rat anococcygeus muscles. *Journal of Physiology (London)*, **309**, 55–64.

Gillespie, J. S. & Sheng, H. (1990). The effects of pyrogallol and hydroquinone on the response to NANC nerve stimulation in the rat anococcygeus and the bovine retractor penis muscles. *British Journal of Pharmacology*, **99**, 194–6.

Girard, P., Sercombe, R., Sercombe, C., Le Lem, G., Seylaz, J. & Potier, P. (1995). A new synthetic flavonoid protects endothelium-derived relaxing factor-induced relaxation in rabbit arteries *in vitro*: evidence for superoxide scavenging. *Biochemical Pharmacology*, **49**, 1533–9.

Giugliano, D., Ceriello, A. & Paolisso, G. (1995). Diabetes mellitus, hypertension, and cardiovascular disease: which role for oxidative stress? *Metabolism*, **44**, 363–8.

Goetz, R. M., Morano, I., Calovini, T., Studer, R. & Holtz, J. (1994). Increased expression of endothelial constitutive nitric oxide synthase in rat aorta during pregnancy. *Biochemical and Biophysical Research Communications*, **205**, 905–10.

Goldring, C. E. P., Narayanan, R., Lagadec, P. & Jeannin, J.-F. (1995). Transcriptional inhibition of the inducible nitric oxide synthase gene by competitive binding of NF-κB/rel proteins. *Biochemical and Biophysical Research Communications*, **209**, 73–9.

Goligorsky, M. S., Tsukahara, H., Magazine, H., Andersen, T. T., Malik, A. B. & Bahou, W. F. (1994). Termination of endothelin signaling: role of nitric oxide. *Journal of Cell Physiology*, **158**, 485–94.

González-Hernández, T., González-González, B., Mantolán-Sarmiento, B., Ferres-Torres, R. & Meyer, G. (1994). Transient NADPH-diaphorase activity in motor nuclei of the foetal human brain stem. *NeuroReport*, **5**, 758–60.

Gorfine, S. R. (1995). Treatment of benign anal disease with topical nitroglycerin. *Diseases of the Colon and Rectum*, **38**, 453–6.

Gorsky, L. D., Förstermann, U., Ishii, K. & Murad, F. (1990). Production of an EDRF-like activity in the cytosol of N1E–115 neuroblastoma cells. *FASEB Journal*, **4**, 1494–500.

Graham, A. M. & Sneddon, P. (1993). Evidence for nitric oxide as an inhibitory neurotransmitter in rabbit isolated anococcygeus. *European Journal of Pharmacology*, **237**, 93–9.

Graham, R. C. & Karnowsky, M. J. (1966). The early stages of absorption of injected horseradish peroxidase in the proximal tubules of the mouse kidney: ultrastructural cytochemistry by a new technique. *Journal of Histochemistry and Cytochemistry*, **14**, 219–302.

Green, L. C., Ruiz de Luzuriaga, K., Wagner, D. A., Rand, W., Istfan, N., Young, V. R. & Tannenbaum, S. R. (1981). Nitrate biosynthesis in man. *Proceedings of the National Academy of Science USA*, **78**, 7764–8.

Greenberg, S. S., Wilcox, D. E. & Rubanyi, G. M. (1990). Endothelium-derived relaxing factor released from canine femoral artery by acetylcholine cannot be identified as free nitric oxide by electron paramagnetic resonance spectroscopy. *Circulation Research*, **67**, 1446–52.

Greenberg, S. S., Xie, J., Kolls, J., Mason, C. & Didier, P. (1995a). Rapid induction of mRNA for nitric oxide synthase II in rat alveolar macrophages by intratracheal administration of *Mycobacterium tuberculosis* and *Mycobacterium avium*. *Proceedings of the Society for Experimental Biology and Medicine*, **209**, 46–53.

Greenberg, S. S., Xie, J., Spitzer, J. J., Wang, J.-F., Lancaster, J., Grisham, M. B., Powers, D. R. & Giles, T. D. (1995b). Nitro containing L-arginine analogs interfere with assays for nitrate and nitrite. *Life Sciences*, **57**, 1949–61.

Greensmith, L., Hasan, H. I. & Vrbová, G. (1994). Nerve injury increases the susceptibility of motoneurons to *N*-methyl-D-aspartate-induced neurotoxicity in the developing rat. *Neuroscience*, **58**, 727–33.

Grider, J. R., Murthy, K. S., Jin, J. G. & Makhlouf, G. M. (1992). Stimulation of nitric oxide from muscle cells by VIP: prejunctional enhancement of VIP release. *American Journal of Physiology*, **262**, G774–8.

Griffiths, M. J. D., Messent, M., MacAllister, R. J. & Evans, T. W. (1993).

Aminoguanidine selectively inhibits inducible nitric oxide synthase. *British Journal of Pharmacology*, **110**, 963–8.

Grisham, M. B., Specian, R. D. & Zimmerman, T. E. (1994). Effects of nitric oxide synthase inhibition on the pathophysiology observed in a model of chronic granulomatous colitis. *Journal of Pharmacology and Experimental Therapeutics*, **271**, 1114–21.

Groscavage, J. M., Wilk, S. & Ignarro, L. J. (1995). Serine and cysteine proteinase inhibitors prevent nitric oxide production by activated macrophages by interfering with transcription of the inducible NO synthase gene. *Biochemical and Biophysical Research Communications*, **215**, 721–9.

Gross, S. S. & Levi, R. (1992). Tetrahydrobiopterin synthesis: an absolute requirement for cytokine-induced nitric oxide generation by vascular smooth muscle. *Journal of Biological Chemistry*, **267**, 25722–9.

Grozdanovic, Z., Mayer, B., Baumgarten, H. G. & Bruning, G. (1994). Nitric oxide synthase-containing nerve fibers and neurons in the genital tract of the female mouse. *Cell and Tissue Research*, **275**, 355–60.

Gruetter, C. A., Barry, B. K., McManara, D. B., Gruetter, D. Y., Kadowitz, P. J. & Ignarro, L. J. (1979). Relaxation of bovine coronary artery and activation of coronary arterial guanylate cyclase by nitric oxide, nitroprusside and a carcinogenic nitrosamine. *Journal of Cyclic Nucleotide Research*, **5**, 211–24.

Grundy, D., Gharib Naseri, M. K. & Hutson, D. (1993). Role of nitric oxide and vasoactive intestinal polypeptide in vagally mediated relaxation of the gastric corpus in the anaesthetized ferret. *Journal of the Autonomic Nervous System*, **43**, 241–6.

Gryglewski, R. J., Palmer, R. M. & Moncada, S. (1986). Superoxide anion is involved in the breakdown of endothelium-derived vascular relaxing factor. *Nature*, **320**, 454–6.

Gustafsson, B. I. & Delbro, D. S. (1993). Tonic inhibition of small intestinal motility by nitric oxide. *Journal of the Autonomic Nervous System*, **44**, 179–87.

Gustafsson, L. E., Wiklund, C. U., Wiklund, N. P., Persson, M. G. & Moncada, S. (1990). Modulation of autonomic neuroeffector transmission by nitric oxide in guinea pig ileum. *Biochemical and Biophysical Research Communications*, **173**, 106–10.

Haberny, K. A., Pou, S. & Eccles, C. U. (1992). Potentiation of quinolate-induced hippocampal lesions by inhibition of NO synthesis. *Neuroscience Letters*, **146**, 187–90.

Hakim, M. A., Hirooka, Y., Coleman, M. J., Bennett, M. R. & Dampney, R. A. (1995). Evidence for a critical role of nitric oxide in the tonic excitation of rabbit renal sympathetic preganglionic neurones. *Journal of Physiology (London)*, **482**, 401–7.

Haley, J. E., Dickenson, A. H. & Schachter, M. (1992a). Electrophysiological evidence for a role of nitric oxide in prolonged chemical nociception in the rat. *Neuropharmacology*, **31**, 251–8.

Haley, J. E., Malen, P. L. & Chapman, P. F. (1993). Nitric oxide synthase inhibitors block long-term potentiation induced by weak but not strong tetanic stimulation at physiological brain temperatures in rat hippocampal slices. *Neuroscience Letters*, **160**, 85–8.

Haley, J. E., Wilcox, G. L. & Chapman, P. F. (1992b). The role of nitric oxide in hippocampal long-term potentiation. *Neuron*, **8**, 211–16.

Hamid, Q., Springall, D. R., Riveros-Moreno, V., Chanez, P., Howarth, P., Redington, A., Bousquet, J., Godard, P., Holgate, S. & Polak, J. M. (1993). Induction of nitric oxide synthase in asthma. *Lancet*, **342**, 1510–13.

Hammer, B., Parker, W. D. & Bennett, J. P. (1993). NMDA receptors increase OH radicals *in vivo* by using nitric oxide synthase and protein kinase C. *NeuroReport*, **5**, 72–4.

Hamon, C. G., Cutler, P. & Blair, J. A. (1989). Tetrahydrobiopterin metabolism in the streptozotocin induced diabetic state in rats. *Clinica Chimica Acta*, **181**, 249–53.

Han, K., Shimoni, Y. & Giles, W. R. (1994). An obligatory role for nitric oxide in autonomic control of mammalian heart rate. *Journal of Physiology (London)*, **476**, 309–14.

Hanazawa, T., Konno, A., Kaneko, T., Tanaka, K., Ohshima, H., Esumi, H. & Chiba, T. (1994). Nitric oxide synthase-immunoreactive nerve fibers in the nasal mucosa of the rat. *Brain Research*, **657**, 7–13.

Hanson, S. R., Hutsell, T. C., Keefer, L. K., Mooradian, D. L. & Smith, D. J. (1995). Nitric oxide donors: a continuing opportunity in drug design. *Advances in Pharmacology*, **34**, 383–98.

Hansson, G. K., Geng, Y.-J., Holm, J., Hårdhammar, P., Wennmalm, Å. & Jennische, E. (1994). Arterial smooth muscle cells express nitric oxide synthase in response to endothelial injury. *Journal of Experimental Medicine*, **180**, 733–8.

Harrison, D. G., Minor, R. L., Guerra, R., Quillen, J. E. & Selke, F. W. (1991). Endothelial dysfunction in atherosclerosis. In *Cardiovascular Significance of Endothelium-Derived Vasoactive Factors*, ed. G. M. Rubanyi, pp. 263–80. Mount Kisco: Futura.

Hartell, N. A. (1994). cGMP acts within cerebellar Purkinje cells to produce long term depression via mechanisms involving PKC and PKG. *NeuroReport*, **5**, 833–6.

Hashikawa, T., Leggio, M. G., Hattori, R. & Yui, Y. (1994). Nitric oxide synthase immunoreactivity colocalized with NADPH-diaphorase histochemistry in monkey cerebral cortex. *Brain Research*, **641**, 341–9.

Hashimoto, S., Kigoshi, S. & Muramatsu, I. (1993). Nitric oxide-dependent and -independent neurogenic relaxation of isolated dog urethra. *European Journal of Pharmacology*, **231**, 209–14.

Hassall, C. J., Saffrey, M. J., Belai, A., Hoyle, C. H. V., Moules, E. W., Moss, J., Schmidt, H. H. H. W., Murad, F., Förstermann, U. & Burnstock, G. (1992). Nitric oxide synthase immunoreactivity and NADPH-diaphorase activity in a subpopulation of intrinsic neurones of the guinea-pig heart. *Neuroscience Letters*, **143**, 65–8.

Hassall, C. J., Saffrey, M. J. & Burnstock, G. (1993). Expression of NADPH-diaphorase activity by guinea-pig paratracheal neurones. *NeuroReport*, **4**, 49–52.

Hata, F., Ishii, T., Kanada, A., Yamano, N., Kataoka, T., Takeuchi, T. & Yagasaki, O. (1990). Essential role of nitric oxide in descending inhibition in the rat proximal colon. *Biochemical and Biophysical Research Communications*, **172**, 1400–6.

Hatchett, R. J., Gryglewski, R. J., Mlochowski, J., Zembowicz, A. & Radziszewski, W. (1994). Carboxyebselen a potent and selective inhibitor of endothelial nitric oxide synthase. *Journal of Physiology and Pharmacology*, **45**, 55–67.

Hattori, R., Inoue, R., Sase, K., Eizawa, H., Kosuga, K., Aoyama, T., Masayasu, H., Kawai, C., Sasayama, S. & Yui, Y. (1994). Preferential inhibition of inducible nitric oxide synthase by ebselen. *European Journal of Pharmacology*, **267**, R1–2.

Hayashi, T., Fukuto, J. M., Ignarro, L. J. & Chaudhuri, G. (1992). Basal release of nitric oxide from aortic rings is greater in female rabbits than in male rabbits: implications for atherosclerosis. *Proceedings of the National Academy of Science USA*, **89**, 11259–63.

Hebb, D. O. (1949). *The Organization of Behaviour.* New York: Wiley.

Hecker, M., Mülsch, A., Bassenge, E., Förstermann, U. & Busse, R. (1994). Subcellular localization and characterization of nitric oxide synthase(s) in endothelial cells: physiological implications. *Biochemical Journal*, **299**, 247–52.

Heller, B., Bürkle, A., Radons, J., Fengler, E., Jalowy, A., Müller, M., Burkart, V. & Kolb, H. (1994). Analysis of oxygen radical toxicity in pancreatic islets at the single cell level. *Biological Chemistry Hoppe-Seyler*, **375**, 597–602.

Henry, Y., Lepoivre, M., Drapier, J. C., Ducrocq, C., Boucher, J. L. & Guissani, A. (1993). EPR characterization of molecular targets for NO in mammalian cells and organelles. *FASEB Journal*, **7**, 1124–34.

Herdegen, T., Rudiger, S., Mayer, B., Bravo, R. & Zimmermann, M. (1994). Expression of nitric oxide synthase and colocalisation with Jun, Fos and Krox transcription factors in spinal cord neurons following noxious stimulation of the rat hindpaw. *Brain Research and Molecular Brain Research*, **22**, 245–58.

Hess, D. T., Patterson, S. I., Smith, D. S. & Skene, J. H. P. (1993). Neuronal growth cone collapse and inhibition of protein fatty acylation by nitric oxide. *Nature*, **366**, 562–5.

Hess, R., Scarpelli, D. R. & Pearse, A. G. E. (1958). Cytochemical localization of pyridine nucleotide-linked dehydrogenases. *Nature*, **181**, 1531–2.

Hevel, J. M. & Marletta, M. A. (1994). Nitric-oxide synthase assays. *Methods in Enzymology*, **233**, 250–8.

Heym, C., Braun, B., Klimaschewski, L. & Kummer, W. (1995). Chemical codes of sensory neurons innervating the guinea-pig adrenal gland. *Cell and Tissue Research*, **279**, 169–81.

Heym, C., Colombo Benckmann, M. & Mayer, B. (1994). Immunohistochemical demonstration of the synthesis enzyme for nitric oxide and of comediators in neurons and chromaffin cells of the human adrenal medulla. *Anatomischer Anzeiger*, **176**, 11–16.

Hibbs, J. B. J., Taintor, R. R. & Vavrin, Z. (1987). Macrophage cytotoxicity: role for L-arginine deiminase and imino nitrogen oxidation to nitrite. *Science*, **235**, 473–6.

Hibbs, J. B. J., Taintor, R. R., Vavrin, Z. & Rachlin, E. M. (1988). Nitric oxide: a cytotoxic activated macrophage effector molecule. *Biochemical and Biophysical Research Communications*, **157**, 87–94.

Higashi, Y., Oshima, T., Ozono, R., Watanabe, M., Matsuura, H. & Kajiyama, G. (1995). Effects of L-arginine infusion on renal hemodynamics in patients with mild essential hypertension. *Hypertension*, **25**, 898–902.

Hiki, K., Hattori, R., Kawai, C. & Yui, Y. (1992). Purification of insoluble nitric oxide synthase from rat cerebellum. *Journal of Biochemistry*, **111**, 556–8.

Hill, C. E. & Gould, D. J. (1995). Modulation of sympathetic vasoconstriction by sensory nerves and nitric oxide in rat irideal arterioles. *Journal of Pharmacology and Experimental Therapeutics*, **273**, 918–26.

Hirano, T. (1991). Differential pre- and postsynaptic mechanisms for synaptic potentiation and depression between a granule cell and a Purkinje cell in rat cerebellar culture. *Synapse*, **7**, 321–3.

Hirata, K., Kuroda, R., Sakoda, T., Katayama, M., Inoue, N., Suematsu, M., Kawashima, S. & Yokoyama, M. (1995). Inhibition of endothelial nitric oxide synthase activity by protein kinase C. *Hypertension*, **25**, 180–5.

Hirsch, J. A. & Gibson, G. E. (1984). Selective alteration of neurotransmitter release by low oxygen *in vitro*. *Neurochemical Research*, **9**, 1039–49.

Hisa, Y., Uno, T., Tadaki, N., Umehara, K., Okamura, H. & Ibata, Y. (1995). NADPH-diaphorase and nitric oxide synthase in the canine superior cervical ganglion. *Cell and Tissue Research*, **279**, 629–31.

Hishikawa, K., Nakaki, T., Marumo, T., Suzuki, H., Kato, R. & Saruto, T. (1995). Up-regulation of nitric oxide synthase by estradiol in human aortic endothelial cells. *FEBS Letters*, **360**, 291–3.

Hobbs, A. J., Tucker, J. F. & Gibson, A. (1991). Differentiation by hydroquinone of relaxations induced by exogenous and endogenous nitrates in non-vascular smooth muscle: role of superoxide anions. *British Journal of Pharmacology*, **104**, 645–50.

Hoffman, R. A., Gleixner, S. L., Ford, H. R., Schattenfroh, N. C. & Simmons, R. L. (1995). Administration of a nitric oxide (NO) synthase inhibitor attenuates lethal graft versus host disease (GvHD). *Endothelium*, **3** (Suppl), S117

Hogaboam, C. M., Jacobson, K., Collins, S. M. & Blennerhassett, M. G. (1995). The selective beneficial effects of nitric oxide inhibition in experimental colitis. *American Journal of Physiology*, **268**, G673–84.

Hohler, B., Mayer, B. & Kummer, W. (1994). Nitric oxide synthase in the rat carotid body and carotid sinus. *Cell and Tissue Research*, **276**, 559–64.

Hökfelt, T., Ceccatelli, S., Gustafsson, L., Hulting, A.-L., Verge, V., Villar, M., Xu, X.-J., Xu, Z.-Q., Wiesenfeld-Hallin, Z. & Zhang, X. (1994). Plasticity of NO synthase expression in the nervous and endocrine systems. *Neuropharmacology*, **33**, 1221–7.

Holmquist, F., Hedlund, H. & Andersson, K. E. (1992). Characterization of inhibitory neurotransmission in the isolated corpus cavernosum from rabbit and man. *Jour-*

nal of Physiology (London), **449**, 295–311.

Holscher, C. & Rose, S. P. (1992). An inhibitor of nitric oxide synthesis prevents memory formation in the chick. *Neuroscience Letters*, **145**, 165–7.

Holscher, C. & Rose, S. P. (1993). Inhibiting synthesis of the putative retrograde messenger nitric oxide results in amnesia in a passive avoidance task in the chick. *Brain Research*, **619**, 189–94.

Holzer, P., Wachter, C., Heinemann, A., Jocic, M., Lippe, I. T. & Herbert, M. K. (1995). Sensory nerves, nitric oxide and NANC vasodilatation. *Archives Internationales de Pharmacodynamie et Therapie*, **329**, 67–79.

Hope, B. T., Michael, G. J., Knigge, K. M. & Vincent, S. R. (1991). Neuronal NADPH diaphorase is a nitric oxide synthase. *Proceedings of the National Academy of Sciences USA*, **88**, 2811–4.

Hope, B. T. & Vincent, S. R. (1989). Histochemical characterization of neuronal NADPH-diaphorase. *Journal of Histochemistry and Cytochemistry*, **37**, 653–61.

Hoyle, C. H. V. & Burnstock, G. (1989). Galanin-like immunoreactivity in enteric neurons of the human colon. *Journal of Anatomy*, **166**, 23–33.

Hoyle, C. H. V. & Burnstock, G. (1995). Criteria for defining enteric neurotransmitters. In *Handbook of Methods in Gastrointestinal Pharmacology*, ed. T. S. Gaginella, pp. 123–40. Boca Raton, FL: CRC Press.

Hoyle, C. H. V., Kamm, M. A., Burnstock, G. & Lennard-Jones, J. E. (1990). Enkephalins modulate inhibitory neuromuscular transmission in circular muscle of human colon via δ-opioid receptors. *Journal of Physiology (London)*, **431**, 465–78.

Hoyle, C. H. V., Stones, R. W. S., Robson, T., Whitely, K. & Burnstock, G. (1996). Innervation of the vasculature and microvasculature of the human vagina by NOS- and neuropeptide-containing nerves. *Journal of Anatomy*, **188**, 633–44.

Hoyt, K. R., Tang, L. H., Aizenman, E. & Reynolds, I. J. (1992). Nitric oxide modulates NMDA-induced increases in intracellular Ca^{2+} in cultured rat forebrain neurons. *Brain Research*, **592**, 310–16.

Hrabák, A., Bajor, T. & Temesi, A. (1994). Comparison of substrate and inhibitor specificity of arginase and nitric oxide (NO) synthase for arginine analogues and related compounds in murine and rat macrophages. *Biochemical and Biophysical Research Communications*, **198**, 206–12.

Hsu, S. M., Raine, L. & Fanger, H. (1981). Use of avidin-biotin-peroxidase complex (ABC) in immunoperoxidase techniques: a comparison between ABC and unlabeled antibody (PAP) procedures. *Journal of Histochemistry and Cytochemistry*, **29**, 577–80.

Huang, P. L., Dawson, T. M., Bredt, D. S., Snyder, S. H. & Fishman, M. C. (1993). Targeted disruption of the neuronal nitric oxide synthase gene. *Cell*, **75**, 1273–86.

Huang, P. L., Huang, Z. H., Moskowitz, M., Bevan, J. & Fishman, M. C. (1995). Targeted disruption of the endothelial nitric oxide synthase gene. *Endothelium*, **3** (Suppl), S5.

Huang, Z., Huang, P. L., Panahian, N., Dalkara, T., Fishman, M. C. & Moskowitz, M. A. (1994). Effects of cerebral ischemia in mice deficient in neuronal nitric oxide synthase. *Science*, **265**, 1883–5.

Hughes, S. R. & Brain, S. D. (1994). Nitric oxide-dependent release of vasodilator quantities of calcitonin gene-related peptide from capsaicin-sensitive nerves in rabbit skin. *British Journal of Pharmacology*, **111**, 425–30.

Hugo Sachs Elektronik (1995). *Research Equipment for Pharmacology and Physiology*, I.1: *Organ baths and Apparatus for Isolated Organs and Tissues.* March-Hugstetten, Germany: Hugo Sachs Elektronik.

Hyun, J., Komori, Y., Chaudhuri, G., Ignarro, L. J. & Fukuto, J. M. (1995). The protective effect of tetrahydrobiopterin on the nitric oxide-mediated inhibition of purified nitric oxide synthase. *Biochemical and Biophysical Research Communications*, **206**, 380–6.

Iadecola, C., Beitz, A. J., Renno, W., Xu, X., Mayer, B. & Zhang, F. (1993a). Nitric oxide synthase-containing neural processes on large cerebral arteries and cerebral microvessels. *Brain Research*, **606**, 148–55.

Iadecola, C., Zhang, F. & Xu, X. (1993b). Role of nitric oxide synthase-containing vascular nerves in cerebrovasodilation elicited from cerebellum. *American Journal of Physiology*, **264**, R738–46.

Ialenti, A., Ianaro, A., Moncada, S. & Di Rosa, M. (1992). Modulation of acute inflammation by endogenous nitric oxide. *European Journal of Pharmacology*, **211**, 177–82.

Ichimori, K., Ishida, H., Fukahori, M., Nakazawa, H. & Murakami, E. (1994). Practical nitric oxide measurement employing a nitric oxide-selective electrode. *Review of Scientific Instruments*, **65**, 1–5.

Ichinose, F., Huang, P. L. & Zapol, W. M. (1995). Effects of targeted neuronal nitric oxide synthase gene disruption and nitro^{G}-L-arginine methylester on the threshold for isoflurane anesthesia. *Laboratory Investigation*, **83**, 101–8.

Ignarro, L. J. (1992). Pharmacological, biochemical, and chemical evidence that EDRF is NO or a labile nitroso precursor. In *Endothelial Regulation of Vascular Tone*, eds. U. S. Ryan & G. M. Rubanyi, pp. 37–49. New York: Marcel Dekker Inc.

Ignarro, L. J., Buga, G. M. & Chaudhuri, G. (1988a). EDRF generation and release from perfused bovine pulmonary artery and vein. *European Journal of Pharmacology*, **149**, 79–88.

Ignarro, L. J., Buga, G. M., Wood, K. S., Byrns, R. E. & Chaudhuri, G. (1987a). Endothelium-derived relaxing factor produced and released from artery and vein is nitric oxide. *Proceedings of the National Academy of Sciences USA*, **84**, 9265–9.

Ignarro, L. J., Bush, P. A., Buga, G. M., Wood, K. S., Fukuto, J. M. & Rajfer, J. (1990). Nitric oxide and cyclic GMP formation upon electrical field stimulation cause relaxation of corpus cavernosum smooth muscle. *Biochemical and Biophysical Research Communications*, **170**, 843–50.

Ignarro, L. J., Byrns, R. E., Buga, G. M. & Wood, K. S. (1987b). Endothelium-derived relaxing factor from pulmonary artery and vein possesses pharmacologic and chemical properties identical to those of nitric oxide radical. *Circulation Research*, **61**, 866–79.

Ignarro, L. J., Byrns, R. E., Buga, G. M., Wood, K. S. & Chaudhuri, G. (1988b). Pharmacological evidence that endothelium-derived relaxing factor is nitric oxide: use of pyrogallol and superoxide dismutase to study endothelium-dependent and nitric oxide-elicited vascular smooth muscle relaxation. *Journal of Pharmacology and Experimental Therapeutics*, **244**, 181–9.

Ignarro, L. J., Lippton, H., Edwards, J. C., Baricos, W. H., Hyman, A. L., Kadowitz, P. J. & Gruetter, C. A. (1981). Mechanism of vascular smooth muscle relaxation by organic nitrates, nitrites, nitroprusside and nitric oxide: evidence for the involvement of S-nitrosothiols as active intermediates. *Journal of Pharmacology and Experimental Therapeutics*, **218**, 739–49.

Ikeda, M., Morita, I., Murota, S., Sekiguchi, F., Yuasa, T. & Miyatake, T. (1993). Cerebellar nitric oxide synthase activity is reduced in nervous and Purkinje cell degeneration mutants but not in climbing fiber-lesioned mice. *Neuroscience Letters*, **155**, 148–50.

Irie, K., Muraki, T., Furukawa, K. & Nomoto, T. (1991). L-N^G-Nitro-arginine inhibits nicotine-induced relaxation of isolated rat duodenum. *European Journal of Pharmacology*, **202**, 285–8.

Irikura, K., Huang, P. L., Ma, J., Lee, W. S., Dalkara, T., Fishman, M. C., Dawson, T. M., Snyder, S. H. & Moskowitz, M. A. (1995). Cerebrovascular alterations in mice lacking neuronal nitric oxide synthase gene expression. *Proceedings of the National Academy of Science USA*, **92**, 6823–7.

Ishii, K., Sheng, H., Warner, T. D., Förstermann, U. & Murad, F. (1991). A simple and sensitive bioassay method for detection of EDRF with RFL–6 rat lung fibroblasts. *American Journal of Physiology*, **261**, H598–603.

Ito, M. (1989). Long-term depression. *Annual Reviews of Neuroscience*, **12**, 85–102.

Ito, M. & Karachot, L. (1990). Messengers mediating long-term desensitization in cerebellar Purkinje cells. *NeuroReport*, **1**, 129–32.

Iyengar, R., Stuehr, D. J. & Marletta, M. A. (1987). Macrophage synthesis of nitrite, nitrate, and *N*-nitrosamines: precursors and role of the respiratory burst. *Proceedings of the National Academy of Science USA*, **84**, 6369–73.

Jacobs, P., Radzioch, D. & Stevenson, M. M. (1995). Nitric oxide expression in the spleen, but not in the liver, correlates with resistance to blood-stage malaria in mice. *Journal of Immunology*, **155**, 5306–13.

James, S. L. (1995). Role of nitric oxide in parasitic infections. *Microbiological Reviews*, **59**, 533–47.

Jenkinson, K. M., Reid, J. J. & Rand, M. J. (1995). Hydroxocobalamin and haemoglobin differentiate between exogenous and neuronal nitric oxide in the rat gastric fundus. *European Journal of Pharmacology*, **275**, 145–52.

Jia, L., Bonaventura, C. & Stamler, J. S. (1995). *S*-Nitrosohemoglobin: a new transport function of arterial erythrocytes involved in regulation of blood pressure. *Endothelium*, **3** (Suppl), S6.

Jiang, H., Stewart, C. A. & Leu, R. W. (1995). Tumor-derived factor synergizes with IFN-γ and LPS, IL-2 or TNF-α to promote macrophage synthesis of TNF-α and TNF receptors for autocrine induction of nitric oxide synthase and enhanced nitric oxide-mediated tumor cytotoxicity. *Immunobiology*, **192**, 321–42.

Jin, J. G., Murthy, K. S., Grider, J. R. & Makhlouf, G. M. (1993). Activation of distinct cAMP- and cGMP-dependent pathways by relaxant agents in isolated gastric muscle cells. *American Journal of Physiology*, **264**, G470–7.

Jordan, M. L., Rominski, B., Jaquins-Gerstl, A., Geller, D. & Hoffman, R. A. (1995). Regulation of inducible nitric oxide production by intracellular calcium. *Surgery*, **118**, 138–45.

Jun, C.-D., Choi, B.-M., Kim, H.-M. & Chung, H.-T. (1995). Involvement of protein kinase C during taxol-induced activation of murine peritoneal macrophages. *Journal of Immunology*, **154**, 6541–7.

Kader, A., Frazzini, V. I., Solomon, R. A. & Trifiletti, R. R. (1993). Nitric oxide production during focal cerebral ischemia in rats. *Stroke*, **24**, 1709–16.

Kajekar, R., Moore, P. K. & Brain, S. D. (1995). Essential role for nitric oxide in neurogenic inflammation in rat cutaneous microcirculation: evidence for an endothelium-independent mechanism. *Circulation Research*, **76**, 441–7.

Kalb, R. G. & Agostini, J. (1993). Molecular evidence for nitric oxide-mediated motor neuron development. *Neuroscience*, **57**, 1–8.

Kalina, M., Plapinger, R. E., Hoshino, Y. & Seligman, A. M. (1972). Nonosmiophilic tetrazolium salts that yield osmiophilic lipophilic formazans for ultrastructural localization of dehdrogenase activity. *Journal of Histochemistry and Cytochemistry*, **20**, 685–95.

Kanada, A., Hata, F., Suthamnatpong, N., Maehara, T., Ishii, T., Takeuchi, T. & Yagasaki, O. (1992). Key roles of nitric oxide and cyclic GMP in nonadrenergic and noncholinergic inhibition in rat ileum. *European Journal of Pharmacology*, **216**, 287–92.

Kanai, A. J., Strauss, H. C., Truskey, G. A., Crews, A. L., Grunfeld, S. & Malinski, T. (1995). Shear stress induces ATP-independent transient nitric oxide release from vascular endothelial cells, measured directly with a porphyrinic microsensor. *Circulation Research*, **77**, 284–93.

Kaneto, H., Fujii, J., Seo, H. G., Suzuki, K., Nakamura, M., Tatsumi, H. & Taniguchi, N. (1995). Apoptotic cell death triggered by nitric oxide in pancreatic β-cells. *Endothelium*, **3** (Suppl), S46.

Kannan, M. S. & Johnson, D. E. (1992). Nitric oxide mediates the neural nonadrenergic, noncholinergic relaxation of pig tracheal smooth muscle. *American Journal of Physiology*, **262**, L511–14.

Kannan, M. S. & Johnson, D. E. (1995). Modulation of nitric oxide-dependent relaxation of pig tracheal smooth muscle by inhibitors of guanylyl cyclase and calcium activated potassium channels. *Life Sciences*, **56**, 2229–38.

Kasakov, L., Cellek, S. & Moncada, S. (1995). Characterization of nitrergic neurotransmission during short-and long-term electrical stimulation of the rabbit anococcygeus muscle. *British Journal of Pharmacology*, **115**, 1149–54.

Kaufman, H. S., Shermak, M. A., May, C. A., Pitt, H. A. & Lillemoe, K. D. (1993). Nitric oxide inhibits resting sphincter of Oddi activity. *American Journal of Surgery*, **165**, 74–80.

Kauppinen, R. A., McMahon, H. T. & Nicholls, D. G. (1988). Ca^{2+}-dependent and Ca^{2+}-independent glutamate release, energy status and cytosolic free Ca^{2+} concentration in isolated nerve terminals following metabolic inhibition: possible relevance to hypoglycaemia and anoxia. *Neuroscience*, **27**, 175–82.

Kawabata, A. & Takagi, H. (1994). The dual role of L-arginine in nociceptive processing in the brain: involvement of nitric oxide and kyotorphin. In *Nitric Oxide: Roles in Neuronal Communication and Neurotoxicity*, eds. H. Takagi, N. Toda & R. D. Hawkins, pp. 115–25. Tokyo: Japan Scientific Societies Press.

Kawahara, H., Blackshaw, L. A., Nisyrios, V. & Dent, J. (1994). Transmitter mechanisms in vagal afferent-induced reduction of lower oesophageal sphincter (LOS) pressure in the rat. *Journal of the Autonomic Nervous System*, **49**, 69–80.

Keaney, J. F., Simon, D. I., Stamler, J. S., Jaraki, O., Scharfstein, J., Vita, J. A. & Loscalzo, J. (1993). NO forms an adduct with serum albumin that has endothelium-derived relaxing factor-like properties. *Journal of Clinical Investigation*, **91**, 1582–9.

Keef, K. D., Du, C., Ward, S. M., McGregor, B. & Sanders, K. M. (1993). Enteric inhibitory neural regulation of human colonic circular muscle: role of nitric oxide. *Gastroenterology*, **105**, 1009–16.

Keef, K. D., Shuttleworth, C. W., Xue, C., Bayguinov, O., Publicover, N. G. & Sanders, K. M. (1994). Relationship between nitric oxide and vasoactive intestinal polypeptide in enteric inhibitory neurotransmission. *Neuropharmacology*, **33**, 1303–14.

Kelm, M., Feelisch, M., Krebber, T., Deussen, A., Motz, W. & Strauer, B. E. (1995). Role of nitric oxide in the regulation of coronary vascular tone in hearts from hypertensive rats: maintenance of nitric oxide-forming capacity and increased basal production of nitric oxide. *Hypertension*, **25**, 186–93.

Kennedy, C. (1984). *Regulation of Vascular Tone by Purine Nucleosides and Nucleotides*. PhD Thesis, University of London.

Keranen, U., Vanhatalo, S., Kiviluoto, T., Kivilaakso, E. & Soinila, S. (1995). Co-localization of NADPH diaphorase reactivity and vasoactive intestinal polypeptide in human colon. *Journal of the Autonomic Nervous System*, **54**, 177–83.

Kerwin, J. F., Lancaster, J. R. & Feldman, P. L. (1995). Nitric oxide: a new paradigm for second messengers. *Journal of Medicinal Chemistry*, **38**, 4343–62.

Kharitonov, S. A., Yates, D. H. & Barnes, P. J. (1996). Inhaled glucocorticoids decrease nitric oxide in exhaled air of asthmatic patients. *American Journal of Respiratory and Critical Care Medicine*, **153**, 454–57.

Kharitonov, S. A., Yates, D., Springall, D. R., Buttery, L., Polak, J., Robbins, R. A. & Barnes, P. J. (1995). Exhaled nitric oxide is increased in asthma. *Chest*, **107** (Suppl), 156S–7S.

Kim, H., Lee, H. S., Chang, K. T., Ko, T. H., Baek, K. J. & Kwon, N. S. (1995a). Chloromethyl ketones block induction of nitric oxide synthase in murine macrophages by preventing activation of nuclear factor-κB. *Journal of Immunology*, **154**, 4741–8.

Kim, N., Azadzoi, K. M., Goldstein, I. & Saenz de Tejada, I. (1991). A nitric oxide-like factor mediates nonadrenergic-noncholinergic neurogenic relaxation of penile corpus cavernosum smooth muscle. *Journal of Clinical Investigation*, **88**, 112–18.

Kim, Y.-M., Bergonia, H. & Lancaster, J. R. (1995b). Nitrogen oxide-induced autoprotection in isolated rat hepatocytes. *FEBS Letters*, **374**, 228–32.

Kirkeby, H. J., Svane, D., Poulsen, J., Tottrup, A., Forman, A. & Andersson, K. E. (1993). Role of the L-arginine/nitric oxide pathway in relaxation of isolated human penile cavernous tissue and circumflex veins. *Acta Physiologica Scandinavica*, **149**, 385–92.

Kishimoto, J., Keverne, E. B., Hardwick, J. & Emson, P. C. (1993). Localization of nitric oxide synthase in the mouse olfactory and vomeronasal system: a histochemical, immunological and *in situ* hybridization study. *European Journal of Neuroscience*, **5**, 1684–94.

Kitchener, P. D. & Diamond, J. (1993). Distribution and colocalization of choline acetyltransferase immunoreactivity and NADPH diaphorase reactivity in neurons within the medial septum and diagonal band of Broca in the rat basal forebrain. *Journal of Comparative Neurology*, **335**, 1–15.

Kitto, K. F., Haley, J. E. & Wilcox, G. L. (1992). Involvement of nitric oxide in spinally mediated hyperalgesia in the mouse. *Neuroscience Letters*, **148**, 1–5.

Klatt, P., Schmidt, K. & Mayer, B. (1992). Brain nitric oxide synthase is a haemoprotein. *Biochemical Journal*, **288**, 15–17.

Klemm, P., Thiemermann, C., Winklmaier, G., Martorana, P. A. & Henning, R. (1995). Effects of nitric oxide synthase inhibition combined with nitric oxide inhalation in a porcine model of endotoxic shock. *British Journal of Pharmacology*, **114**, 363–8.

Klimaschewski, L., Kummer, W., Mayer, B., Couraud, J. Y., Preissler, U., Philippin, B. & Heym, C. (1992). Nitric oxide synthase in cardiac nerve fibers and neurons of rat and guinea pig heart. *Circulation Research*, **71**, 1533–7.

Knispel, H. H., Goessl, C. & Beckmann, R. (1992). Nitric oxide mediates neurogenic relaxation induced in rabbit cavernous smooth muscle by electric field stimulation. *Urology*, **40**, 471–6.

Knowles, R. G. & Moncada, S. (1994). Nitric oxide synthases in mammals. *Biochemical Journal*, **298**, 249–58.

Knowles, R. G., Palacios, M., Palmer, R. M. & Moncada, S. (1989). Formation of nitric oxide from L-arginine in the central nervous system: a transduction mechanism for stimulation of the soluble guanylate cyclase. *Proceedings of the National Academy of Sciences USA*, **86**, 5159–62.

Knudsen, M. A., Glavind, E. B. & Tottrup, A. (1995). Transmitter interactions in rabbit internal anal sphincter. *American Journal of Physiology*, **269**, G232–9.

Knudsen, M. A., Svane, D. & Tottrup, A. (1992). Action profiles of nitric oxide,

S-nitroso-L-cysteine, SNP, and NANC responses in opossum lower esophageal sphincter. *American Journal of Physiology*, **262**, G840–6.

Kobayashi, H., O'Brian, D. S. & Puri, P. (1995). Immunochemical characterization of neural cell adhesion molecule (NCAM), nitric oxide synthase, and neurofilament protein expression in pyloric muscle of patients with pyloric stenosis. *Journal of Paediatric Gastroenterology and Nutrition*, **20**, 319–25.

Kobzik, L., Bredt, D. S., Lowenstein, C. J., Drazen, J., Gaston, B., Sugarbaker, D. & Stamler, J. S. (1993). Nitric oxide synthase in human and rat lung: immunocytochemical and histochemical localization. *American Journal of Respiratory Cell Molecular Biology*, **9**, 371–7.

Kobzik, L., Reid, M. B., Bredt, D. S. & Stamler, J. S. (1994). Nitric oxide in skeletal muscle. *Nature*, **372**, 546–8.

Kolb, H. & Kolb-Bachofen, V. (1992). Type 1 (insulin-dependent) diabetes mellitus and nitric oxide. *Diabetologia*, **35**, 796–7.

Kollegger, H., McBean, G. J. & Tipton, K. F. (1993). Reduction of striatal *N*-methyl-D-aspartate toxicity by inhibition of nitric oxide synthase. *Biochemical Pharmacology*, **45**, 260–4.

Koller, A., Huang, A., Sun, D. & Kaley, G. (1995). Exercise training augments flow-dependent dilation in rat skeletal muscle arterioles: role of endothelial nitric oxide and prostaglandins. *Circulation Research*, **76**, 544–50.

Komarov, A., Mattson, D., Jones, M. M., Singh, P. K. & Lai, C.-S. (1993). *In vivo* spin trapping of nitric oxide in mice. *Biochemical and Biophysical Research Communications*, **195**, 1191–8.

Komori, Y., Wallace, G. C. & Fukuto, J. M. (1994). Inhibition of purified nitric oxide synthase from rat cerebellum and macrophage by L-arginine analogs. *Archives in Biochemistry and Biophysics*, **315**, 213–18.

Korenaga, R., Ando, J., Tsuboi, H., Yang, W., Sakuma, I., Toyo-oka, T. & Kamiya, A. (1994). Laminar flow stimulates ATP- and shear stress-dependent nitric oxide production in cultured bovine endothelial cells. *Biochemical and Biophysical Research Communications*, **198**, 213–19.

Kortezova, N., Velkova, V., Mizhorkova, Z., Bredy Dobreva, G., Vizi, E. S. & Papasova, M. (1994). Participation of nitric oxide in the nicotine-induced relaxation of the cat lower esophageal sphincter. *Journal of the Autonomic Nervous System*, **50**, 73–8.

Kossel, A., Bonhoeffer, T. & Bolz, J. (1990). Non-Hebbian synapses in rat visual cortex. *NeuroReport*, **1**, 115–18.

Kowaluk, E. A. & Fung, H. L. (1990). Spontaneous liberation of nitric oxide cannot account for *in vitro* vascular relaxation by *S*-nitrosothiols. *Journal of Pharmacology and Experimental Therapeutics*, **255**, 1256–64.

Krammer, H. J., Karahan, S. T., Mayer, B., Zhang, M. & Kühnel, W. (1993). Distribution of nitric oxide synthase immunoreactive neurons in the submucosal plexus of the porcine small intestine. *Annals of Anatomy*, **175**, 225–30.

Kröncke, K.-D., Fehsel, K. & Kolb-Bachofen, V. (1995). Inducible nitric oxide synthase and its product nitric oxide, a small molecule with complex biological activities. *Biological Chemistry Hoppe-Seyler*, **376**, 327–43.

Kubes, P. & Granger, D. N. (1992). Nitric oxide modulates microvascular permeability. *American Journal of Physiology*, **262**, H611–15.

Kuchan, M. J. & Frangos, J. A. (1994). Role of calcium and calmodulin in flow-induced nitric oxide production in endothelial cells. *American Journal of Physiology*, **266**, C628–36.

Kulkarni, A. P., Getchell, T. V. & Getchell, M. L. (1994). Neuronal nitric oxide synthase is localized in extrinsic nerves regulating perireceptor processes in the chemosensory nasal mucosae of rats and humans. *Journal of Comparative Neurology*, **345**, 125–38.

Kumari, K., Umar, S., Bansal, V. & Sahib, M. J. (1991). Inhibition of diabetes-associated complications by nucleophilic compounds. *Diabetes*, **40**, 1079–84.

Kummer, W., Fischer, A., Mundel, P., Mayer, B., Hoba, B., Philippin, B. & Preissler, U. (1992). Nitric oxide synthase in VIP-containing vasodilator nerve fibres in the guinea-pig. *NeuroReport*, **3**, 653–5.

Kummer, W. & Mayer, B. (1993). Nitric oxide synthase-immunoreactive axons innervating the guinea-pig lingual artery: an ultrastructural immunohistochemical study using elastic brightfield imaging. *Histochemistry*, **99**, 175–9.

Kunz, D., Walker, G., Eberhardt, W. & Pfeilschifter, J. (1996). Molecular nechanisms of dexamethasone inhibition of nitric oxide synthase expression in interleukin 1-stimulated mesangial cells: evidence for the involvement of transcriptional and posttranscriptional regulation. *Proceedings of the National Academy of Science USA*, **93**, 255–9.

Kuonen, D. R., Kemp, M. C. & Roberts, P. J. (1988). Demonstration and biochemical characterisation of rat brain NADPH-dependent diaphorase. *Journal of Neurochemistry*, **50**, 1017–25.

Kurenni, D. E., Thurlow, G. A., Turner, R. W., Moroz, L. L., Sharkey, K. A. & Barnes, S. (1995). Nitric oxide synthase in tiger salamander retina. *Journal of Comparative Neurology*, **361**, 525–36.

Kurose, I., Miura, S., Saito, H., Tada, S., Fukumura, D., Higuchi, H. & Ishii, H. (1995). Rat Kupffer cell-derived nitric oxide modulates induction of lymphokine-activated killer cell. *Gastroenterology*, **109**, 1958–68.

Kwon, G., Corbett, J. A., Rodi, C. P., Sullivan, P. & McDaniel, M. L. (1995). Interleukin-1-induced nitric oxide synthase expression by rat pancreatic b-cells: evidence for the involvement of nuclear factor B in the signaling mechanism. *Endocrinology*, **136**, 4790–5.

Lancaster, J. R. (1992). Nitric oxide in cells. *American Scientist*, **80**, 248–59.

Lancaster, J. R. (1994). Simulation of the diffusion and reaction of endogenously produced nitric oxide. *Proceedings of the National Academy of Sciences USA*, **91**, 8137–41.

Lancaster, J. R., Langrehr, J. M., Bergonia, H. A., Murase, N., Simmons, R. L. & Hoffman, R. A. (1992). EPR detection of heme and nonheme iron-containing protein nitrosylation by nitric oxide during rejection of rat heart allograft. *Journal of Biological Chemistry*, **267**, 10994–8.

Lang, D., Smith, J. A. & Lewis, M. J. (1993). Induction of a calcium-independent NO synthase by hypercholesterolaemia in the rabbit. *British Journal of Pharmacology*, **108**, 290–2.

Langenstroer, P. & Pieper, G. M. (1992). Regulation of spontaneous EDRF release in diabetic rat aorta by oxygen free radicals. *American Journal of Physiology*, **263**, H257–65.

Langrehr, J. M., Hoffman, R. A., Lancaster, J. R. & Simmons, R. L. (1993). Nitric oxide: a new endogenous immunomodulator. *Transplantation*, **55**, 1205–12.

Langrehr, J. M., Müller, A. R., Bergonia, H. A., Jacobs, T. D., Lee, T. K., Schraut, W. H., Lancaster, J. R., Hoffman, R. A. & Simmons, R. L. (1992). Detection of NO by EPR spectroscopy during rejection and graft-versus-host disease following small-bowel transplantation. *Surgery*, **112**, 395–402.

Lassen, L. H., Thomsen, L. L., Iversen, H. K. & Olesen, J. (1995). Histamine-1 receptor blockade does not prevent nitroglycerin induced migraine. *Endothelium*, **3** (Suppl), S78.

Laszio, F., Evans, S. M. & Whittle, B. J. R. (1995). Aminoguanidine inhibits both constitutive and inducible nitric oxide synthase isoforms in rat intestinal microvasculature *in vivo*. *European Journal of Pharmacology*, **272**, 169–75.

Lawrence, E. & Brain, S. D. (1992). Altered microvascular reactivity to endothelin-1, endothelin-3 and N^G-nitro-L-arginine methyl ester in streptozotocin-induced diabetes mellitus. *British Journal of Pharmacology*, **106**, 1035–40.

Lee, J. G., Wein, A. J. & Levin, R. M. (1994). Comparative pharmacology of the male and female rabbit bladder neck and urethra: involvement of nitric oxide. *Pharmacology*, **48**, 250–9.

Lee, J.-H., Price, R. H., Williams, F. G., Mayer, B. & Beitz, A. J. (1993). Nitric oxide synthase is found in some spinothalamic neurons and in neuronal processes that appose spinal neurons that express Fos induced by noxious stimulation. *Brain Research*, **608**, 324–33.

Lefebvre, R. A. (1993). Non-adrenergic non-cholinergic neurotransmission in the proximal stomach. *General Pharmacology*, **24**, 257–66.

Lefebvre, R. A., Baert, E. & Barbier, A. J. (1992a). Influence of N^G-nitro-L-arginine on non-adrenergic non-cholinergic relaxation in the guinea-pig gastric fundus. *British Journal of Pharmacology*, **106**, 173–9.

Lefebvre, R. A., De Vriese, A. & Smits, G. J. (1992b). Influence of vasoactive intestinal polypeptide and N^G-nitro-L-arginine methyl ester on cholinergic neurotransmission in the rat gastric fundus. *European Journal of Pharmacology*, **221**, 235–42.

Lefebvre, R. A., Hasrat, J. & Gobert, A. (1994). Influence of N^G-nitro-L-arginine methyl ester on vagally induced gastric relaxation in the anaesthetized rat. *British Journal of Pharmacology*, **105**, 315–20.

Lefer, A. M. & Lefer, D. J. (1993). Pharmacology of the endothelium in ischemia-reperfusion and circulatory shock. *Annual Reviews in Pharmacology and Toxicol-*

ogy, **33**, 71–90.

Lefer, D. J., Nakanishi, K., Johnston, W. E. & Vinten-Johansen, J. (1993). Anti neutrophil and myocardial protecting actions of a novel nitric oxide donor after acute myocardial ischemia and reperfusion in dogs. *Circulation*, **88**, 2337–50.

Lei, S. Z., Pan, Z. H., Aggarwal, S. K., Chen, H. S., Hartman, J., Sucher, N. J. & Lipton, S. A. (1992). Effect of nitric oxide production on the redox modulatory site of the NMDA receptor-channel complex. *Neuron*, **8**, 1087–99.

Leone, A. M., Francis, P. L., Rhodes, P. & Moncada, S. (1994a). A rapid and simple method for the measurement of nitrite and nitrate in plasma by high performance capillary electrophoresis. *Biochemical and Biophysical Research Communications*, **200**, 951–7.

Leone, A. M., Wiklund, N. P., Hokfelt, T., Brundin, L. & Moncada, S. (1994b). Release of nitric oxide by nerve stimulation in the human urogenital tract. *NeuroReport*, **5**, 733–6.

Lepoivre, M., Chenais, B., Yapo, A., Lemaire, G., Thelander, L. & Tenu, J. P. (1990). Alterations of ribonucleotide reductase activity following induction of the nitrite-generating pathway in adenocarcinoma cells. *Journal of Biological Chemistry*, **265**, 14 143–9.

Lerner Natoli, M., Rondouin, G., de Bock, F. & Bockaert, J. (1992). Chronic NO synthase inhibition fails to protect hippocampal neurones against NMDA toxicity. *NeuroReport*, **3**, 1109–12.

Lewis, M. J. & Smith, J. A. (1992). Factors regulating the release of endothelium-derived relaxing factor. In *Endothelial Regulation of Vascular Tone*, eds. U. S. Ryan & G. M. Rubanyi, pp. 139–54. New York: Marcel Dekker Inc.

Lewis, R. S. & Deen, W. M. (1994). Kinetics of the reaction of nitric oxide with oxygen in aqueous solutions. *Chemical Research in Toxicology*, **7**, 568–74.

Li, C. G. & Rand, M. J. (1989). Evidence for a role of nitric oxide in the neurotransmitter system mediating relaxation of the rat anococcygeus muscle. *Clinical and Experimental Pharmacology and Physiology*, **16**, 933–8.

Li, C. G. & Rand, M. J. (1990). Nitric oxide and vasoactive intestinal polypeptide mediate non-adrenergic, non-cholinergic inhibitory transmission to smooth muscle of the rat gastric fundus. *European Journal of Pharmacology*, **191**, 303–9.

Li, C. G. & Rand, M. J. (1991). Evidence that part of the NANC relaxant response of guinea-pig trachea to electrical field stimulation is mediated by nitric oxide. *British Journal of Pharmacology*, **102**, 91–4.

Li, C. G. & Rand, M. J. (1993). Effects of hydroxocobalamin and haemoglobin on NO-mediated relaxations in the rat anococcygeus muscle. *Clinical and Experimental Pharmacology and Physiology*, **20**, 633–40.

Li, Z. S. & Furness, J. B. (1993). Nitric oxide synthase in the enteric nervous system of the rainbow trout, *Salmo gairdneri*. *Archives of Histology and Cytology*, **56**, 185–93.

Li, Z. S., Murphy, S., Furness, J. B., Young, H. M. & Campbell, G. (1993). Relationships between nitric oxide synthase, vasoactive intestinal peptide and substance P immunoreactivities in neurons of the amphibian intestine. *Journal of the Autonomic Nervous System*, **44**, 197–206.

Li, Z. S., Young, H. M. & Furness, J. B. (1994). Nitric oxide synthase in neurons of the gastrointestinal tract of an avian species, *Coturnix coturnix*. *Journal of Anatomy*, **184**, 261–72.

Li, Z. S., Young, H. M. & Furness, J. B. (1995). Do vasoactive intestinal peptide (VIP)-and nitric oxide synthase-immunoreactive terminals synapse exclusively with VIP cell bodies in the submucous plexus of the guinea-pig ileum? *Cell and Tissue Research*, **281**, 485–91.

Liao, J. K., Shin, W. S., Lee, W. Y. & Clark, S. L. (1995). Oxidized low-density lipoprotein decreases the expression of endothelial nitric oxide synthase. *Journal of Biological Chemistry*, **270**, 319–24.

Liepe, B. A., Stone, C., Koistinaho, J. & Copenhagen, D. R. (1994). Nitric oxide synthase in Muller cells and neurons of salamander and fish retina. *Journal of Neuroscience*, **14**, 7641–54.

Lin, J.-Y., Seguin, R., Keller, K. & Chadee, K. (1995). Transforming growth factor-1 primes macrophages for enhanced expression of the nitric oxide synthase gene for nitric oxide-dependent cytotoxicity against *Entamoeba histolytica*. *Immunology*, **85**, 400–7.

Lincoln, J. & Burnstock, G. (1990). Neural-endothelial interactions in control of local blood flow. In *The Endothelium: An Introduction to Current Research*, ed. J. B. Warren, pp. 21–32. New York: Wiley-Liss.

Lincoln, J., Crowe, R., Blacklay, P. F., Pryor, J. P., Lumley, J. S. P. & Burnstock, G. (1987). Changes in VIPergic, cholinergic and adrenergic innervation of human penile tissue in diabetic and non-diabetic impotent males. *Journal of Urology*, **137**, 1053–9.

Lincoln, J., Crowe, R. & Burnstock, G. (1991a). Neuropeptides and impotence. In *Impotence: Diagnosis and Management of Male Erectile Dysfunction*, eds. R. S. Kirby, C. Carson & G. D. Webster, pp. 3–18. Oxford: Butterworth-Heinemann Ltd.

Lincoln, J., Hoyle, C. H. V. & Burnstock, G. (1995). Transmission: nitric oxide. In *Autonomic Neuroeffector Mechanisms*, eds. G. Burnstock & C. H. V. Hoyle, pp. 509–39. Reading: Harwood Academic.

Lincoln, J. & Messersmith, W. A. (1995). Conditions required for the measurement of nitric oxide synthase activity in a myenteric plexus/smooth muscle preparation from the rat ileum. *Journal of Neuroscience Methods*, **59**, 191–7.

Lincoln, J., Messersmith, W. A., Belai, A. & Burnstock, G. (1993). The effects of streptozotocin-induced diabetes and ganglioside treatment on the activities of the enzymes involved in nitric oxide synthesis in the rat ileum myenteric plexus. *Diabetes Medicine*, **10** (Suppl 3), P71.

Lincoln, J., Ralevic, V. & Burnstock, G. (1991b). Neurohumoral substances and the endothelium. In *Cardiovascular Significance of Endothelium-Derived Vasoactive Factors*, ed. G. M. Rubanyi, pp. 83–110. Mount Kisco: Futura.

Linden, D. J. & Connor, J. A. (1991). Participation of postsynaptic PKC in cerebellar long-term depression in culture. *Science*, **254**, 1656–9.

Linden, D. J. & Connor, J. A. (1993). Cellular mechanisms of long-term depression in the cerebellum. *Current Opinions in Neurology*, **3**, 401–6.

Lindsay, R. M., Smith, W., Rossiter, S. P., McIntyre, M. A., Williams, B. C. & Baird, J. D. (1995). *N*-Nitro-L-arginine methyl ester reduces the incidence of IDDM in BB/E rats. *Diabetes*, **44**, 365–8.

Ling, L., Karius, D. R., Fiscus, R. R. & Speck, D. F. (1992). Endogenous nitric oxide required for an integrative respiratory function in the cat brain. *Journal of Neurophysiology*, **68**, 1910–12.

Lipton, S. A., Choi, Y.-B., Pan, Z.-H., Sizheng, Z. L., Chen, H.-S. V., Sucher, N. J., Loscalzo, J., Singel, D. J. & Stamler, J. S. (1993). A redox-based mechanism for the neuroprotective and neurodestructive effects of nitric oxide and related nitroso-compounds. *Nature*, **364**, 626–32.

Lipton, S. A., Singel, D. J. & Stamler, J. S. (1994). Nitric oxide in the central nervous system. *Progress in Brain Research*, **103**, 359–64.

Lipton, S. A. & Stamler, J. S. (1994). Actions of redox-related congeners of nitric oxide at the NMDA receptor. *Neuropharmacology*, **33**, 1229–33.

Liu, J. & Sessa, W. C. (1994). Identification of covalently bound amino-terminal myristic acid in endothelial nitric oxide synthase. *Journal of Biological Chemistry*, **269**, 11691–4.

Liu, L. & Barajas, L. (1993). Nitric oxide synthase immunoreactive neurons in the rat kidney. *Neuroscience Letters*, **161**, 145–8.

Liu, S., Adcock, I. M., Old, R. W., Barnes, P. J. & Evans, T. W. (1993). Lipopolysaccharide treatment *in vivo* induces widespread tissue expression of inducible nitric oxide synthase mRNA. *Biochemical and Biophysical Research Communications*, **196**, 1208–13.

Liu, S. F., Crawley, D. E., Evans, T. W. & Barnes, P. J. (1992a). Endothelium-dependent nonadrenergic, noncholinergic neural relaxation in guinea pig pulmonary artery. *Journal of Pharmacology and Experimental Therapeutics*, **260**, 541–8.

Liu, S. F., Crawley, D. E., Rohde, J. A., Evans, T. W. & Barnes, P. J. (1992b). Role of nitric oxide and guanosine 3′,5′-cyclic monophosphate in mediating nonadrenergic, noncholinergic relaxation in guinea-pig pulmonary arteries. *British Journal of Pharmacology*, **107**, 861–6.

Llewellyn Smith, I. J., Song, Z. M., Costa, M., Bredt, D. S. & Snyder, S. H. (1992). Ultrastructural localization of nitric oxide synthase immunoreactivity in guinea-pig enteric neurons. *Brain Research*, **577**, 337–42.

Loder, P. B., Kamm, M. A., Nicholls, R. J. & Phillips, R. K. (1994). 'Reversible chemical sphincterotomy' by local application of glyceryl trinitrate. *British Journal of Surgery*, **81**, 1386–9.

Loesch, A., Belai, A. & Burnstock, G. (1993). Ultrastructural localization of NADPH-diaphorase and colocalization of nitric oxide synthase in endothelial cells of the rabbit aorta. *Cell and Tissue Research*, **274**, 539–45.

Loesch, A., Belai, A. & Burnstock, G. (1994). An ultrastructural study of NADPH-diaphorase and nitric oxide synthase in the perivascular nerves and vascular endothelium of the rat basilar artery. *Journal of Neurocytology*, **23**, 49–59.

Loesch, A., Belai, A., Lincoln, J. & Burnstock, G. (1986). Enteric nerves in diabetic rats: electron microscopic evidence for neuropathy of vasoactive intestinal polypeptide-containing fibres. *Acta Neuropathologica*, **70**, 161–8.

Loesch, A. & Burnstock, G. (1995). Ultrastructural localization of nitric oxide synthase and endothelin in coronary and pulmonary arteries of newborn rats. *Cell and Tissue Research*, **279**, 475–83.

Loesch, A. & Burnstock, G. (1996). Ultrastructural study of perivascular nerve fibres and endothelial cells of the rat basilar artery immunolabelled with monoclonal antibodies to neuronal and endothelial nitric oxide synthase. *Journal of Neurocytology*, **25**, 525–34.

Lopes, M. C., Cardoso, S. A., Schousboe, A. & Carvalho, A. P. (1994). Amino acids differentially inhibit the L-[^{3}H]arginine transport and nitric oxide synthase in rat brain synaptosomes. *Neuroscience Letters*, **181**, 1–4.

Loskove, J. A. & Frishman, W. H. (1995). Nitric oxide donors in the treatment of cardiovascular and pulmonary diseases. *American Heart Journal*, **129**, 604–13.

Lowenstein, C. J., Dinerman, J. L. & Snyder, S. H. (1994). Nitric oxide: a physiologic messenger. *Annals of Internal Medicine*, **120**, 227–37.

Lowenstein, C. J. & Snyder, S. H. (1992). Nitric oxide, a novel biological messenger. *Cell*, **70**, 705–7.

Lowenstein, C. J. & Snyder, S. H. (1994). Purification, cloning and expression of nitric-oxide synthase. *Methods in Enzymology*, **233**, 264–9.

Lunn, R. J. (1995). Inhaled nitric oxide therapy. *Mayo Clinical Proceedings*, **70**, 247–55.

Luo, D., Das, S. & Vincent, S. R. (1995). Effects of methylene blue and LY83583 on neuronal nitric oxide synthase and NADPH-diaphorase. *European Journal of Pharmacology*, **290**, 247–51.

Luo, D., Knezevich, S. & Vincent, S. R. (1993). *N*-Methyl-D-aspartate-induced nitric oxide release: an *in vivo* microdialysis study. *Neuroscience*, **57**, 897–900.

Lüscher, T. F. (1993). Platelet-vessel wall interaction: role of nitric oxide, prostaglandins and endothelins. *Baillières Clinical Haematology*, **6**, 609–27.

Lüscher, T. F. & Noll, G. (1994). Endothelium dysfunction in the coronary circulation. *Journal of Cardiovascular Pharmacology*, **24** (Suppl 3), S16–26.

Lüscher, T. F., Vanhoutte, P. M., Boulanger, C., Dohi, Y. & Bühler, F. R. (1991). Endothelial dysfunction in hypertension. In *Cardiovascular Significance of Endothelium-Derived Vasoactive Factors*, ed. G. M. Rubanyi, pp. 199–221. Mount Kisco: Futura.

Luth, H. J., Hedlich, A., Hilbig, H., Winkelmann, E. & Mayer, B. (1994). Morphological analyses of NADPH-diaphorase/nitric oxide synthase positive structures in human visual cortex. *Journal of Neurocytology*, **23**, 770–82.

Lynn, R. B., Sankey, S. L., Chakder, S. & Rattan, S. (1995). Colocalization of NADPH-diaphorase staining and VIP immunoreactivity in neurons in opossum internal anal sphincter. *Digestive Diseases and Sciences*, **40**, 781–91.

Lyons, C. R., Orloff, G. J. & Cunningham, J. M. (1992). Molecular cloning and functional expression of an inducible nitric oxide synthase from a macrophage cell line. *Journal of Biological Chemistry*, **267**, 6370–4.

MacAllister, R. J., Fickling, S. A., Whitley, G. S. J. & Vallance, P. (1994a). Metabolism of methylguanidines by human vasculature; implications for the regulation of nitric oxide synthesis. *British Journal of Pharmacology*, **112**, 43–8.

MacAllister, R. J., Whitley, G. S. J. & Vallance, P. (1994b). Effects of guanidino and uremic compounds on nitric oxide pathways. *Kidney International*, **45**, 737–42.

MacMicking, J. D., Nathan, C., Hom, G., Chartrain, N., Fletcher, D. S., Trumbauer, M., Stevens, K., Xie, Q.-W., Sokoi, K., Hutchinson, N., Chen, H. & Mudgett, J. S. (1995). Altered responses to bacterial infection and endotoxic shock in mice lacking inducible nitric oxide synthase. *Cell*, **81**, 641–50.

MacNaul, K. L. & Hutchinson, N. I. (1993). Differential expression of iNOS and cNOS mRNA in human vascular smooth muscle cells and endothelial cells under normal and inflammatory conditions. *Biochemical and Biophysical Research Communications*, **196**, 1330–4.

Maggi, C. A., Barbanti, G., Turini, D. & Giuliani, S. (1991). Effect of N^G-monomethyl L-arginine (L-NMMA) and N^G-nitro L-arginine (L-NOARG) on non-adrenergic non-cholinergic relaxation in the circular muscle of the human ileum. *British Journal of Pharmacology*, **103**, 1970–2.

Malinski, T., Radomski, M. W., Taha, Z. & Moncada, S. (1993b). Direct electrochemical measurement of nitric oxide released from human platelets. *Biochemical and Biophysical Research Communications*, **194**, 960–5.

Malinski, T. & Taha, Z. (1992). Nitric oxide release from a single cell measured *in situ* by a porphyrinic-based microsensor. *Nature*, **358**, 676–8.

Malinski, T., Taha, Z., Grunfeld, S., Patton, S., Kapturczak, M. & Tomboulian, P. (1993a). Diffusion of nitric oxide in the aorta wall monitored *in situ* by porphyrinic microsensors. *Biochemical and Biophysical Research Communications*, **193**, 1076–82.

Malmberg, A. B. & Yaksh, T. L. (1993). Spinal nitric oxide synthesis inhibition blocks NMDA-induced thermal hyperalgesia and produces antinociception in the formalin test in rats. *Pain*, **54**, 291–300.

Manilow, R., Schulman, R. & Tsien, R. W. (1989). Inhibition of postsynaptic PKC or CAMKII blocks induction but not expression of LTP. *Science*, **245**, 862–6.

Manilow, R. & Tsien, R. W. (1990). Presynaptic enhancemant shown by whole-cell recordings of long-term potentiation in hippocampal slices. *Nature*, **346**, 177–80.

Maqbool, A., Batten, T. F. & McWilliam, P. N. (1995). Co-localization of neurotransmitter immunoreactivities in putative nitric oxide synthesizing neurones of the cat brain stem. *Journal of Chemical Neuroanatomy*, **8**, 191–206.

Maragos, C. M., Morley, D., Wink, D. A., Dunams, T. M., Saavedra, J. E., Hoffman, A., Bove, A. A., Isaac, L., Hrabie, J. A. & Keefer, L. K. (1991). Complexes of .NO with nucleophiles as agents for the controlled biological release of nitric oxide. Vasorelaxant effects. *Journal of Medicinal Chemistry*, **34**, 3242–7.

Marklund, S. & Marklund, G. (1974). Involvement of the superoxide anion radical in the autoxidation of pyrogallol and a convenient assay for superoxide dismutase. *European Journal of Biochemistry*, **47**, 469–74.

Marletta, M. A. (1989). Nitric oxide: biosynthesis and biological significance. *Trends in Biochemical Sciences*, **14**, 488–92.

Marletta, M. A. (1993). Nitric oxide synthase structure and mechanism. *Journal of Biological Chemistry*, **268**, 12231–4.

Marletta, M. A. (1994). Nitric oxide synthase: aspects concerning structure and catalysis. *Cell*, **78**, 927–30.

Marletta, M. A., Yoon, P. S., Iyengar, R., Leaf, C. D. & Wishnok, J. S. (1988). Macrophage oxidation of L-arginine to nitrite and nitrate: nitric oxide is an intermediate. *Biochemistry*, **27**, 8706–11.

Marsden, P. A., Schappert, K. T., Chen, H. S., Flowers, M., Sundell, C. L., Wilcox, J. N. & Michel, T. (1992). Molecular cloning and characterization of human endothelial nitric oxide synthase. *FEBS Letters*, **307**, 287–93.

Martin, W., Gillespie, J. S. & Gibson, I. F. (1993). Actions and interactions of N^G-substituted analogues of L-arginine on NANC neurotransmission in the bovine retractor penis and rat anococcygeus muscles. *British Journal of Pharmacology*, **108**, 242–7.

Martin, W., McAllister, K. H. & Paisley, K. (1994). NANC neurotransmission in the bovine retractor penis muscle is blocked by superoxide anion following inhibition of superoxide dismutase with diethyldithiocarbamate. *Neuropharmacology*, **33**, 1293–301.

Martin, W., Smith, J. A., Lewis, M. J. & Henderson, A. H. (1988). Evidence that inhibitory factor extracted from bovine retractor penis is nitrite, whose acid-activated derivative is stabilized nitric oxide. *British Journal of Pharmacology*, **93**, 579–86.

Martin, W., Villani, G. M., Jothianandan, D. & Furchgott, R. F. (1985). Selective blockade of endothelium-dependent and glyceryl trinitrate-induced relaxation by hemoglobin and by methylene blue in the rabbit aorta. *Journal of Pharmacology and Experimental Therapeutics*, **232**, 708–16.

Mascolo, N., Izzo, A. A., Barbato, F. & Capasso, F. (1993). Inhibitors of nitric oxide synthetase prevent castor-oil-induced diarrhoea in the rat. *British Journal of Pharmacology*, **108**, 861–4.

Matini, P., Faussone Pellegrini, M. S., Cortesini, C. & Mayer, B. (1995). Vasoactive intestinal polypeptide and nitric oxide synthase distribution in the enteric plexuses of the human colon: an histochemical study and quantitative analysis. *Histochemistry and Cell Biology*, **103**, 415–23.

Matsumoto, T., Mitchell, J. A., Schmidt, H. H. H. W., Kohlhaas, K. L., Warner, T. D., Förstermann, U. & Murad, F. (1992). Nitric oxide synthase in ferret brain: localization and characterization. *British Journal of Pharmacology*, **107**, 849–52.

Matsumoto, T., Nakane, M., Pollock, J. S., Kuk, J. E. & Förstermann, U. (1993). A correlation between soluble brain nitric oxide synthase and NADPH-diaphorase activity is only seen after exposure of the tissue to fixative. *Neuroscience Letters*, **155**, 61–4.

Matsuoka, A., Stuehr, D. J., Olson, J. S., Clark, P. & Ikeda-Saito, M. (1994). L-Arginine and calmodulin regulation of the heme iron reactivity in neuronal nitric oxide synthase. *Journal of Biological Chemistry*, **269**, 20335–9.

Matsuoka, I., Giuili, G., Poyard, M., Stengel, D., Parma, J., Guellaen, G. & Hanoune, J. (1992). Localization of adenylyl and guanylyl cyclase in rat brain by *in situ* hybridization: comparison with calmodulin mRNA distribution. *Journal of Neuroscience*, **12**, 3350–60.

Mayer, B., Brunner, F. & Schmidt, K. (1993). Inhibition of nitric oxide synthesis by methylene blue. *Biochemical Pharmacology*, **45**, 367–74.

Mayer, B., Klatt, P., Bohme, E. & Schmidt, K. (1992). Regulation of neuronal nitric oxide and cyclic GMP formation by Ca^{2+}. *Journal of Neurochemistry*, **59**, 2024–9.

Mayer, B., Klatt, P., Werner, E. R. & Schmidt, K. (1994). Molecular mechanisms of inhibition of porcine brain nitric oxide synthase by the antinociceptive drug 7-nitro-indazole. *Neuropharmacology*, **33**, 1253–9.

Mayhan, W. G. (1992). Role of nitric oxide in modulating permeability of hamster cheek pouch in response to adenosine 5′-diphosphate and bradykinin. *Inflammation*, **16**, 295–305.

McCall, T. B., Feelisch, M., Palmer, R. M. J. & Moncada, S. (1991). Identification of *N*-iminoethyl-L-ornithine as an irreversible inhibitor of nitric oxide synthase in phagocytic cells. *British Journal of Pharmacology*, **102**, 234–8.

McKirdy, H. C., Marshall, R. W. & Taylor, B. A. (1993). Control of the human ileocaecal junction: an *in vitro* analysis of adrenergic and non-adrenergic non-cholinergic mechanisms. *Digestion*, **54**, 200–6.

McKirdy, H. C., McKirdy, M. L., Lewis, M. J. & Marshall, R. W. (1992). Evidence for involvement of nitric oxide in the non-adrenergic non-cholinergic (NANC) relaxation of human lower oesophageal sphincter muscle strips. *Experimental Physiology*, **77**, 509–11.

McLaren, A., Li, C. G. & Rand, M. J. (1993). Mediators of nicotine-induced relaxations of the rat gastric fundus. *Clinical and Experimental Pharmacology and Physiology*, **20**, 451–7.

McLennan, H. (1970). *Synaptic Transmission*, 2nd edn. Philadelphia: W. B. Saunders.

McNamara, D. B., Bedi, B., Aurora, H., Tena, L., Ignarro, L. J., Kadowitz, P. J. & Akers, D. L. (1993). L-Arginine inhibits balloon catheter-induced intimal hyperplasia. *Biochemical and Biophysical Research Communications*, **193**, 291–6.

McQuillan, L. P., Leung, G. K., Marsden, P. A., Kostyk, S. K. & Kourembanas, S. (1994). Hypoxia inhibits expression of eNOS via transcriptional and posttranscriptional mechanisms. *American Journal of Physiology*, **267**, H1921–7.

McVeigh, G. E., Brennan, G. M., Johnston, G. D., McDermott, B. J., McGrath, L. T., Henry, W. R., Andrews, J. W. & Hayes, J. R. (1992). Impaired endothelium-dependent and independent vasodilation in patients with Type 2 (non-insulin-dependent) diabetes mellitus. *Diabetologia*, **35**, 771–6.

Mearin, F., Mourelle, M., Guarner, F., Salas, A., Riveros-Moreno, V., Moncada, S. & Malagelada, J. R. (1993). Patients with achalasia lack nitric oxide synthase in the gastro-oesophageal junction. *European Journal of Clinical Investigation*, **23**, 724–8.

Melillo, G., Musso, T., Sica, A., Taylor, L. S., Cox, G. W. & Varesio, L. (1995). A hypoxia-responsive element mediates a novel pathway of activation of the inducible nitric oxide synthase promoter. *Journal of Experimental Medicine*, **182**, 1683–93.

Melis, M. R., Stancampiano, R. & Argiolas, A. (1994). Prevention by N^G-nitro-L-arginine methyl ester of apomorphine-and oxytocin-induced penile erection and yawning: site of action in the brain. *Pharmacology and Biochemistry of Behaviour*, **48**, 799–804.

Melis, M. R., Stancampiano, R. & Argiolas, A. (1995a). Role of nitric oxide in penile erection and yawning induced by 5-HT1c receptor agonists in male rats. *Naunyn Schmiedeberg's Archives of Pharmacology*, **351**, 439–45.

Melis, M. R., Stancampiano, R., Lai, C. & Argiolas, A. (1995b). Nitroglycerin-induced penile erection and yawning in male rats: mechanism of action in the brain. *Brain Research Bulletin*, **36**, 527–31.

Meller, S. T., Cummings, C. P., Traub, R. J. & Gebhart, G. F. (1994). The role of nitric oxide in the development and maintenance of the hyperalgesia produced by intraplantar injection of carrageenan in the rat. *Neuroscience*, **60**, 367–74.

Meller, S. T. & Gebhart, G. F. (1993). Nitric oxide (NO) and nociceptive processing in the spinal cord. *Pain*, **52**, 127–36.

Meller, S. T. & Gebhart, G. F. (1994a). The role of nitric oxide in spinal nociceptive processing. In *Nitric Oxide: Roles in Neuronal Communication and Neurotoxicity*, eds. H. Takagi, N. Toda & R. D. Hawkins, pp. 103–14. Tokyo: Japan Scientific Societies Press.

Meller, S. T. & Gebhart, G. F. (1994b). Spinal mediators of hyperalgesia. *Drugs*, **47** (Suppl 5), 10–20.

Meller, S. T., Pechman, P. S., Gebhart, G. F. & Maves, T. J. (1992). Nitric oxide mediates the thermal hyperalgesia produced in a model of neuropathic pain in the rat. *Neuroscience*, **50**, 7–10.

Menon, N. K., Wolf, A., Zehetgruber, M. & Bing, R. J. (1989). An improved chemiluminescence assay suggests non nitric oxide-mediated action of lysophosphatidylcholine and acetylcholine. *Proceedings of the Society for Experimental Biology and Medicine*, **191**, 316–19.

Michel, T. & Lamas, S. (1992). Molecular cloning of constitutive endothelial nitric oxide synthase: evidence for a family of related genes. *Journal of Cardiovascular Pharmacology*, **20** (Suppl 12), S45–9.

Milano, S., Arcoleo, F., Dieli, M., D'Agostino, R., D'Agostino, P., de Nucci, G. & Cillari, E. (1995). Prostanglandin E_2 regulates inducible nitric oxide synthase in the murine macrophage cell line J774. *Prostaglandins*, **49**, 105–15.

Miles, A. M., Bohle, D. S., Glassbrenner, P. A., Hansert, B., Wink, D. A. & Grisham, M. B. (1996). Modulation of superoxide-dependent oxidation and hydroxylation reactions by nitric oxide. *Journal of Biological Chemistry*, **271**, 40–7.

Miller, M. J. S., Sadowska-Krowicka, H., Chotinaruemol, S., Kakkis, J. L. & Clark, D. A. (1993). Amelioration of chronic ileitis by nitric oxide synthase inhibition. *Journal of Pharmacology and Experimental Therapeutics*, **264**, 11–16.

Minami, T., Nishihara, I., Ito, S., Sakamoto, K., Hyodo, M. & Hayaishi, O. (1995). Nitric oxide mediates allodynia induced by intrathecal administration of prostaglandin E_2 or prostaglandin $F_2\alpha$ in conscious mice. *Pain*, **61**, 285–90.

Minor, R. L. J., Myers, P. R., Guerra, R. J., Bates, J. N. & Harrison, D. G. (1990). Diet-induced atherosclerosis increases the release of nitrogen oxides from rabbit aorta. *Journal of Clinical Investigation*, **86**, 2109–16.

Mirza, U. A., Chait, B. T. & Landers, H. M. (1995). Monitoring reactions of nitric oxide with peptides and proteins by electrospray ionization-mass spectrometry. *Journal of Biological Chemistry*, **270**, 17185–8.

Misko, T. P., Moore, W. M., Kasten, T. P., Nickols, G. A., Corbett, J. A., Tilton, R. G., McDaniel, M. L., Williamson, J. R. & Currie, M. G. (1993a). Selective inhibition of the inducible nitric oxide synthase by aminoguanidine. *European Journal of Pharmacology*, **233**, 119–25.

Misko, T. P., Schilling, R. J., Salvemini, D., Moore, W. M. & Currie, M. G. (1993b). A fluorometric assay for the measurement of nitrite in biological samples. *Analytical Biochemistry*, **214**, 11–16.

Mitchell, H. H., Shonle, H. A. & Grindley, H. S. (1916). The origin of the nitrates in the urine. *Journal of Biological Chemistry*, **24**, 461–90.

Mitchell, J. A., Sheng, H., Förstermann, U. & Murad, F. (1991). Characterization of nitric oxide synthases in non-adrenergic non-cholinergic nerve containing tissue from the rat anococcygeus muscle. *British Journal of Pharmacology*, **104**, 289–91.

Mitrovic, B., Ignarro, L. J., Montestruque, S., Smoll, A. & Merrill, J. E. (1994). Nitric oxide as a potential pathological mechanism in demyelination: its differential effects on primary glial cells *in vitro*. *Neuroscience*, **61**, 575–85.

Mittal, C. K. & Murad, F. (1977). Activation of guanylate cyclase by superoxide dismutase and hydroxyl radical: a physiological regulator of guanosine 3′,5′-monophosphate formation. *Proceedings of the National Academy of Sciences USA*, **74**, 4360–4.

Mizhorkova, Z., Kortezova, N., Bredy Dobreva, G. & Papasova, M. (1994). Role of nitric oxide in mediating non-adrenergic non-cholinergic relaxation of the cat ileocecal sphincter. *European Journal of Pharmacology*, **265**, 77–82.

Modin, A., Weitzberg, E. & Lundberg, J. M. (1994). Nitric oxide regulates peptide release from parasympathetic nerves and vascular reactivity to vasoactive intestinal polypeptide *in vivo*. *European Journal of Pharmacology*, **261**, 185–97.

Modolell, M., Corraliza, I. M., Link, F., Soler, G. & Eichmann, K. (1995). Reciprocal regulation of the nitric oxide synthase/arginase balance in mouse bone marrow-derived macrophages by TH1 and TH2 cytokines. *European Journal of Immunology*, **25**, 1101–4.

Molina, R., Hidalgo, A. & Garcia de Boto, M. J. (1992). Influence of mechanical endothelium removal techniques and conservation conditions on rat aorta responses. *Methods and Findings in Experimental Clinical Pharmacology*, **14**, 91–6.

Moncada, S. (1992). The 1991 Ulf von Euler Lecture: the L-arginine:nitric oxide pathway. *Acta Physiologica Scandinavica*, **145**, 201–27.

Moncada, S. & Higgs, A. (1993). The L-arginine-nitric oxide pathway. *New England Journal of Medicine*, **329**, 2002–12.

Moncada, S., Palmer, R. M. J. & Higgs, E. A. (1991). Nitric oxide: physiology, pathophysiology, and pharmacology. *Pharmacological Reviews*, **43**, 109–42.

Moore, P. K., Al Swayeh, O. A., Chong, N. W., Evans, R. A. & Gibson, A. (1990). L-N^G-Nitro arginine (L-NOARG), a novel, L-arginine-reversible inhibitor of endothelium-dependent vasodilatation *in vitro*. *British Journal of Pharmacology*, **99**, 408–12.

Moore, P. K., Babbedge, R. C., Wallace, P., Gaffen, Z. A. & Hart, S. L. (1993a). 7-Nitro indazole, an inhibitor of nitric oxide synthase, exhibits anti-nociceptive activity in the mouse without increasing blood pressure. *British Journal of Pharmacology*, **108**, 296–7.

Moore, P. K., Oluyomi, A. O., Babbedge, R. C., Wallace, P. & Hart, S. L. (1991). L-N^G-Nitro arginine methyl ester exhibits antinociceptive activity in the mouse. *British Journal of Pharmacology*, **102**, 198–202.

Moore, P. K., Wallace, P., Gaffen, Z., Hart, S. L. & Babbedge, R. C. (1993b). Characterization of the novel nitric oxide synthase inhibitor 7-nitro indazole and related indazoles: antinociceptive and cardiovascular effects. *British Journal of Pharmacology*, **110**, 219–24.

Moore, P. K., Webber, R. K., Jerome, G. M., Tjoeng, F. S., Misko, T. P. & Currie, M. G. (1994). L-N^6-(1-Iminoethyl)lysine: a selective inhibitor of inducible nitric oxide synthase. *Journal of Medicinal Chemistry*, **37**, 3886–8.

Morgan, C. V. J., Babbedge, R. C., Gaffen, Z., Wallace, P., Hart, S. L. & Moore, P. K. (1992). Synergistic anti-nociceptive effect of L-N^G-nitro arginine methyl ester (L-NAME) and flurbiprofen in the mouse. *British Journal of Pharmacology*, **106**, 493–7.

Morley, D. & Keefer, L. K. (1993). Nitric oxide/nucleophile complexes: a unique class of nitric oxide-based vasodilators. *Journal of Cardiovascular Pharmacology*, **22** (Suppl 7), S3–9.

Morley, D., Maragos, C. M., Zhang, X. Y., Boignon, M., Wink, D. A. & Keefer, L. K. (1993). Mechanism of vascular relaxation induced by the nitric oxide (NO)/nucleophile complexes, a new class of NO-based vasodilators. *Journal of Cardiovascular Pharmacology*, **21**, 670–6.

Moro, M. A., Russell, R. J., Cellek, S., Lizasoain, I., Su, Y., Darley-Usmar, V. M., Radomski, M. W. & Moncada, S. (1996). cGMP mediates the vascular and platelet actions of nitric oxide: confirmation using an inhibitor of the soluble guanylyl cyclase. *Proceedings of the National Academy of Sciences USA*, **93**, 1480–5.

Morris, J. L. (1993). Co-transmission from autonomic vasodilator neurons supplying the guinea pig uterine artery. *Journal of the Autonomic Nervous System*, **42**, 11–21.

Moskowitz, M. A., Buzzi, M. G., Sakas, D. E. & Linnik, M. D. (1989). Pain mechanisms underlying vascular headaches. *Reviews in Neurology (Paris)*, **145**, 181–93.

Moss, D. W., Wei, X., Liew, F. Y., Moncada, S. & Charles, I. G. (1995). Enzymatic characterisation of recombinant murine inducible nitric oxide synthase. *European Journal of Pharmacology*, **289**, 41–8.

Mountcastle, V. B. & Sastre, A. (1980). Synaptic transmission. In *Medical Physiology*, 14th edn, ed. V. B. Mountcastle, pp. 184–223. St. Louis, MO: CV Mosby.

Mourelle, M., Guarner, F., Moncada, S. & Malagelada, J. R. (1993). The arginine/nitric oxide pathway modulates sphincter of Oddi motor activity in guinea pigs and rabbits. *Gastroenterology*, **105**, 1299–305.

Mufson, E. J. & Brandabur, M. M. (1994). Sparing of NADPH-diaphorase striatal neurons in Parkinson's and Alzheimer's diseases. *NeuroReport*, **5**, 705–8.

Mugge, A., Elwell, J. H., Peterson, T. E. & Harrison, D. G. (1991). Release of intact endothelium-derived relaxing factor depends on endothelial superoxide dismutase activity. *American Journal of Physiology*, **260**, C219–25.

Muller, U. (1994). Ca^{2+}/calmodulin-dependent nitric oxide synthase in *Apis mellifera* and *Drosophila melanogaster. European Journal of Neuroscience*, **6**, 1362–70.

Muller, U. & Bicker, G. (1994). Calcium-activated release of nitric oxide and cellular distribution of nitric oxide-synthesizing neurons in the nervous system of the locust. *Journal of Neuroscience*, **14**, 7521–8.

Mullins, M. E., Sondheimer, N. J., Huang, Z., Singel, D. J., Stamler, J., Huang, P. L., Fishman, M. C., Jensen, F. E., Lipton, S. A. & Moskowitz, M. A. (1995). Spin-trapping NO in nNOS-deficient mice: indications for stroke therapy. *Endothelium*, **3** (Suppl), S4.

Mülsch, A. (1994). Nitrogen monoxide transport mechanisms. *Arzneimittel-Forschung*, **44** (Suppl 3A), 408–11.

Mulvany, M. J. & Halpern, W. (1976). Mechanical properties of vascular smooth muscle cells *in situ*. *Nature*, **260**, 617–19.

Mundel, P., Bachmann, S., Bader, M., Fischer, A., Kummer, W., Mayer, B. & Kriz, W. (1992). Expression of nitric oxide synthase in kidney macula densa cells. *Kidney International*, **42**, 1017–19.

Murad, F., Mittal, C. K., Arnold, W. P., Katsuki, S. & Kimura, H. (1978). Guanylate cyclase: activation by azide, nitro compounds, nitric oxide, and hydroxyl radical and inhibition by hemoglobin and myoglobin. *Advances in Cyclic Nucleotide Research*, **9**, 145–58.

Murphy, M. E. & Noack, E. (1994). Nitric oxide assay using hemoglobin method. *Methods in Enzymology*, **233**, 240–50.

Murphy, S., Simmons, M. L., Agullo, L., Garcia, A., Feinstein, D. L., Galea, E., Reis, D. J., Minc-Golomb, D. & Schwartz, J. P. (1993). Synthesis of nitric oxide in CNS glial cells. *Trends in Neuroscience*, **16**, 323–8.

Murray, J., Du, C., Ledlow, A., Bates, J. N. & Conklin, J. L. (1991). Nitric oxide: mediator of nonadrenergic noncholinergic responses of opossum esophageal muscle. *American Journal of Physiology*, **261**, G401–6.

Murray, J. A., Ledlow, A., Launspach, J., Evans, D., Loveday, M. & Conklin, J. L. (1995). The effects of recombinant human hemoglobin on esophageal motor functions in humans. *Gastroenterology*, **109**, 1241–8.

Murrell, W. (1879). Nitroglycerin as a remedy for angina pectoris. *Lancet*, **i**, 80–1, 113–115, 151–152, 225–227.

Myers, P. R., Guerra, R. J. & Harrison, D. G. (1989). Release of NO and EDRF from cultured bovine aortic endothelial cells. *American Journal of Physiology*, **256**, H1030–7.

Myers, P. R., Minor, R. L. J., Guerra, R. J., Bates, J. N. & Harrison, D. G. (1990). Vasorelaxant properties of the endothelium-derived relaxing factor more closely resemble *S*-nitrosocysteine than nitric oxide. *Nature*, **345**, 161–3.

Myers, P. R., Wright, T. F., Tanner, M. A. & Ostlund, R. E. J. (1994). The effects of native LDL and oxidized LDL on EDRF bioactivity and nitric oxide production in vascular endothelium. *Journal of Laboratory and Clinical Medicine*, **124**, 672–83.

Nachlas, M. M., Wayler, D. G. & Seligano, A. M. (1958). A histochemical method for the demonstration of diphosphopyridine diaphorase. *Journal of Biophysical and Biochemical Cytology*, **4**, 29–38.

Nagafuji, T., Matsui, T., Koide, T. & Asano, T. (1992). Blockade of nitric oxide formation by N^{ω}-nitro-L-arginine mitigates ischemic brain edema and subsequent cerebral infarction in rats. *Neuroscience Letters*, **147**, 159–62.

Nagafuji, T., Sugiyama, M., Matsui, T. & Koide, T. (1993). A narrow therapeutical window of a nitric oxide synthase inhibitor against transient ischemic brain injury. *European Journal of Pharmacology*, **248**, 325–8.

Naka, Y., Chowdhury, N. C., Oz, M. C., Smith, C. R., Yano, O. J., Michler, R. E., Stern, D. M. & Pinsky, D. J. (1995). Nitroglycerin maintains graft vascular homeostasis and enhances preservation in an orthotopic rat lung transplant model. *Journal of Thoracic and Cardiovascular Surgery*, **109**, 206–10.

Nakane, M., Klinghofer, V., Kuk, J. E., Donnelly, J. L., Budzik, G. P., Pollock, J. S., Basha, F. & Carter, G. W. (1995). Novel potent and selective inhibitors of inducible nitric oxide synthase. *Molecular Pharmacology*, **47**, 831–4.

Nakane, M., Mitchell, J., Förstermann, U. & Murad, F. (1991). Phosphorylation by calcium calmodulin-dependent protein kinase II and protein kinase C modulates the activity of nitric oxide synthase. *Biochemical and Biophysical Research Communications*, **180**, 1396–402.

Nakane, M., Schmidt, H. H. H. W., Pollock, J. S., Förstermann, U. & Murad, F. (1993). Cloned human brain nitric oxide synthase is highly expressed in skeletal muscle. *FEBS Letters*, **316**, 175–80.

Nakos, G. & Gossrau, R. (1994). When NADPH diaphorase (NADPHd) works in the presence of formaldehyde, the enzyme appears to visualize selectively cells with constitutive nitric oxide synthase (NOS). *Acta Histochemica*, **96**, 335–43.

Narayanan, K., Spack, L., McMillan, K., Kilbourn, R. G., Hayward, M. A., Siler Masters, B. S. & Griffith, O. W. (1995). *S*-Alkyl-L-thiocitrullines: potent stereoselective inhibitors of nitric oxide synthase with strong pressor activity *in vivo*. *Journal of Biological Chemistry*, **270**, 11103–10.

Natanson, C., Hoffman, W. D., Suffredini, A. F., Eichacker, P. Q. & Danner, R. L. (1994). Selected treatment strategies for septic shock based on proposed mechanisms of pathogenesis. *Annals of Internal Medicine*, **120**, 771–83.

Nathan, C. (1992). Nitric oxide as a secretory product of mammalian cells. *FASEB Journal*, **6**, 3051–64.

Nathan, C. (1995). Inducible nitric oxide synthase: regulation subserves function. *Current Topics in Microbiology and Immunology*, **196**, 1–4.

Nathan, C. & Xie, Q.-W. (1994a). Regulation of biosynthesis of nitric oxide. *Journal of Biological Chemistry*, **269**, 13725–18.

Nathan, C. & Xie, Q.-W. (1994b). Nitric oxide synthases: roles, tolls, and controls. *Cell*, **78**, 915–18.

Nathanson, J. A. & McKee, M. (1995). Identification of an extensive system of nitric oxide-producing cells in the ciliary muscle and outflow pathway of the human eye. *Investigative Ophthalmology and Visual Science*, **36**, 1765–73.

Natuzzi, E. S., Ursell, P. C., Harrison, M., Buscher, C. & Riemer, R. K. (1993). Nitric oxide synthase activity in the pregnant uterus decreases at parturition. *Biochemical and Biophysical Research Communications*, **194**, 1–8.

Nava, E., Palmer, R. M. J. & Moncada, S. (1991). Inhibition of nitric oxide synthesis in septic shock: how much is beneficial? *Lancet*, **338**, 1555–7.

Needleman, P. (1973). Oral nitrates: efficacious? *Annals of Internal Medicine*, **78**, 457–8.

Nelson, R. J., Demas, G. E., Huang, P. L., Fishman, M. C., Dawson, V. L., Dawson, T. M. & Snyder, S. H. (1995). Behavioural abnormalities in male mice lacking neuronal nitric oxide synthase. *Nature*, **378**, 383–6.

Nemade, R. V., Lewis, A. I., Zuccarello, M. & Keller, J. T. (1995). Immunohistochemical localization of endothelial nitric oxide synthase in vessels of the dura mater of the Sprague-Dawley rat. *Neuroscience Letters*, **197**, 78–80.

Niimi, Y., Azuma, H. & Hirakawa, K. (1994). Repeated endothelial removal augments intimal thickening and attenuates EDRF release. *American Journal of Physiology*, **266**, H1348–56.

Nishida, K., Harrison, D. G., Navas, J. P., Fisher, A. A., Dockery, S. P., Uematsu, M., Nerem, R. M., Alexander, R. W. & Murphy, T. J. (1992). Molecular cloning and characterization of the constitutive bovine aortic endothelial cell nitric oxide synthase. *Journal of Clinical Investigation*, **90**, 2092–6.

Nishizawa, O., Kawahara, T., Shimoda, N., Suzuki, K., Fujieda, N., Kudo, T., Suzuki, T., Noto, H., Harada, T. & Tsuchida, S. (1992). Effect of methylene blue on the vesicourethral function in the rats. *Tohoku Journal of Experimental Medicine*, **168**, 621–2.

Noack, E., Kubitzek, D. & Kojda, G. (1992). Spectrophotometric determination of nitric oxide using hemoglobin. *Neuroprotocols*, **1**, 133–9.

Noris, M., Morigi, M., Donadelli, R., Aiello, S., Foppolo, M., Todeschini, M., Orisio, S., Remuzzi, G. & Remuzzi, A. (1995). Nitric oxide synthesis by cultured endothelial cells is modulated by flow conditions. *Circulation Research*, **76**, 536–43.

Nozaki, K., Moskowitz, M. A., Maynard, K. I., Koketsu, N., Dawson, T. M., Bredt, D. S. & Snyder, S. H. (1993). Possible origins and distribution of immunoreactive nitric oxide synthase-containing nerve fibers in cerebral arteries. *Journal of Cerebral Blood Flow and Metabolism*, **13**, 70–9.

Nunoshiba, T., DeRojas-Walker, T., Tannenbaum, S. R. & Demple, B. (1995). Roles of nitric oxide in inducible resistance of *Escherichia coli* to activated murine macrophages. *Infection and Immunity*, **63**, 794–8.

Nussler, A. K. & Billiar, T. R. (1993). Inflammation, immunoregulation, and inducible nitric oxide synthase. *Journal of Leukocyte Biology*, **54**, 171–8.

Ny, L., Alm, P., Ekstrom, P., Hannibal, J., Larsson, B. & Andersson, K. E. (1994). Nitric oxide synthase-containing, peptide-containing, and acetylcholinesterase-positive nerves in the cat lower oesophagus. *Histochemical Journal*, **26**, 721–33.

O'Brien, A. J., Young, H. M., Povey, J. M. & Furness, J. B. (1995). Nitric oxide synthase is localized predominantly in the Golgi apparatus and cytoplasmic vesicles of vascular endothelial cells. *Histochemistry and Cell Biology*, **103**, 221–5.

O'Dell, T. J., Hawkins, R. D., Kandel, E. R. & Arancio, O. (1991). Tests of roles of two diffusible substances in long-term potentiation: evidence for nitric oxide as a possible early retrograde messenger. *Proceedings of the National Academy of Sciences USA*, **88**, 11 285–9.

O'Dell, T. J., Huang, P. L., Dawson, T. M., Dinerman, J. L., Snyder, S. H., Kandel, E. R. & Fishman, M. C. (1994). Endothelial NOS and the blockade of LTP by NOS inhibitors in mice lacking neuronal NOS. *Science*, **265**, 542–6.

O'Kelly, T., Brading, A. & Mortensen, N. (1993). Nerve mediated relaxation of the human internal anal sphincter: the role of nitric oxide. *Gut*, **34**, 689–93.

Ogura, T., Yokoyama, T., Fujisawa, H., Kurashima, Y. & Esumi, H. (1993). Structural diversity of neuronal nitric oxide synthase mRNA in the nervous system. *Biochemical and Biophysical Research Communications*, **193**, 1014–22.

Ohara, Y., Peterson, T. E. & Harrison, D. G. (1993). Hypercholesterolemia increases endothelial superoxide anion production. *Journal of Clinical Investigation*, **91**, 2546–51.

Ohara, Y., Sayegh, H. S., Yamin, J. J. & Harrison, D. G. (1995). Regulation of endothelial constitutive nitric oxide synthase by protein kinase C. *Hypertension*, **25**, 415–20.

Ohta, A., Takagi, H., Matsui, T., Hamai, Y., Iida, S. & Esumi, H. (1993). Localization of nitric oxide synthase-immunoreactive neurons in the solitary nucleus and ventrolateral medulla oblongata of the rat: their relation to catecholaminergic neurons. *Neuroscience Letters*, **158**, 33–5.

Okamura, T., Enokibori, M. & Toda, N. (1993). Neurogenic and non-neurogenic relaxations caused by nicotine in isolated dog superficial temporal artery. *Journal of Pharmacology and Experimental Therapeutics*, **266**, 1416–21.

Okamura, T. & Toda, N. (1994). Mechanism underlying nicotine-induced relaxation in dog saphenous arteries. *European Journal of Pharmacology*, **263**, 85–91.

Okamura, T., Yoshida, K. & Toda, N. (1995). Nitroxidergic innervation in dog and monkey renal arteries. *Hypertension*, **25**, 1090–5.

Olesen, J., Thomsen, L. L. & Iversen, H. (1994). Nitric oxide is a key molecule in migraine and other vascular headaches. *Trends in Pharmacological Sciences*, **15**, 149–53.

Oliveira, R. B., Matsuda, N. M., Antoniolli, A. R. & Ballejo, G. (1992). Evidence for the involvement of nitric oxide in the electrically induced relaxations of human lower esophageal sphincter and distal pylorus. *Brazilian Journal of Medical and Biological Research*, **25**, 853–5.

Oliver, J. A. (1992). Endothelium-derived relaxing factor contributes to the regulation of endothelial permeability. *Journal of Cell Physiology*, **151**, 506–11.

Olken, N. M., Osawa, Y. & Marletta, M. A. (1994). Characterization of the inactivation of nitric oxide synthase by N^G-methyl-L-arginine: evidence for heme loss. *Biochemistry*, **33**, 14784–91.

Omar, H. A., Cherry, P. D., Mortelliti, M. P., Burke Wolin, T. & Wolin, M. S. (1991). Inhibition of coronary artery superoxide dismutase attenuates endothelium-dependent and -independent nitrovasodilator relaxation. *Circulation Research*, **69**, 601–8.

Oshima, H. & Bartsch, H. (1994). Chronic infections and inflammatory processes as cancer risk factors: possible role of nitric oxide in carcinogenesis. *Mutation Research*, **305**, 253–64.

Ostholm, T., Holmqvist, B. I., Alm, P. & Ekstrom, P. (1994). Nitric oxide synthase in the CNS of the Atlantic salmon. *Neuroscience Letters*, **168**, 233–7.

Pacelli, R., Wink, D. A., Cook, J. A., Krishna, M. C., DeGraff, W., Friedman, N., Tsokos, M., Samuni, A. & Mitchell, J. B. (1995). Nitric oxide potentiates hydrogen peroxide-induced killing of *Escherichia coli*. *Journal of Experimental Medicine*, **182**, 1469–79.

Paisley, K. & Martin, W. (1996). Blockade of nitrergic transmission by hydroquinone, hydroxocobalamin and carboxy-PTIO in bovine retractor penis: role of superoxide anion. *British Journal of Pharmacology*, **117**, 1633–8.

Palmer, R. M. J., Ashton, D. S. & Moncada, S. (1988). Vascular endothelial cells synthesize nitric oxide from L-arginine. *Nature*, **333**, 664–6.

Palmer, R. M. J., Ferrige, A. G. & Moncada, S. (1987). Nitric oxide release accounts for the biological activity of endothelium-derived relaxing factor. *Nature*, **327**, 524–6.

Papka, R. E., McCurdy, J. R., Williams, S. J., Mayer, B., Marson, L. & Platt, K. B. (1995a). Parasympathetic preganglionic neurons in the spinal cord involved in uterine innervation are cholinergic and nitric oxide-containing. *Anatomical Record*, **241**, 554–62.

Papka, R. E., McNeill, D. L., Thompson, D. & Schmidt, H. H. H. W. (1995b). Nitric oxide nerves in the uterus are parasympathetic, sensory, and contain neuropeptides. *Cell and Tissue Research*, **279**, 339–49.

Parlani, M., Conte, B. & Manzini, S. (1993). Nonadrenergic, noncholinergic inhibitory control of the rat external urethral sphincter: involvement of nitric oxide. *Journal of Pharmacology and Experimental Therapeutics*, **265**, 713–19.

Patel, S. B., Clayden, G. S., Burleigh, D. E., Ward, H. C. & Ward, J. P. R. (1995). The internal anal sphincter in children with severe idiopathic constipation: a possible therapeutic role for nitric oxide. *Endothelium*, **3** (Suppl), S13.

Paterson, W. G., Anderson, M. A. & Anand, N. (1992). Pharmacological characterization of lower esophageal sphincter relaxation induced by swallowing, vagal efferent nerve stimulation, and esophageal distention. *Canadian Journal of Physiology and Pharmacology*, **70**, 1011–15.

Paton, W. D. M. (1958). Central and synaptic transmission in the nervous system (pharmacological aspects). *Annual Reviews of Physiology*, **20**, 431–70.

Pauletzki, J. G., Sharkey, K. A., Davison, J. S., Bomzon, A. & Shaffer, E. A. (1993). Involvement of L-arginine-nitric oxide pathways in neural relaxation of the sphincter of Oddi. *European Journal of Pharmacology*, **232**, 263–70.

Pearse, A. G. E. (1972). *Histochemistry: Theoretical and Applied*, third edn. Edinburgh: Churchill Livingstone.

Pearson, J. D. (1994). Endothelial cell function and thrombosis. *Baillières Clinical Haematology*, **7**, 441–52.

Peiró, C., Redondo, J., Rodríguez-Martínez, M. A., Angulo, J., Marín, J. & Sánchez-Ferrer, C. F. (1995). Influence of endothelium on cultured vascular smooth muscle cell proliferation. *Hypertension*, **25**, 748–51.

Peredo, H. A. & Enero, M. A. (1993). Effect of endothelium removal on basal and muscarinic cholinergic stimulated rat mesenteric vascular bed prostanoid synthesis. *Prostaglandins Leukotrienes and Essential Fatty Acids*, **48**, 373–8.

Perez, M. T., Larsson, B., Alm, P., Andersson, K. E. & Ehinger, B. (1995). Localisation of neuronal nitric oxide synthase-immunoreactivity in rat and rabbit retinas. *Experimental Brain Research*, **104**, 207–17.

Persson, K., Alm, P., Johansson, K., Larsson, B. & Andersson, K. E. (1993). Nitric oxide synthase in pig lower urinary tract: immunohistochemistry, NADPH diaphorase histochemistry and functional effects. *British Journal of Pharmacology*, **110**, 521–30.

Persson, K., Alm, P., Johansson, K., Larsson, B. & Andersson, K. E. (1995). Co-existence of nitrergic, peptidergic and acetylcholine esterase-positive nerves in the pig lower urinary tract. *Journal of the Autonomic Nervous System*, **52**, 225–36.

Persson, K. & Andersson, K. E. (1992). Nitric oxide and relaxation of pig lower urinary tract. *British Journal of Pharmacology*, **106**, 416–22.

Persson, K., Igawa, Y., Mattiasson, A. & Andersson, K. E. (1992). Effects of inhibition of the L-arginine/nitric oxide pathway in the rat lower urinary tract *in vivo* and *in vitro*. *British Journal of Pharmacology*, **107**, 178–84.

Peterson, D. A., Peterson, D. C., Archer, S. & Weir, E. K. (1992). The non specificity of specific nitric oxide synthase inhibitors. *Biochemical and Biophysical Research*

Communications, **187**, 797–801.

Petros, A., Lamb, G., Leone, A., Moncada, S., Bennett, D. & Vallance, P. (1994). Effects of a nitric oxide synthase inhibitor in humans with septic shock. *Cardiovascular Research*, **28**, 34–9.

Peunova, N. & Enikolopov, G. (1993). Amplification of calcium-induced gene transcription by nitric oxide in neuronal cells. *Nature*, **364**, 450–3.

Pickard, R. S., King, P., Zar, M. A. & Powell, P. H. (1994). Corpus cavernosal relaxation in impotent men. *British Journal of Urology*, **74**, 485–91.

Pickard, R. S., Powell, P. H. & Zar, M. A. (1991). The effect of inhibitors of nitric oxide biosynthesis and cyclic GMP formation on nerve-evoked relaxation of human cavernosal smooth muscle. *British Journal of Pharmacology*, **104**, 755–9.

Pickard, R. S., Powell, P. H. & Zar, M. A. (1995). Nitric oxide and cyclic GMP formation following relaxant nerve stimulation in isolated human corpus cavernosum. *British Journal of Urology*, **75**, 516–22.

Pinsky, D. J., Oz, M. C., Koga, S., Taha, Z., Broekman, M. J., Marcus, A. J., Liao, H., Naka, Y., Brett, J., Cannon, P. J., Nowygrod, R., Malinski, T. & Stern, D. M. (1994). Cardiac preservation is enhanced in a heterotopic rat transplant model by supplementing the nitric oxide pathway. *Journal of Clinical Investigation*, **93**, 2291–7.

Pipili-Synetos, E., Sakkoula, E., Haralabopoulos, G., Andriopoulou, P., Peristeris, P. & Maragoudakis, M. E. (1994). Evidence that nitric oxide is an endogenous antiangiogenic mediator. *British Journal of Pharmacology*, **111**, 894–902.

Poeggel, G., Muller, M., Seidel, I., Rechardt, L. & Bernstein, H. G. (1992). Histochemistry of guanylate cyclase, phosphodiesterase, and NADPH-diaphorase (nitric oxide synthase) in rat brain vasculature. *Journal of Cardiovascular Pharmacology*, **20** (Suppl 12), S76–9.

Pollock, J. S., Förstermann, U., Mitchell, J. A., Warner, T. D., Schmidt, H. H. H. W., Nakane, M. & Murad, F. (1991). Purification and characterization of particulate and endothelium-derived relaxing factor synthase from cultured and native bovine aortic endothelial cells. *Proceedings of the National Academy of Science USA*, **88**, 10480–4.

Prabhakar, N. R., Kumar, G. K., Chang, C. H., Agani, F. H. & Haxhiu, M. A. (1993). Nitric oxide in the sensory function of the carotid body. *Brain Research*, **625**, 16–22.

Preiksaitis, H. G., Tremblay, L. & Diamant, N. E. (1994). Nitric oxide mediates inhibitory nerve effects in human esophagus and lower esophageal sphincter. *Digestive Diseases and Sciences*, **39**, 770–5.

Preiser, J.-C., Lejeune, P., Roman, A., Carlier, E., De Backer, D., Leeman, M., Kahn, R. J. & Vincent, J.-L. (1995). Methylene blue administration in septic shock: a clinical trial. *Critical Care Medicine*, **23**, 259–64.

Price, R. H., Mayer, B. & Beitz, A. J. (1993). Nitric oxide synthase neurons in rat brain express more NMDA receptor mRNA than non-NOS neurons. *NeuroReport*, **4**, 807–10.

Pullen, A. H. & Humphreys, P. (1995). Diversity in localization of nitric oxide synthase antigen and NADPH-diaphorase histochemical staining in sacral somatic motor neurones of the cat. *Neuroscience Letters*, **196**, 33–6.

Qiu, H. Y., Henrion, D. & Levy, B. I. (1994). Alterations in flow-dependent vasomotor tone in spontaneously hypertensive rats. *Hypertension*, **24**, 474–9.

Radomski, M. W. & Moncada, S. (1993). Regulation of vascular homeostasis by nitric oxide. *Thrombosis and Haemostasis*, **70**, 36–41.

Radomski, M. W., Palmer, R. M. J. & Moncada, S. (1987a). Endogenous nitric oxide inhibits human platelet adhesion to vascular endothelium. *Lancet*, **2**, 1057–8.

Radomski, M. W., Palmer, R. M. J. & Moncada, S. (1987b). Comparative pharmacology of endothelium-derived relaxing factor, nitric oxide and prostacyclin in platelets. *British Journal of Pharmacology*, **92**, 181–7.

Radomski, M. W., Rees, D. D., Dutra, A. & Moncada, S. (1992). *S*-Nitroso-glutathione inhibits platelet activation *in vitro* and *in vivo*. *British Journal of Pharmacology*, **107**, 745–9.

Radomski, M. W., Vallance, P., Whitley, G., Foxwell, N. & Moncada, S. (1993). Platelet adhesion to human vascular endothelium is modulated by constitutive and cytokine induced nitric oxide. *Cardiovascular Research*, **27**, 1380–2.

Rajanayagam, M. A., Li, C. G. & Rand, M. J. (1993). Differential effects of hydroxocobalamin on NO-mediated relaxations in rat aorta and anococcygeus muscle. *British Journal of Pharmacology*, **108**, 3–5.

Rajfer, J., Aronson, W. J., Bush, P. A., Dorey, F. J. & Ignarro, L. J. (1992). Nitric oxide as a mediator of relaxation of the corpus cavernosum in response to nonadrenergic, noncholinergic neurotransmission. *New England Journal of Medicine*, **326**, 90–4.

Ralevic, V. & Burnstock, G. (1991). Effects of purines and pyrimidines on the rat mesenteric arterial bed. *Circulation Research*, **69**, 1583–90.

Ralevic, V. & Burnstock, G. (1993). *Neural-Endothelial Interactions in the Control of Local Vascular Tone.* Boca Raton, FL: CRC Press.

Ralevic, V., Dikranian, K. & Burnstock, G. (1995). Long-term sensory denervation does not modify endothelial function or endothelial substance P and nitric oxide synthase in rat mesenteric arteries. *Journal of Vascular Research*, **32**, 320–7.

Ralevic, V., Hoyle, C. H. V., Goss Sampson, M. A., Milla, P. J. & Burnstock, G. (1996). Effect of chronic vitamin E deficiency on sensory-motor vasodilatation and sympathetic vasoconstriction in the mesenteric arterial bed of the rat *in vitro*. *Journal of Physiology (London)*, **490**, 181–9.

Ralevic, V., Kristek, F., Hudlicka, O. & Burnstock, G. (1989). A new protocol for removal of the endothelium from the perfused rat hind-limb preparation. *Circulation Research*, **64**, 1190–6.

Ramagopal, M. V. & Leighton, H. J. (1989). Effects of N^G-monomethyl-L-arginine on field stimulation-induced decreases in cytosolic Ca^{2+} levels and relaxation in the rat anococcygeus muscle. *European Journal of Pharmacology*, **174**, 297–9.

Rand, M. J. (1992). Nitrergic transmission: nitric oxide as a mediator of non-adrenergic, non-cholinergic neuro-effector transmission. *Clinical and Experimental Pharmacology and Physiology*, **19**, 147–69.

Rand, M. J. & Li, C. G. (1992). Effects of argininosuccinic acid on nitric oxide-mediated relaxations in rat aorta and anococcygeus muscle. *Clinical and Experimental Pharmacology and Physiology*, **19**, 331–4.

Rand, M. J. & Li, C. G. (1993). Differential effects of hydroxocobalamin on relaxations induced by nitrosothiols in rat aorta and anococcygeus muscle. *European Journal of Pharmacology*, **241**, 249–54.

Rattan, S. & Chakder, S. (1992). Role of nitric oxide as a mediator of internal anal sphincter relaxation. *American Journal of Physiology*, **262**, G107–12.

Rattan, S., Rosenthal, G. J. & Chakder, S. (1995). Human recombinant hemoglobin (rHb1.1) inhibits nonadrenergic noncholinergic (NANC) nerve-mediated relaxation of internal anal sphincter. *Journal of Pharmacology and Experimental Therapeutics*, **272**, 1211–16.

Rattan, S., Sarkar, A. & Chakder, S. (1992). Nitric oxide pathway in rectoanal inhibitory reflex of opossum internal anal sphincter. *Gastroenterology*, **103**, 43–50.

Rattan, S. & Thatikunta, P. (1993). Role of nitric oxide in sympathetic neurotransmission in opossum internal anal sphincter. *Gastroenterology*, **105**, 827–36.

Regidor, J., Edvinsson, L. & Divac, I. (1993). NOS neurones lie near branchings of cortical arteriolae. *NeuroReport*, **4**, 112–14.

Rengasamy, A. & Johns, R. A. (1993). Regulation of nitric oxide synthase by nitric oxide. *Molecular Pharmacology*, **44**, 124–8.

Rengasamy, A., Xue, C. & Johns, R. A. (1994). Immunohistochemical demonstration of a paracrine role of nitric oxide in bronchial function. *American Journal of Physiology*, **267**, L704–11.

Ribbons, K. A., Zhang, X.-J., Thompson, J. H., Greenberg, S. S., Moore, W. M., Kornmeier, C. M., Currie, M. G., Lerche, N., Blanchard, J., Clark, D. A. & Miller, M. J. S. (1995). Potential role of nitric oxide in a model of chronic colitis in rhesus macaques. *Gastroenterology*, **108**, 705–11.

Ribeiro, M. O., Antunes, E., de Nucci, G., Lovisolo, S. M. & Zatz, R. (1992). Chronic inhibition of nitric oxide synthesis: a new model of arterial hypertension. *Hypertension*, **20**, 298–303.

Ringheim, G. E. & Pan, J. (1995). Particulate and soluble forms of the inducible nitric oxide synthase are distinguishable at the amino terminus in RAW 264.7 macrophage cells. *Biochemical and Biophysical Research Communications*, **210**, 711–16.

Robinson, L. J., Busconi, L. & Michel, T. (1995). Agonist-modulated palmitoylation of endothelial nitric oxide synthase. *Journal of Biological Chemistry*, **270**, 995–8.

Rocha, M., Krüger, A., Van Rooijen, N., Schirrmacher, V. & Umansky, V. (1995). Liver endothelial cells participate in T-cell dependent host resistance to lymphoma metastasis by production of nitric oxide *in vivo*. *International Journal of Cancer*, **63**, 405–11.

Rochelle, L. G., Morana, S. J., Kruszyna, H., Russell, M. A., Wilcox, D. E. & Smith, R. P. (1995). Interactions between hydroxocobalamin and nitric oxide (NO): evidence for a redox reaction between NO and reduced cobalamin and reversible NO binding to oxidized cobalamin. *Journal of Pharmacology and Experimental Therapeutics*, **275**, 48–52.

Rodrigo, J., Springall, D. R., Uttenthal, O., Bentura, M. L., Abadia Molina, F., Riveros Moreno, V., Martinez Murillo, R., Polak, J. M. & Moncada, S. (1994). Localization of nitric oxide synthase in the adult rat brain. *Philosophical Transactions of the Royal Society of London, Series B, Biological Sciences*, **345**, 175–221.

Roitt, I. (1994). *Essential Immunology*, 8th edn. London: Blackwell Scientific Publications.

Roman, L. J., Sheta, E. A., Martasek, P., Gross, S. S., Liu, Q. & Masters, B. S. S. (1995). A novel system for expressing high levels of functional nitric oxide synthase in *Escherichia coli*. *Endothelium*, **3** (Suppl), S21.

Ross, G., Chaudhuri, G., Ignarro, L. J. & Chyu, K. Y. (1991). Acetylcholine vasodilation of resistance vessels *in vivo* may not entirely depend on newly synthesized nitric oxide. *European Journal of Pharmacology*, **195**, 291–3.

Rubanyi, G. M. (1993). The role of the endothelium in cardiovascular homeostasis and diseases. *Journal of Cardiovascular Pharmacology*, **22** (Suppl 4), S1–14.

Rubanyi, G. M. & Vanhoutte, P. M. (1986). Superoxide anions and hyperoxia inactivate endothelium-derived relaxing factor. *American Journal of Physiology*, **250**, H822–7.

Rubin, L. L., Hall, D. E., Porter, S., Barbu, K., Cannon, C., Horner, H. C., Janatpour, M., Liaw, C. W., Manning, K., Morales, J., Tanner, L. I., Tomaselli, K. J. & Bard, F. (1991). A cell culture model of the blood brain barrier. *Journal of Cell Biology*, **115**, 1725–35.

Ruderman, N. B., Williamson, J. R. & Brownlee, M. (1992). Glucose and diabetic vascular disease. *FASEB Journal*, **6**, 2905–14.

Rueff, A., Patel, I. A., Urban, L. & Dray, A. (1994). Regulation of bradykinin sensitivity in peripheral sensory fibres of the neonatal rat by nitric oxide and cyclic GMP. *Neuropharmacology*, **33**, 1139–45.

Ruggeri, Z. M. (1993). Mechanisms of shear-induced platelet adhesion and aggregation. *Thrombosis and Haemostasis*, **70**, 119–23.

Saenz de Tejada, I., Goldstein, I. & Krane, R. J. (1988). Local control of penile erection: nerves, smooth muscle and endothelium. *Urologic Clinics of North America*, **15**, 9–15.

Saffrey, M. J., Hassall, C. J., Hoyle, C. H. V., Belai, A., Moss, J., Schmidt, H. H. H. W., Förstermann, U., Murad, F. & Burnstock, G. (1992). Colocalization of nitric oxide synthase and NADPH-diaphorase in cultured myenteric neurones. *NeuroReport*, **3**, 333–6.

Saffrey, M. J., Hassall, C. J., Moules, E. W. & Burnstock, G. (1994). NADPH diaphorase and nitric oxide synthase are expressed by the majority of intramural neurons in the neonatal guinea pig urinary bladder. *Journal of Anatomy*, **185**, 487–95.

Saito, S., Kidd, G. J., Trapp, B. D., Dawson, T. M., Bredt, D. S., Wilson, D. A., Traystman, R. J., Snyder, S. H. & Hanley, D. F. (1994). Rat spinal cord neurons contain nitric oxide synthase. *Neuroscience*, **59**, 447–56.

Salas, E., Moro, M. A., Askew, S., Hodson, H. F., Butler, A. R., Radomski, M. W. & Moncada, S. (1994). Comparative pharmacology of analogues of *S*-nitroso-*N*-acetyl-DL-penicillamine on human platelets. *British Journal of Pharmacology*, **112**, 1071–6.

Salazar, F. J. & Llinás, M. T. (1996). Role of nitric oxide in the control of sodium excretion. *News in Physiological Science*, **11**, 62–7.

Salter, M., Knowles, R. G. & Moncada, S. (1991). Widespread tissue distribution, species distribution and changes in activity of Ca^{2+}-dependent and Ca^{2+}-independent nitric oxide synthases. *FEBS Letters*, **291**, 145–9.

Sancesario, G., Morello, M., Massa, R., Fusco, F. & Bernardi, G. (1993). NADPH diaphorase activity is inhibited by EDTA in neurons but not in choroid plexus epithelium. *Neuroscience Letters*, **158**, 101–4.

Sarih, M., Souvannavong, V. & Adam, A. (1993). Nitric oxide synthase induces macrophage death by apoptosis. *Biochemical and Biophysical Research Communications*, **191**, 503–8.

Sato, K., Miyakawa, K., Takeya, M., Hattori, R., Yui, Y., Sunamoto, M., Ichimori, Y., Ushio, Y. & Takahashi, K. (1995). Immunohistochemical expression of inducible nitric oxide synthase (iNOS) in reversible endotoxic shock studied by a novel monoclonal antibody against rat iNOS. *Journal of Leukocyte Biology*, **57**, 36–44.

Satoh, K., Arai, R., Ikemoto, K., Narita, M., Nagai, T., Ohshima, H. & Kitahama, K. (1995). Distribution of nitric oxide synthase in the central nervous system of *Macaca fuscata:* subcortical regions. *Neuroscience*, **66**, 685–96.

Schemann, M., Schaaf, C. & Mader, M. (1995). Neurochemical coding of enteric neurons in the guinea pig stomach. *Journal of Comparative Neurology*, **353**, 161–78.

Schilling, K., Schmidt, H. H. H. W. & Baader, S. L. (1994). Nitric oxide synthase expression reveals compartments of cerebellar granule cells and suggests a role for mossy fibers in their development. *Neuroscience*, **59**, 893–903.

Schmidt, H. H. H. W. (1992). NO, CO and OH: endogenous soluble guanylyl cyclase-activating factors. *FEBS Letters*, **307**, 102–7.

Schmidt, H. H. H. W., Baeblich, S. E., Zernikow, B. C., Klein, M. M. & Böhme, E. (1990). L-Arginine and arginine analogues: effects on isolated blood vessels and cultured endothelial cells. *British Journal of Pharmacology*, **101**, 145–51.

Schmidt, H. H. H. W., Gagne, G. D., Nakane, M., Pollock, J. S., Miller, M. F. & Murad, F. (1992a). Mapping of neural nitric oxide synthase in the rat suggests frequent co-localization with NADPH diaphorase but not with soluble guanylyl cyclase, and novel paraneural functions for nitrinergic signal transduction. *Journal of Histochemistry and Cytochemistry*, **40**, 1439–56.

Schmidt, H. H. H. W., Hofmann, H., Shutenko, Z. S., Cunningham, D. D. & Feelisch, M. (1995). No nitric oxide from nitric oxide synthase. *Endothelium*, **3** (Suppl), S2.

Schmidt, H. H. H. W., Lohman, S. M. & Walter, U. (1993). The nitric oxide and cGMP signal transduction system: regulation and mechanism of action. *Biochimica et Biophysica Acta*, **1178**, 153–75.

Schmidt, H. H. H. W., Pollock, J. S., Nakane, M., Förstermann, U. & Murad, F. (1992b). Ca^{2+}/calmodulin-regulated nitric oxide synthases. *Cell Calcium*, **13**, 427–34.

Schmidt, H. H. H. W., Pollock, J. S., Nakane, M., Gorsky, L. D., Förstermann, U. & Murad, F. (1991). Purification of a soluble isoform of guanylyl cyclase-activating-factor synthase. *Proceedings of the National Academy of Science USA*, **88**, 365–9.

Schmidt, H. H. H. W., Seifert, R. & Böhme, E. (1989a). Formation and release of nitric oxide from human neutrophils and HL–60 cells induced by chemotactic peptide, platelet activating factor and leukotriene B_4. *FEBS Letters*, **244**, 357–60.

Schmidt, H. H. H. W., Wilke, P., Evers, B. & Böhme, E. (1989b). Enzymatic formation of nitrogen oxides from L-arginine in bovine brain cytosol. *Biochemical and Biophysical Research Communications*, **165**, 284–91.

Schmidt, K., Klatt, P. & Mayer, B. (1994). Reaction of peroxynitrite with oxyhaemoglobin: interference with photometrical determination of nitric oxide. *Biochemical Journal*, **301**, 645–7.

Schmidt, K., Werner, E. R., Mayer, B., Wachter, H. & Kukovetz, W. R. (1992). Tetrahydrobiopterin-dependent formation of endothelium-derived relaxing factor (nitric oxide) in aortic endothelial cells. *Biochemical Journal*, **281**, 297–300.

Schoedon, G., Schneemann, M., Blau, N., Edgell, C.-J. S. & Schaffner, A. (1993). Modulation of human endothelial cell tetrahydrobiopterin synthesis by activating and deactivating cytokines: new perspectives on endothelium-derived relaxing factor. *Biochemical and Biophysical Research Communications*, **196**, 1343–8.

Schuman, E. M. & Madison, D. V. (1991). A requirement for the intercellular messenger nitric oxide in long-term potentiation. *Science*, **254**, 1503–6.

Schuman, E. M. & Madison, D. V. (1994a). Nitric oxide and synaptic function. *Annual Reviews of Neuroscience*, **17**, 153–83.

Schuman, E. M. & Madison, D. V. (1994b). Locally distributed synaptic potentiation in the hippocampus. *Science*, **263**, 532–6.

Schuman, E. M., Meffert, M. K., Schulman, H. & Madison, D. V. (1992). A potential role for an ADP-ribosyltransferase in hippocampal long-term potentiation. *Society of Neuroscience Abstracts*, **18**, 761.

Scott, T. R. & Bennett, M. R. (1993). The effect of nitric oxide on the efficacy of synaptic transmission through the chick ciliary ganglion. *British Journal of Pharmacology*, **110**, 627–32.

Sekizawa, K., Fukushima, T., Ikarashi, Y., Maruyama, Y. & Sasaki, H. (1993). The role of nitric oxide in cholinergic neurotransmission in rat trachea. *British Journal of Pharmacology*, **110**, 816–20.

Semple, J. W., Speck, E. R., Milev, Y. P., Blanchette, V. & Freedman, J. (1995). Indirect allorecognition of platelets by T helper cells during platelet transfusions correlates with anti-major histocompatibility complex antibody and cytotoxic T lymohocyte formation. *Blood*, **86**, 805–12.

Sessa, W. C. (1994). The nitric oxide synthase family of proteins. *Journal of Vascular Research*, **31**, 131–43.

Sessa, W. C., Barber, C. M. & Lynch, K. R. (1993). Mutation of *N*-myristoylation site converts endothelial cell nitric oxide synthase from a membrane to a cytosolic protein. *Circulation Research*, **72**, 921–4.

Sessa, W. C., García-Cardeña, G., Liu, J., Keh, A., Pollock, J. S., Bradley, J., Thiru, S., Braverman, I. M. & Desai, K. M. (1995). The Golgi association of endothelial nitric oxide synthase is necessary for the synthesis of NO. *Endothelium*, **3** (Suppl), S8.

Sessa, W. C., Harrison, J. K., Barber, C. M., Zeng, D., Durieux, M. E., D'Angelo, D. D., Lynch, K. R. & Peach, M. J. (1992). Molecular cloning and expression of a cDNA encoding endothelial cell nitric oxide synthase. *Journal of Biological Chemistry*, **267**, 15274–6.

Sessa, W. C., Pritchard, K., Seyedi, N., Wang, J. & Hintze, T. H. (1994). Chronic exercise in dogs increases coronary vascular nitric oxide production and endothelial cell nitric oxide synthase gene expression. *Circulation Research*, **74**, 349–53.

Sexton, A. J., Loesch, A., Turmaine, M., Miah, S. & Burnstock, G. (1995). Nitric oxide and human umbilical vessels: pharmacological and immunohistochemical studies. *Placenta*, **16**, 277–88.

Shastry, B. S. (1994). More to learn from gene knockouts. *Molecular and Cellular Biochemistry*, **136**, 171–82.

Sheffler, L. A., Wink, D. A., Melillo, G. & Cox, G. W. (1995). Exogenous nitric oxide regulates IFN-γ plus lipopolysaccharide-induced nitric oxide synthase expression in mouse macrophages. *Journal of Immunology*, **155**, 886–94.

Sheng, H. (1994). Effect of LY 83583 on the response to NANC nerve stimulation in rat anococcygeus and bovine retractor penis muscles. *Journal of Pharmacology*, **104**, 137P.

Sheng, H., Schmidt, H. H. H. W., Nakane, M., Mitchell, J. A., Pollock, J. S., Förstermann, U. & Murad, F. (1992). Characterization and localization of nitric oxide synthase in non-adrenergic non-cholinergic nerves from bovine retractor penis muscles. *British Journal of Pharmacology*, **106**, 768–73.

Sherman, M. P., Griscavage, J. M. & Ignarro, L. J. (1992). Nitric oxide-mediated neuronal injury in multiple sclerosis. *Medical Hypotheses*, **39**, 143–6.

Shew, R. L., Papka, R. E., McNeill, D. L. & Yee, J. A. (1993). NADPH-diaphorase-positive nerves and the role of nitric oxide in CGRP relaxation of uterine contraction. *Peptides*, **14**, 637–41.

Shibuki, K. (1990). An electrochemical microprobe for detecting nitric oxide release in brain tissue. *Neuroscience Research*, **9**, 69–76.

Shibuki, K. & Okada, D. (1991). Endogenous nitric oxide release required for long-term synaptic depression in the cerebellum. *Nature*, **349**, 326–8.

Shibuta, S., Mashimoto, T., Ohara, A., Zhang, P. & Yoshiya, I. (1995). Intracerebroventricular administration of a nitric oxide-releasing compound, NOC-18, produces thermal hyperalgesia in rats. *Neuroscience Letters*, **187**, 103–6.

Shimamura, K., Fujikawa, A., Toda, N. & Sunano, S. (1993). Effects of N^{G}-nitro-L-arginine on electrical and mechanical responses to stimulation of non-adrenergic, non-cholinergic inhibitory nerves in circular muscle of the rat gastric fundus.

European Journal of Pharmacology, **231**, 103–9.

Shimokawa, H., Aarhus, L. L. & Vanhoutte, P. M. (1987). Porcine coronary arteries with regenerated endothelium have a reduced endothelium-dependent responsiveness to aggregating platelets and serotonin. *Circulation Research*, **61**, 256–70.

Shimosegawa, T. & Toyota, T. (1994). NADPH-diaphorase activity as a marker for nitric oxide synthase in neurons of the guinea pig respiratory tract. *American Journal of Respiratory and Critical Care Medicine*, **150**, 1402–10.

Shuttleworth, C. W., Murphy, R. & Furness, J. B. (1991). Evidence that nitric oxide participates in non-adrenergic inhibitory transmission to intestinal muscle in the guinea-pig. *Neuroscience Letters*, **130**, 77–80.

Shuttleworth, C. W., Sanders, K. M. & Keef, K. D. (1993a). Inhibition of nitric oxide synthesis reveals non-cholinergic excitatory neurotransmission in the canine proximal colon. *British Journal of Pharmacology*, **109**, 739–47.

Shuttleworth, C. W., Xue, C., Ward, S. M., de Vente, J. & Sanders, K. M. (1993b). Immunohistochemical localization of 3′,5′-cyclic guanosine monophosphate in the canine proximal colon: responses to nitric oxide and electrical stimulation of enteric inhibitory neurons. *Neuroscience*, **56**, 513–22.

Singer, I. I., Kawka, D. W., Scott, S., Weidner, J., Mumford, R. & Stenson, W. F. (1995). Inducible nitric oxide synthase and nitrotyrosine are localized in damaged intestinal epithelium during human inflammatory bowel disease (IBD). *Endothelium*, **3** (Suppl), S105.

Sladek, S. M., Regenstein, A. C., Lykins, D. & Roberts, J. M. (1993). Nitric oxide synthase activity in pregnant rabbit uterus decreases on the last day of pregnancy. *American Journal of Obstetrics and Gynecology*, **169**, 1285–91.

Smet, P. J., Edyvane, K. A., Jonavicius, J. & Marshall, V. R. (1994). Colocalization of nitric oxide synthase with vasoactive intestinal peptide, neuropeptide Y, and tyrosine hydroxylase in nerves supplying the human ureter. *Journal of Urology*, **152**, 1292–6.

Sneddon, P. & Graham, A. (1992). Role of nitric oxide in the autonomic innervation of smooth muscle. *Journal of Autonomic Pharmacology*, **12**, 445–56.

Snyder, S. H. & Bredt, D. S. (1992). Biological roles of nitric oxide. *Scientific American*, **266**, 28–35.

Sobrevia, L., Yudilevich, D. L. & Mann, G. E. (1995). Diabetes and hyperglycaemia induced activation of the human endothelial cell L-arginine transporter and NO synthase: inhibitory effects of insulin in diabetes. *Endothelium*, **3** (Suppl), S31.

Soediono, P. & Burnstock, G. (1994). Contribution of ATP and nitric oxide to NANC inhibitory transmission in rat pyloric sphincter. *British Journal of Pharmacology*, **113**, 681–6.

Solodkin, A., Traub, R. J. & Gebhart, G. F. (1992). Unilateral hindpaw inflammation produces a bilateral increase in NADPH-diaphorase histochemical staining in the rat lumbar spinal cord. *Neuroscience*, **51**, 495–9.

Son, K. & Kim, Y.-M. (1995). *In vivo* cisplatin-exposed macrophages increase immunostimulant-induced nitric oxide synthesis for tumor cell killing. *Cancer Re-*

search, **55**, 5524–7.

Song, Z. M., Brookes, S. J. & Costa, M. (1993). NADPH-diaphorase reactivity in nerves supplying the rat anococcygeus muscle. *Neuroscience Letters*, **158**, 221–4.

Sosunov, A. A., Hassall, C. J., Loesch, A., Turmaine, M. & Burnstock, G. (1995). Ultrastructural investigation of nitric oxide synthase-immunoreactive nerves associated with coronary blood vessels of rat and guinea-pig. *Cell and Tissue Research*, **280**, 575–82.

Southam, E. & Garthwaite, J. (1991). Intercellular action of nitric oxide in adult rat cerebellar slices. *NeuroReport*, **2**, 658–60.

Southam, E. & Garthwaite, J. (1993). The nitric oxide-cyclic GMP signalling pathway in rat brain. *Neuropharmacology*, **32**, 1267–77.

Southan, G. J., Szabó, C. & Thiemermann, C. (1995). Isothioureas: potent inhibitors of nitric oxide synthases with variable isoform selectivity. *British Journal of Pharmacology*, **114**, 510–16.

Soyombo, A. A., Thurston, V. J. & Newby, A. C. (1993). Endothelial control of vascular smooth muscle proliferation in an organ culture of human saphenous vein. *European Heart Journal*, **14** (Suppl I), 201–6.

Spessert, R. & Layes, E. (1994). Fixation conditions affect the intensity but not the pattern of NADPH-diaphorase staining as a marker for neuronal nitric oxide synthase in rat olfactory bulb. *Journal of Histochemistry and Cytochemistry*, **42**, 1309–15.

Spink, J., Cohen, J. & Evans, T. J. (1995). The cytokine responsive vascular smooth muscle cell enhancer of inducible nitric oxide synthase: activation by nuclear factor-κB. *Journal of Biological Chemistry*, **270**, 29 541–7.

Spitsin, S. V., Koprowski, H. & Michaels, F. H. (1995). Characterization and functional analysis of the human inducible nitric oxide synthase gene promoter. *Endothelium*, **3** (Suppl), S50.

Stamler, J. S. (1994). Redox signaling: nitrolysation and related target interactions of nitric oxide. *Cell*, **78**, 931–6.

Stamler, J. S., Simon, D. I., Osborne, J. A., Mullins, M. E., Jaraki, O., Michel, T., Singel, D. J. & Loscalzo, J. (1992). *S*-Nitrosylation of proteins with nitric oxide: synthesis and characterization of biologically active compounds. *Proceedings of the National Academy of Science USA*, **89**, 444–8.

Stanboli, A. & Morin, A. M. (1994). Nitric oxide synthase in cerebrovascular endothelial cells is inhibited by brefeldin A. *Neuroscience Letters*, **171**, 209–12.

Stark, M. E., Bauer, A. J. & Szurszewski, J. H. (1991). Effect of nitric oxide on circular muscle of the canine small intestine. *Journal of Physiology (London)*, **444**, 743–61.

Stefanovic-Racic, M., Stadler, J. & Evans, C. H. (1993). Nitric oxide and arthritis. *Arthritis and Rheumatism*, **36**, 1036–44.

Struck, A. T., Hogg, N., Thomas, J. P. & Kalyanaraman, B. (1995). Nitric oxide donor compounds inhibit the toxicity of oxidized low-density lipoprotein to endothelial cells. *FEBS Letters*, **361**, 291–4.

Stuehr, D. J., Cho, H. J., Soo Kwon, N., Weise, M. F. & Nathan, C. F. (1991a).

Purification and characterization of the cytokine-induced macrophage nitric oxide synthase: an FAD- and FMN-containing flavoprotein. *Proceedings of the National Academy of Science USA*, **88**, 7773–7.

Stuehr, D. J. & Ikeda-Saito, M. (1992). Spectral characterization of brain and macrophage nitric oxide synthases. Cytochrome P-450-like hemeproteins that contain a flavin semiquinone radical. *Journal of Biological Chemistry*, **267**, 20547–50.

Stuehr, D. J., Kwon, N. S., Nathan, C., Griffith, O., Feldman, P. & Wiseman, J. (1991b). N^{ω}-Hydroxy-L-arginine is an intermediate in the biosynthesis of nitric oxide from L-arginine. *Journal of Biological Chemistry*, **266**, 6259–63.

Stuehr, D. J. & Marletta, M. A. (1985). Mammalian nitrate biosynthesis: mouse macrophages produce nitrite and nitrate in response to *Escherichia coli* lipopolysaccharide. *Proceedings of the National Academy of Science USA*, **82**, 7738–42.

Stuehr, D. J. & Nathan, C. F. (1989). Nitric oxide: a macrophage product responsible for cytostasis and respiratory inhibition in tumour target cells. *Journal of Experimental Medicine*, **169**, 1543–55.

Sup, S. J., Gordon Green, B. & Grant, S. K. (1994). 2-Iminobiotin is an inhibitor of nitric oxide synthases. *Biochemical and Biophysical Research Communications*, **204**, 962–8.

Szabó, C., Southan, G. J. & Thiemermann, C. (1994a). Beneficial effects and improved survival in rodent models of septic shock with *S*-methylisothiourea sulfate, a potent and selective inhibitor of inducible nitric oxide synthase. *Proceedings of the National Academy of Science USA*, **91**, 12472–6.

Szabó, C., Southan, G. J., Thiemermann, C. & Vane, J. R. (1994b). The mechanism of the inhibitory effect of polyamines on the induction of nitric oxide synthase: role of aldehyde metabolites. *British Journal of Pharmacology*, **113**, 757–66.

Szatkowski, M., Barbour, B. & Attwell, D. (1990). Non-vesicular release of glutamate from glial cells by reversed electrogenic glutamate uptake. *Nature*, **348**, 443–6.

Takenaga, M., Kawasaki, H., Wada, A. & Eto, T. (1995). Calcitonin gene-related peptide mediates acetylcholine-induced endothelium-independent vasodilation in mesenteric resistance blood vessels of the rat. *Circulation Research*, **76**, 935–41.

Tam, F. S. & Hillier, K. (1992). The role of nitric oxide in mediating non-adrenergic non-cholinergic relaxation in longitudinal muscle of human taenia coli. *Life Sciences*, **51**, 1277–84.

Tamm, E. R., Flugel Koch, C., Mayer, B. & Lutjen Drecoll, E. (1995). Nerve cells in the human ciliary muscle: ultrastructural and immunocytochemical characterization. *Investigative Ophthalmology and Visual Science*, **36**, 414–26.

Tanaka, K., Hassall, C. J. & Burnstock, G. (1993a). Distribution of intracardiac neurones and nerve terminals that contain a marker for nitric oxide, NADPH-diaphorase, in the guinea-pig heart. *Cell and Tissue Research*, **273**, 293–300.

Tanaka, K., Ohshima, H., Esumi, H. & Chiba, T. (1993b). Direct synaptic contacts of nitric oxide synthase-immunoreactive nerve terminals on the neurons of the intracardiac ganglia of the guinea pig. *Neuroscience Letters*, **158**, 67–70.

Tanaka, M., Yoshida, S., Yano, M. & Hanaoka, F. (1994). Roles of endogenous nitric oxide in cerebellar cortical development in slice cultures. *NeuroReport*, **5**, 2049–52.

Tannenbaum, S. R., Fett, D., Young, V. R., Land, P. D. & Bruce, W. R. (1978). Nitrite and nitrate are formed by endogenous synthesis in the human intestine. *Science*, **200**, 1487–9.

Tannenbaum, S. R., Tamir, S., de Rojas-Walker, T. & Wishnok, J. S. (1994). DNA damage and cytotoxicity caused by nitric oxide. In *Nitrosamines and Related N-Nitroso Compounds*. Washington: American Chemical Society.

Tay, S. S. & Moules, E. W. (1994). NADPH-diaphorase is colocalized with nitric oxide synthase and vasoactive intestinal polypeptide in rat pancreatic neurons in culture. *Archives of Histology and Cytology*, **57**, 253–7.

Tay, S. S., Moules, E. W. & Burnstock, G. (1994). Colocalisation of NADPH-diaphorase with nitric oxide synthase and vasoactive intestinal polypeptide in new-born pancreatic neurons. *Journal of Anatomy*, **184**, 545–52.

Teale, D. M. & Atkinson, A. M. (1994). L-Canavanine restores blood pressure in a rat model of endotoxic shock. *European Journal of Pharmacology*, **271**, 87–92.

Telfer, J. F., Lyall, F., Norman, J. E. & Cameron, I. T. (1995). Identification of nitric oxide synthase in human uterus. *Human Reproduction*, **10**, 19–23.

Terenghi, G., Riveros Moreno, V., Hudson, L. D., Ibrahim, N. B. & Polak, J. M. (1993). Immunohistochemistry of nitric oxide synthase demonstrates immunoreactive neurons in spinal cord and dorsal root ganglia of man and rat. *Journal of Neurological Sciences*, **118**, 34–7.

Terenzi, F., Casado, M., Martin-Sanz, P. & Boscá, L. (1995). Epidermal growth factor inhibits cytokine-dependent nitric oxide synthase expression in hepatocytes. *FEBS Letters*, **368**, 193–6.

Tesfamariam, B. (1994). Free radicals in diabetic endothelial cell dysfunction. *Free Radical Biology and Medicine*, **16**, 383–91.

Tesfamariam, B., Palacino, J. J., Weisbrod, R. M. & Cohen, R. A. (1993). Aldose reductase inhibition restores endothelial cell function in diabetic rabbit aorta. *Journal of Cardiovascular Pharmacology*, **21**, 205–11.

Thiemermann, C., Ruetten, H., Wu, C.-C. & Vane, J. R. (1995). Inhibitors of nitric oxide synthase in the multiple organ failure (MOF) syndrome caused by endotoxin. *Endothelium*, **3** (Suppl), S98.

Thomas, E. & Pearse, A. G. E. (1964). The solitary active cells. Histochemical demonstration of damage-resistant nerve cells with a TPN-diaphorase reaction. *Acta Neuropathologica*, **3**, 238–49.

Thomsen, L. L., Miles, D. W., Happerfield, L., Bobrow, L. G., Knowles, R. G. & Moncada, S. (1995). Nitric oxide synthase activity in human breast cancer. *British Journal of Cancer*, **72**, 41–4.

Thornbury, K. D., Hollywood, M. A. & McHale, N. G. (1992). Mediation by nitric oxide of neurogenic relaxation of the urinary bladder neck muscle in sheep. *Journal of Physiology (London)*, **451**, 133–44.

Thornbury, K. D., Ward, S. M., Dalziel, H. H., Carl, A., Westfall, D. P. & Sanders, K. M.

(1991). Nitric oxide and nitrosocysteine mimic nonadrenergic, noncholinergic hyperpolarization in canine proximal colon. *American Journal of Physiology*, **261**, G553–7.

Thune, A., Delbro, D. S., Nilsson, B., Friman, S. & Svanvik, J. (1995). Role of nitric oxide in motility and secretion of the feline hepatobiliary tract. *Scandinavian Journal of Gastroenterology*, **30**, 715–20.

Tilton, R. G., Chang, K., Hasan, K. S., Smith, S. R., Petrash, J. M., Misko, T. P., Moore, W. M., Currie, M. G., Corbett, J. A., McDaniel, M. L. *et al.* (1993). Prevention of diabetic vascular dysfunction by guanidines. Inhibition of nitric oxide synthase versus advanced glycation end-product formation. *Diabetes*, **42**, 221–32.

Timmermans, J. P., Barbiers, M., Scheuermann, D. W., Bogers, J. J., Adriaensen, D., Fekete, E., Mayer, B., van Marck, E. A. & De Groodt-Lasseel, M. H. (1994). Nitric oxide synthase immunoreactivity in the enteric nervous system of the developing human digestive tract. *Cell and Tissue Research*, **275**, 235–45.

Toda, N. (1993). Mediation by nitric oxide of neurally-induced human cerebral artery relaxation. *Experientia*, **49**, 51–3.

Toda, N., Ayajiki, K. & Okamura, T. (1993a). Cerebroarterial relaxations mediated by nitric oxide derived from endothelium and vasodilator nerve. *Journal of Vascular Research*, **30**, 61–7.

Toda, N., Ayajiki, K., Yoshida, K., Kimura, H. & Okamura, T. (1993b). Impairment by damage of the pterygopalatine ganglion of nitroxidergic vasodilator nerve function in canine cerebral and retinal arteries. *Circulation Research*, **72**, 206–13.

Toda, N., Baba, H. & Okamura, T. (1990a). Role of nitric oxide in non-adrenergic, non-cholinergic nerve-mediated relaxation in dog duodenal longitudinal muscle strips. *Japanese Journal of Pharmacology*, **53**, 281–4.

Toda, N., Baba, H., Tanobe, Y. & Okamura, T. (1992). Mechanism of relaxation induced by K^+ and nicotine in dog duodenal longitudinal muscle. *Journal of Pharmacology and Experimental Therapeutics*, **260**, 697–701.

Toda, N., Kimura, T., Yoshida, K., Bredt, D. S., Snyder, S. H., Yoshida, Y. & Okamura, T. (1994a). Human uterine arterial relaxation induced by nitroxidergic nerve stimulation. *American Journal of Physiology*, **266**, H1446–50.

Toda, N., Kitamura, Y. & Okamura, T. (1994b). Role of nitroxidergic nerve in dog retinal arterioles *in vivo* and arteries *in vitro*. *American Journal of Physiology*, **266**, H1985–92.

Toda, N., Kitamura, Y. & Okamura, T. (1995a). Functional role of nerve-derived nitric oxide in isolated dog ophthalmic arteries. *Investigative Ophthalmology and Visual Science*, **36**, 563–70.

Toda, N., Minami, Y. & Okamura, T. (1990b). Inhibitory effects of L-N^G-nitro-arginine on the synthesis of EDRF and the cerebroarterial response to vasodilator nerve stimulation. *Life Sciences*, **47**, 345–51.

Toda, N. & Okamura, T. (1990a). Possible role of nitric oxide in transmitting information from vasodilator nerve to cerebroarterial muscle. *Biochemical and Biophysical Research Communications*, **170**, 308–13.

Toda, N. & Okamura, T. (1990b). Mechanism underlying the response to vasodilator nerve stimulation in isolated dog and monkey cerebral arteries. *American Journal of Physiology*, **259**, H1511–17.

Toda, N. & Okamura, T. (1990c). Modification by L-N^G-monomethyl arginine (L-NMMA) of the response to nerve stimulation in isolated dog mesenteric and cerebral arteries. *Japanese Journal of Pharmacology*, **52**, 170–3.

Toda, N. & Okamura, T. (1991a). Reciprocal regulation by putatively nitroxidergic and adrenergic nerves of monkey and dog temporal arterial tone. *American Journal of Physiology*, **261**, H1740–5.

Toda, N. & Okamura, T. (1991b). Role of nitric oxide in neurally induced cerebroarterial relaxation. *Journal of Pharmacology and Experimental Therapeutics*, **258**, 1027–32.

Toda, N. & Okamura, T. (1992). Mechanism of neurally induced monkey mesenteric artery relaxation and contraction. *Hypertension*, **19**, 161–6.

Toda, N. & Okamura, T. (1993). Responses to perivascular nerve stimulation of distal temporal arteries from dogs and monkeys. *Journal of Cardiovascular Pharmacology*, **22**, 744–9.

Toda, N., Tanobe, Y. & Baba, H. (1991a). Suppression by N^G-nitro-L-arginine of relaxations induced by non-adrenergic, non-cholinergic nerve stimulation in dog duodenal longitudinal muscle. *Japanese Journal of Pharmacology*, **57**, 527–34.

Toda, N., Yoshida, K. & Okamura, T. (1991b). Analysis of the potentiating action of N^G-nitro-L-arginine on the contraction of the dog temporal artery elicited by transmural stimulation of noradrenergic nerves. *Naunyn Schmiedeberg's Archives of Pharmacology*, **343**, 221–4.

Toda, N., Yoshida, K. & Okamura, T. (1995b). Involvement of nitroxidergic and noradrenergic nerves in the relaxation of dog and monkey temporal veins. *Journal of Cardiovascular Pharmacology*, **25**, 741–7.

Tojo, A., Gross, S. S., Zhang, L., Tisher, C. C., Schmidt, H. H. H. W., Wilcox, C. S. & Madsen, K. M. (1994). Immunocytochemical localization of distinct isoforms of nitric oxide synthase in the juxtaglomerular apparatus of normal rat kidney. *Journal of the American Society of Nephrologists*, **4**, 1438–47.

Tomimoto, H., Akiguchi, I., Wakita, H., Nakamura, S. & Kimura, J. (1994). Histochemical demonstration of membranous localization of endothelial nitric oxide synthase in endothelial cells of the rat brain. *Brain Research*, **667**, 107–10.

Tominaga, T., Sato, S., Ohnishi, T. & Ohnishi, S. T. (1993). Potentiation of nitric oxide formation following bilateral carotid occlusion and focal cerebral ischemia in the rat: *in vivo* detection of the nitric oxide radical by electron paramagnetic resonance spin trapping. *Brain Research*, **614**, 342–6.

Tottrup, A., Glavind, E. B. & Svane, D. (1992). Involvement of the L-arginine-nitric oxide pathway in internal anal sphincter relaxation. *Gastroenterology*, **102**, 409–15.

Tottrup, A., Knudsen, M. A. & Gregersen, H. (1991a). The role of the L-arginine-nitric oxide pathway in relaxation of the opossum lower oesophageal sphincter. *British Journal of Pharmacology*, **104**, 113–16.

Tottrup, A., Ny, L., Alm, P., Larsson, B., Forman, A. & Andersson, K. E. (1993). The role of the L-arginine/nitric oxide pathway for relaxation of the human lower oesophageal sphincter. *Acta Physiologica Scandinavica*, **149**, 451–9.

Tottrup, A., Svane, D. & Forman, A. (1991b). Nitric oxide mediating NANC inhibition in opossum lower esophageal sphincter. *American Journal of Physiology*, **260**, G385–9.

Tracey, W. R., Nakane, M., Pollock, J. S. & Förstermann, U. (1993). Nitric oxide synthases in neuronal cells, macrophages and endothelium are NADPH diaphorases, but represent only a fraction of total cellular diaphorase activity. *Biochemical and Biophysical Research Communications*, **195**, 1035–40.

Traystman, R. J., Moore, L. E., Helfaer, M. A., Davis, S., Banasiak, K., Williams, M. & Hurn, P. D. (1995). Nitro-L-arginine analogues: dose- and time-related nitric oxide synthase inhibition in brain. *Stroke*, **26**, 864–9.

Trifiletti, R. R. (1992). Neuroprotective effects of N^G-nitro-L-arginine in focal stroke in the 7-day old rat. *European Journal of Pharmacology*, **218**, 197–8.

Trigo-Rocha, F., Aronson, W. J., Hohenfellner, M., Ignarro, L. J., Rajfer, J. & Lue, T. F. (1993). Nitric oxide and cGMP: mediators of pelvic nerve-stimulated erection in dogs. *American Journal of Physiology*, **264**, H419–22.

Triguero, D., Prieto, D. & Garcia Pascual, A. (1993). NADPH-diaphorase and NANC relaxations are correlated in the sheep urinary tract. *Neuroscience Letters*, **163**, 93–6.

Tsao, P. S., Aoki, N., Lefer, D. J., Johnson, G. & Lefer, A. M. (1990). Time course of endothelial dysfunction and myocardial injury during myocardial ischemia and reperfusion in the cat. *Circulation*, **82**, 1402–12.

Tsao, P. S., McEvoy, L. M., Drexler, H., Butcher, E. C. & Cooke, J. P. (1994). Enhanced endothelial adhesiveness in hypercholesterolemia is attenuated by L-arginine. *Circulation*, **89**, 2176–82.

Tsukahara, H., Ende, H., Magazine, H. I., Bahou, W. F. & Goligorsky, M. S. (1994). Molecular and functional characterization of the non-isopeptide-selective ET_B receptor in endothelial cells: receptor coupling to nitric oxide synthase. *Journal of Biological Chemistry*, **269**, 21 778–85.

Tsukahara, H., Gordienko, D. V. & Goligorsky, M. S. (1993). Continuous monitoring of nitric oxide release from human umbilical vein endothelial cells. *Biochemical and Biophysical Research Communications*, **193**, 722–9.

Turk, J., Corbett, J. A., Ramanadham, S., Bohrer, A. & McDaniel, M. L. (1993). Biochemical evidence for nitric oxide formation from streptozotocin in isolated pancreatic islets. *Biochemical and Biophysical Research Communications*, **197**, 1458–64.

Tzeng, E., Billiar, T. R., Robbins, P. D., Loftus, M. & Stuehr, D. J. (1995). Expression of human inducible nitric oxide synthase in a tetrahydrobiopterin (H_4B)-deficient cell line: H_4B promotes assembly of enzyme subunits into an active dimer. *Proceedings of the National Academy of Sciences USA*, **92**, 11771–75.

Urbanics, R., Kapinya, K., Dézsi, L. & Kovách, A. G. B. (1995). Dose dependent effects of the 7-nitro indazole, a neural NOS inhibitor on focal cerebral ischemia. *Endothelium*, **3** (Suppl), S15.

Vallance, P., Leone, A., Calver, A., Collier, J. & Moncada, S. (1992). Accumulation of an endogenous inhibitor of nitric oxide synthesis in chronic renal failure. *Lancet*, **339**, 572–5.

Valtschanoff, J. G., Weinberg, R. J., Kharazia, V. N., Nakane, M. & Schmidt, H. H. H. W. (1993a). Neurons in rat hippocampus that synthesize nitric oxide. *Journal of Comparative Neurology*, **331**, 111–21.

Valtschanoff, J. G., Weinberg, R. J., Kharazia, V. N., Schmidt, H. H. H. W., Nakane, M. & Rustioni, A. (1993b). Neurons in rat cerebral cortex that synthesize nitric oxide: NADPH diaphorase histochemistry, NOS immunocytochemistry, and colocalization with GABA. *Neuroscience Letters*, **157**, 157–61.

Valtschanoff, J. G., Weinberg, R. J. & Rustioni, A. (1992). NADPH diaphorase in the spinal cord of rats. *Journal of Comparative Neurology*, **321**, 209–22.

Van Dam, A. M., Bauer, J., Man, A. K., Marquette, C., Tilders, F. J. & Berkenbosch, F. (1995). Appearance of inducible nitric oxide synthase in the rat central nervous system after rabies virus infection and during experimental allergic encephalomyelitis but not after peripheral administration of endotoxin. *Journal of Neuroscience Research*, **40**, 251–60.

Vanderkooi, J. M., Wright, W. W. & Erecinska, M. (1994). Nitric oxide diffusion coefficients in solutions, proteins and membranes determined by phosphorescence. *Biochimica et Biophysica Acta*, **1207**, 249–54.

Vanderwinden, J. M., Mailleux, P., Schiffmann, S. N., Vanderhaeghen, J. J. & De Laet, M. H. (1992). Nitric oxide synthase activity in infantile hypertrophic pyloric stenosis. *New England Journal of Medicine*, **327**, 511–15.

Vargas, H. M., Ignarro, L. J. & Chaudhuri, G. (1990). Physiological release of nitric oxide is dependent on the level of vascular tone. *European Journal of Pharmacology*, **190**, 393–7.

Varndell, I. M. & Polak, J. M. (1986). Electron microscopical immunocytochemistry. In *Immunocytochemistry: Modern Methods and Applications*, Second edn., eds. J. M. Polak & S. Van Noorden, pp. 146–66. Bristol: Wright.

Venema, R. C., Nishida, K., Alexander, R. W., Harrison, D. G. & Murphy, T. J. (1994). Organization of the bovine gene encoding the endothelial nitric oxide synthase. *Biochimica et Biophysica Acta*, **1218**, 413–20.

Venkova, K. & Krier, J. (1994). A nitric oxide and prostaglandin-dependent component of NANC off-contractions in cat colon. *American Journal of Physiology*, **266**, G40–7.

Verge, V. M., Xu, Z., Xu, X. J., Wiesenfeld Hallin, Z. & Hokfelt, T. (1992). Marked increase in nitric oxide synthase mRNA in rat dorsal root ganglia after peripheral axotomy: *in situ* hybridization and functional studies. *Proceedings of the National Academy of Sciences USA*, **89**, 11617–21.

Verma, A., Hirsch, D. J., Glatt, C. E., Ronnett, G. V. & Snyder, S. H. (1993). Carbon monoxide: a putative neural messenger. *Science*, **259**, 381–4.

Vickerstaff, T. (1950). *The Physical Chemistry of Dyeing.* New York: Interscience.

Villar, M. J., Settembrini, B. P., Hokfelt, T. & Tramezzani, J. H. (1994). NOS is present in the brain of *Triatoma infestans* and is colocalized with CCK. *NeuroReport*, **6**, 81–4.

Vincent, S. R. (1994). Nitric oxide: a radical neurotransmitter in the central nervous system. *Progress in Neurobiology*, **42**, 129–60.

Vincent, S. R. & Hope, B. T. (1992). Neurons that say NO. *Trends in Neuroscience*, **15**, 108–13.

Vincent, S. R., Staines, W. A. & Fibiger, H. C. (1983). Histochemical demonstration of separate populations of somatostatin and cholinergic neurons in the rat striatum. *Neuroscience Letters*, **35**, 111–14.

Vizzard, M. A., Erdman, S. L. & de Groat, W. C. (1993). Localization of NADPH-diaphorase in pelvic afferent and efferent pathways of the rat. *Neuroscience Letters*, **152**, 72–6.

Vizzard, M. A., Erdman, S. L. & de Groat, W. C. (1995). Increased expression of neuronal nitric oxide synthase (NOS) in visceral neurons after nerve injury. *Journal of Neuroscience*, **15**, 4033–45.

Vizzard, M. A., Erdman, S. L., Förstermann, U. & de Groat, W. C. (1994). Differential distribution of nitric oxide synthase in neural pathways to the urogenital organs (urethra, penis, urinary bladder) of the rat. *Brain Research*, **646**, 279–91.

Vladutiu, A. O. (1995). Role of nitric oxide in autoimmunity. *Clinical Immunology and Immunopathology*, **76**, 1–11.

von der Leyen, H. E., Gibbons, G. H., Morishita, R., Lewis, N. P., Zhang, L., Nakajima, M., Kaneda, Y., Cooke, J. P. & Dzau, V. J. (1995). Gene therapy inhibiting neointimal vascular lesion: *in vivo* transfer of endothelial cell nitric oxide synthase gene. *Proceedings of the National Academy of Science USA*, **92**, 1137–41.

Wagner, D. A., Young, V. R. & Tannenbaum, S. R. (1983). Mammalian nitrate biosynthesis: incorporation of $^{15}NH_3$ into nitrate is enhanced by endotoxin treatment. *Proceedings of the National Academy of Science USA*, **80**, 4518–21.

Walker, M. W., Kinter, M. T., Roberts, R. J. & Spitz, D. R. (1995). Nitric oxide-induced cytotoxicity: involvement of cellular resistance to oxidative stress and the role of glutathione in protection. *Pediatric Research*, **37**, 41–9.

Wallace, J. L., Reuter, B. K. & Cirino, G. (1994). Nitric oxide-releasing non-steroidal anti-inflammatory drugs: a novel approach for reducing gastrointestinal toxicity. *Journal of Gastroenterology and Hepatology*, **9** (Suppl 1), S40–4.

Wallis, R. A., Panizzon, K. L., Henry, D. & Wasterlain, C. G. (1993). Neuroprotection against nitric oxide injury with inhibitors of ADP-ribosylation. *NeuroReport*, **5**, 245–8.

Wallis, R. A., Panizzon, K. & Wasterlain, C. G. (1992). Inhibition of nitric oxide synthase protects against hypoxic neuronal injury. *NeuroReport*, **3**, 645–8.

Wang, B.-Y., Singer, A. H., Tsao, P. S., Drexler, H., Kosek, J. & Cooke, J. P. (1994a). Dietary arginine prevents atherogenesis in the coronary artery of the hypercholesterolemic rabbit. *Journal of the American College of Cardiology*, **23**, 452–8.

Wang, J., Stuehr, D. J., Ikeda-Saito, M. & Rousseau, D. L. (1993a). Heme coordination and structure of the catalytic site in nitric oxide synthase. *Journal of Biological Chemistry*, **268**, 22 255–8.

Wang, P. G., Yu, L.-B., Ramirez, J. E., Guo, Z.-M., Li, J. & McGill, A. D. (1995). Chemistry and biochemistry of nitric oxide through synthetic and physical organic approach: a comprehensive scale of transnitrosation potentials of organic compounds. *Endothelium*, **3** (Suppl), S64.

Wang, Z. Z., Bredt, D. S., Fidone, S. J. & Stensaas, L. J. (1993b). Neurons synthesizing nitric oxide innervate the mammalian carotid body. *Journal of Comparative Neurology*, **336**, 419–32.

Wang, Z. Z., Stensaas, L. J., Bredt, D. S., Dinger, B. & Fidone, S. J. (1994b). Localization and actions of nitric oxide in the cat carotid body. *Neuroscience*, **60**, 275–86.

Ward, J. K., Barnes, P. J., Springall, D. R., Abelli, L., Tadjkarimi, S., Yacoub, M. H., Polak, J. M. & Belvisi, M. G. (1995). Distribution of human i-NANC bronchodilator and nitric oxide-immunoreactive nerves. *American Journal of Respiratory Cell Molecular Biology*, **13**, 175–84.

Ward, J. K., Belvisi, M. G., Fox, A. J., Miura, M., Tadjkarimi, S., Yacoub, M. H. & Barnes, P. J. (1993). Modulation of cholinergic neural bronchoconstriction by endogenous nitric oxide and vasoactive intestinal peptide in human airways *in vitro*. *Journal of Clinical Investigation*, **92**, 736–42.

Ward, S. M., Dalziel, H. H., Bradley, M. E., Buxton, I. L., Keef, K. D., Westfall, D. P. & Sanders, K. M. (1992a). Involvement of cyclic GMP in non-adrenergic, non-cholinergic inhibitory neurotransmission in dog proximal colon. *British Journal of Pharmacology*, **107**, 1075–82.

Ward, S. M., Dalziel, H. H., Khoyi, M. A., Westfall, A. S., Sanders, K. M. & Westfall, D. P. (1996). Hyperpolarization and inhibition of contraction mediated by nitric oxide released from enteric inhibitory neurones in guinea-pig taenia coli. *British Journal of Pharmacology*, **118**, 49–56.

Ward, S. M., Dalziel, H. H., Thornbury, K. D., Westfall, D. P. & Sanders, K. M. (1992b). Nonadrenergic, noncholinergic inhibition and rebound excitation in canine colon depend on nitric oxide. *American Journal of Physiology*, **262**, G237–43.

Ward, S. M., McKeen, E. S. & Sanders, K. M. (1992c). Role of nitric oxide in non-adrenergic, non-cholinergic inhibitory junction potentials in canine ileocolonic sphincter. *British Journal of Pharmacology*, **105**, 776–82.

Ward, S. M., Shuttleworth, C. W. & Kenyon, J. L. (1994a). Dorsal root ganglion neurons of embryonic chicks contain nitric oxide synthase and respond to nitric oxide. *Brain Research*, **648**, 249–58.

Ward, S. M., Xue, C. & Sanders, K. M. (1994b). Localization of nitric oxide synthase in canine ileocolonic and pyloric sphincters. *Cell and Tissue Research*, **275**, 513–27.

Ward, S. M., Xue, C., Shuttleworth, C. W., Bredt, D. S., Snyder, S. H. & Sanders, K. M. (1992d). NADPH diaphorase and nitric oxide synthase colocalization in enteric neurons of canine proximal colon. *American Journal of Physiology*, **263**, G277–84.

Wascher, T. C., Toplak, H., Krejs, G. J., Simecek, S., Kukovetz, W. R. & Graier, W. F. (1994). Intracellular mechanisms involved in D-glucose-mediated amplification of agonist-induced Ca^{2+} response and EDRF formation in vascular endothelial cells. *Diabetes*, **43**, 984–91.

Waterman, S. A. & Costa, M. (1994). The role of enteric inhibitory motoneurons in peristalsis in the isolated guinea-pig small intestine. *Journal of Physiology (London)*, **477**, 459–68.

Waterman, S. A., Costa, M. & Tonini, M. (1994). Accommodation mediated by enteric inhibitory reflexes in the isolated guinea-pig small intestine. *Journal of Physiology (London)*, **474**, 539–46.

Watson, N., Maclagan, J. & Barnes, P. J. (1993). Vagal control of guinea pig tracheal smooth muscle: lack of involvement of VIP or nitric oxide. *Journal of Applied Physiology*, **74**, 1964–71.

Way, K. J. & Reid, J. J. (1994). Effect of aminoguanidine on the impaired nitric oxide-mediated neurotransmission in anococcygeus muscle from diabetic rats. *Neuropharmacology*, **33**, 1315–22.

Wei, X., Charles, I. G., Smith, A., Ure, J., Feng, G., Huang, F., Xu, D., Muller, W., Moncada, S. & Liew, F. Y. (1995). Altered immune responses in mice lacking inducible nitric oxide synthase. *Nature*, **375**, 408–11.

Weinberg, J. B., Misukonis, M. A., Shami, P. J., Mason, S. N., Sauls, D. L., Dittman, W. A., Wood, E. R., Smith, G. K., McDonald, B., Bachus, K. E., Haney, A. F. & Granger, D. L. (1995). Human mononuclear phagocyte inducible nitric oxide synthase (iNOS): analysis of iNOS mRNA, iNOS protein, biopterin, and nitric oxide production by blood monocytes and peritoneal macrophages. *Blood*, **86**, 1184–95.

Weiner, C. P., Lizasoain, I., Baylis, S. A., Knowles, R. G., Charles, I. G. & Moncada, S. (1994). Induction of calcium-dependent nitric oxide synthases by sex hormones. *Proceedings of the National Academy of Science USA*, **91**, 5212–16.

Weissman, B. A., Kadar, T., Brandeis, R. & Shapira, S. (1992). N^G-Nitro-L-arginine enhances neuronal death following transient forebrain ischemia in gerbils. *Neuroscience Letters*, **146**, 139–42.

Weitzberg, E., Rudehill, A., Modin, A. & Lundberg, J. M. (1995). Effect of combined nitric oxide inhalation and N^G-nitro-L-arginine infusion in porcine endotoxin shock. *Critical Care Medicine*, **23**, 909–1018.

Wells, D. G., Talmage, E. K. & Mawe, G. M. (1995). Immunohistochemical identification of neurons in ganglia of the guinea pig sphincter of Oddi. *Journal of Comparative Neurology*, **352**, 106–16.

Welsh, N., Eizirik, D. L. & Sandler, S. (1994). Nitric oxide and pancreatic β-cell destruction in insulin dependent diabetes mellitus: don't take NO for an answer. *Autoimmunity*, **18**, 285–90.

Wendland, B., Schweizer, F. E., Ryan, T. A., Nakane, M., Murad, F., Scheller, R. H. & Tsien, R. W. (1994). Existence of nitric oxide synthase in rat hippocampal pyramidal cells. *Proceedings of the National Academy of Sciences USA*, **91**,

2151–5.

Werman, R. (1966). Criteria for identification of a central nervous system transmitter. *Comparative Biochemistry and Physiology*, **18**, 745–66.

Westendorp, R. G. J., Draijer, R., Meinders, A. E. & van Hinsbergh, V. W. M. (1994). Cyclic-GMP-mediated decrease in permeability of human umbilical and pulmonary artery endothelial cell monolayers. *Journal of Vascular Research*, **31**, 42–51.

Wetts, R. & Vaughn, J. E. (1993). Transient expression of β-NADPH diaphorase in developing rat dorsal root ganglia neurons. *Developmental Brain Research*, **76**, 278–82.

White, C. R., Brock, T. A., Chang, L.-Y., Crapo, J., Briscoe, P., Ku, D., Bradley, W. A., Gianturco, S. H., Gore, J., Freeman, B. A. & Tarpey, M. M. (1994). Superoxide and peroxynitrite in atherosclerosis. *Proceedings of the National Academy of Science USA*, **91**, 1044–8.

Wiest, E., Trach, V. & Dammgen, J. (1989). Removal of endothelial function in coronary resistance vessels by saponin. *Basic Research in Cardiology*, **84**, 469–78.

Wiklund, C. U., Wiklund, N. P. & Gustafsson, L. E. (1993a). Modulation of neuroeffector transmission by endogenous nitric oxide: a role for acetylcholine receptor-activated nitric oxide formation, as indicated by measurements of nitric oxide/nitrite release. *European Journal of Pharmacology*, **240**, 235–42.

Wiklund, N. P., Leone, A. M., Gustafsson, L. E. & Moncada, S. (1993b). Release of nitric oxide evoked by nerve stimulation in guinea-pig intestine. *Neuroscience*, **53**, 607–11.

Williams, C. V., Nordquist, D. & McLoon, S. C. (1994). Correlation of nitric oxide synthase expression with changing patterns of axonal projections in the developing visual system. *Journal of Neuroscience*, **14**, 1746–55.

Williams, M. B., Errington, M. L. & Bliss, T. V. P. (1989). Arachidonic acid induces a long-term activity-dependent enhancement of synaptic transmission in the hippocampus. *Nature*, **341**, 739–42.

Williamson, J. R., Chang, K., Frangos, M., Hasan, K. S., Ido, Y., Kawamura, T., Nyengaard, J. R., van den Enden, M., Kilo, C. & Tilton, R. G. (1993). Hyperglycemic pseudohypoxia and diabetic complications. *Diabetes*, **42**, 801–13.

Wink, D. A., Kasprzak, K. S., Maragos, C. M., Elespuru, R. K., Misra, M., Dunams, T. M., Cebula, T. A., Koch, W. H., Andrews, A. W., Allen, J. S. & Keefer, L. K. (1991). DNA deaminating ability and genotoxicity of nitric oxide and its progenitors. *Science*, **254**, 1001–3.

Wolf, G., Wurdig, S. & Schunzel, G. (1992). Nitric oxide synthase in rat brain is predominantly located at neuronal endoplasmic reticulum: an electron microscopic demonstration of NADPH-diaphorase activity. *Neuroscience Letters*, **147**, 63–6.

Wolfe, G. C., MacNaul, K. L., Raju, S., McCauley, E., Weidner, J., Mumford, R.,

Schmidt, J. & Hutchinson, N. I. (1995). Comparison of the calcium and calmodulin requirements of the three human nitric oxide synthase isoforms expressed in baculovirus-infected SF9 cells. *Endothelium*, **3** (Suppl), S21.

Wolff, D. J., Datto, G. A. & Samatovicz, R. A. (1993b). The dual mode of inhibition of calmodulin-dependent nitric-oxide synthase by antifungal imidazole agents. *Journal of Biological Chemistry*, **268**, 9430–6.

Wolff, D. J., Datto, G. A., Samatovicz, R. A. & Tempsick, R. A. (1993a). Calmodulin-dependent nitric-oxide synthase: mechanism of inhibition by imidazole and phenylimidazoles. *Journal of Biological Chemistry*, **268**, 9425–9.

Wolff, D. J. & Lubeskie, A. (1995). Aminoguanidine is an isoform-selective, mechanism-based inactivator of nitric oxide synthase. *Archives in Biochemistry and Biophysics*, **316**, 290–301.

Wolff, D. J., Lubeskie, A. & Umansky, S. (1994). The inhibition of the constitutive bovine endothelial nitric oxide synthase by imidazole and indazole agents. *Archives in Biochemistry and Biophysics*, **314**, 360–6.

Wood, P. L., Emmett, M. R., Rao, T. S., Cler, J., Mick, S. & Iyengar, S. (1990). Inhibition of nitric oxide synthase blocks *N*-methyl-D-aspartate-, quisqualate-, kainate-, harmaline-, and pentylenetetrazole-dependent increases in cerebellar cyclic GMP *in vivo*. *Journal of Neurochemistry*, **55**, 346–8.

Wood, P. L., Emmett, M. R. & Wood, J. A. (1994). Involvement of granule, basket and stellate neurons but not Purkinje or Golgi cells in cerebellar cGMP increases *in vivo*. *Life Sciences*, **54**, 615–20.

Woolf, C. J. & Thompson, S. W. (1991). The induction and maintenance of central sensitization is dependent on *N*-methyl-D-aspartic receptor activation: implications for the treatment of post-injury pain hypersensitivity states. *Pain*, **44**, 293–9.

Worl, J., Mayer, B. & Neuhuber, W. L. (1994a). Nitrergic innervation of the rat esophagus: focus on motor endplates. *Journal of the Autonomic Nervous System*, **49**, 227–33.

Worl, J., Wiesand, M., Mayer, B., Greskotter, K. R. & Neuhuber, W. L. (1994b). Neuronal and endothelial nitric oxide synthase immunoreactivity and NADPH-diaphorase staining in rat and human pancreas: influence of fixation. *Histochemistry*, **102**, 353–64.

Wu, C.-C., Thiemermann, C. & Vane, J. R. (1995). Glibenclamide-induced inhibition of the expression of inducible nitric oxide synthase in cultured macrophages and in the anaesthetized rat. *British Journal of Pharmacology*, **114**, 1273–81.

Wu, G. (1995). Nitric oxide synthesis and the effect of aminoguanidine and N^G-monomethyl-L-arginine on the onset of diabetes in the spontaneously diabetic BB rat. *Diabetes*, **44**, 360–4.

Wu, H. H., Williams, C. V. & McLoon, S. C. (1994a). Involvement of nitric oxide in the elimination of a transient retinotectal projection in development. *Science*, **265**, 1593–6.

Wu, W. (1993). Expression of nitric-oxide synthase (NOS) in injured CNS neurons as shown by NADPH-diaphorase histochemistry. *Experimental Neurology*, **120**, 153–9.

Wu, W. & Li, L. (1993). Inhibition of nitric oxide synthase reduces motoneuron death due to spinal root avulsion. *Neuroscience Letters*, **153**, 121–4.

Wu, W., Li, Y. & Schinco, F. P. (1994b). Expression of c-*jun* and neuronal nitric oxide synthase in rat spinal motoneurons following axonal injury. *Neuroscience Letters*, **179**, 157–61.

Xiao, L., Eneroth, P. H. E. & Qureshi, G. A. (1995). Nitric oxide synthase pathway may mediate human natural killer cell cytotoxicity. *Scandinavian Journal of Immunology*, **42**, 505–11.

Xie, Q.-W., Cho, H. J., Calaycay, J., Mumford, R. A., Swiderek, K. M., Lee, T. D., Ding, A., Troso, T. & Nathan, C. (1992). Cloning and characterization of inducible nitric oxide synthase from mouse macrophages. *Science*, **256**, 225–8.

Xu, J. Y. & Tseng, L. F. (1995). Nitric oxide/cyclic guanosine monophosphate system in the spinal cord differentially modulates intracerebroventricularly administered morphine- and β-endorphin-induced antinociception in the mouse. *Journal of Pharmacology and Experimental Therapeutics*, **274**, 8–16.

Xue, C., Pollock, J., Schmidt, H. H. H. W., Ward, S. M. & Sanders, K. M. (1994a). Expression of nitric oxide synthase immunoreactivity by interstitial cells of the canine proximal colon. *Journal of the Autonomic Nervous System*, **49**, 1–14.

Xue, C., Rengasamy, A., Le Cras, T. D., Koberna, P. A., Dailey, G. C. & Johns, R. A. (1994b). Distribution of NOS in normoxic vs. hypoxic rat lung: upregulation of NOS by chronic hypoxia. *American Journal of Physiology*, **267**, L667–78.

Yamada, K., Noda, Y., Nakayama, S., Komori, Y., Sugihara, H., Hasegawa, T. & Nabeshima, T. (1995). Role of nitric oxide in learning and memory and in monoamine metabolism in the rat brain. *British Journal of Pharmacology*, **115**, 852–8.

Yamamoto, R., Bredt, D. S., Snyder, S. H. & Stone, R. A. (1993b). The localization of nitric oxide synthase in the rat eye and related cranial ganglia. *Neuroscience*, **54**, 189–200.

Yamamoto, T. & Shimoyama, N. (1995). Role of nitric oxide in the development of thermal hyperesthesia induced by sciatic nerve constriction injury in the rat. *Anesthesiology*, **82**, 1266–73.

Yamamoto, T., Shimoyama, N. & Mizuguchi, T. (1993a). Nitric oxide synthase inhibitor blocks spinal sensitization induced by formalin injection into the rat paw. *Anesthesia and Analgesia*, **77**, 886–90.

Yamato, S., Saha, J. K. & Goyal, R. K. (1992). Role of nitric oxide in lower esophageal sphincter relaxation to swallowing. *Life Sciences*, **50**, 1263–72.

Yan, X. X., Jen, L. S. & Garey, L. J. (1993). Parasagittal patches in the granular layer of the developing and adult rat cerebellum as demonstrated by NADPH-diaphorase histochemistry. *NeuroReport*, **4**, 1227–30.

Yang, J., Kawamura, I., Zhu, H. & Mitsuyama, M. (1995). Involvement of natural killer cells in nitric oxide production by spleen cells after stimulation with *Mycobacterium bovis* BCG. *Journal of Immunology*, **155**, 5728–35.

Yang, W., Ando, J., Korenaga, R., Toyo-oka, T. & Kamiya, A. (1994). Exogenous nitric oxide inhibits proliferation of cultured vascular endothelial cells. *Biochemical and Biophysical Research Communications*, **203**, 1160–7.

Yang, Z. & Lüscher, T. F. (1993). Basic cellular mechanisms of coronary bypass graft disease. *European Heart Journal*, **14** (Suppl I), 193–7.

Yokoi, I., Kabuto, H., Habu, H., Inada, K., Toma, J. & Mori, A. (1994). Structure–activity relationships of arginine analogues on nitric oxide synthase activity in the rat brain. *Neuropharmacology*, **33**, 1261–5.

Yoshida, M., Akaike, T., Wada, Y., Sato, K., Ikeda, K., Ueda, S. & Maeda, H. (1994). Therapeutic effects of imidazolineoxyl *N*-oxide against endotoxin shock through its direct nitric oxide-scavenging activity. *Biochemical and Biophysical Research Communications*, **202**, 923–30.

Young, H. M., Furness, J. B., Shuttleworth, C. W., Bredt, D. S. & Snyder, S. H. (1992). Co-localization of nitric oxide synthase immunoreactivity and NADPH diaphorase staining in neurons of the guinea-pig intestine. *Histochemistry*, **97**, 375–8.

Yuan, Y., Granger, H. J., Zawieja, D. C., DeFily, D. V. & Chilian, W. M. (1993). Histamine increases venular permeability via a phospholipase C-NO synthase-guanylate cyclase cascade. *American Journal of Physiology*, **264**, H1734–9.

Zembowicz, A., Hatchett, R. J., Radziszewski, W. & Gryglewski, R. J. (1993). Inhibition of endothelial nitric oxide synthase by ebselen. Prevention by thiols suggests the inactivation by ebselen of a critical thiol essential for the catalytic activity of nitric oxide synthase. *Journal of Pharmacology and Experimental Therapeutics*, **267**, 1112–18.

Zhang, J., Dawson, V. L., Dawson, T. M. & Snyder, S. H. (1994). Nitric oxide activation of poly(ADP-ribose) synthetase in neurotoxicity. *Science*, **263**, 687–9.

Zhang, J. & Snyder, S. H. (1992). Nitric oxide stimulates auto-ADP-ribosylation of glyceraldehyde-3-phosphate dehydrogenase. *Proceedings of the National Academy of Sciences USA*, **89**, 9382–5.

Zhang, J. & Snyder, S. H. (1993). Purification of a nitric oxide-stimulated ADP-ribosylated protein using biotinylated beta-nicotinamide adenine dinucleotide. *Biochemistry*, **32**, 2228–33.

Zhang, J. & Snyder, S. H. (1995). Nitric oxide in the nervous system. *Annual Reviews in Pharmacology and Toxicology*, **35**, 213–33.

Zhang, M. & Vogel, H. J. (1994). Characterization of the calmodulin-binding domain of rat cerebellar nitric oxide synthase. *Journal of Biological Chemistry*, **269**, 981–5.

Zhang, X., Verge, V., Wiesenfeld-Hallin, Z., Ju, G., Bredt, D. S., Snyder, S. H. & Hökfelt, T. (1993). Nitric oxide synthase-like immunoreactivity in lumbar dorsal root ganglia and spinal cord of rat and monkey and effect of peripheral axotomy. *Journal of Comparative Neurology*, **335**, 563–75.

Zhuo, M., Meller, S. T. & Gebhart, G. F. (1993). Endogenous nitric oxide is required for tonic cholinergic inhibition of spinal mechanical transmission. *Pain*, **54**, 71–8.

Zhou, P., Sieve, M. C., Bennett, J., Kwon-Chung, K. J., Tewari, R. P., Gazzinelli, R. T., Sher, A. & Seder, R. A. (1995). IL-12 prevents mortality in mice infected with

Histoplasma capsulatum through induction of IFN-γ. *Journal of Immunology*, **155**, 785–95.

Ziche, M., Morbidelli, L., Masini, E., Amerini, S., Granger, H. J. & Maggi, C. A. (1994). Nitric oxide mediates angiogenesis *in vivo* and endothelial cell growth and migration *in vitro* promoted by substance P. *Journal of Clinical Investigation*, **94**, 2036–44.

Zielasek, J., Jung, S., Gold, R., Liew, F. Y., Toyka, K. V. & Hartung, H. P. (1995). Administration of nitric oxide synthase inhibitors in experimental autoimmune neuritis and experimental autoimmune encephalomyelitis. *Journal of Neuroimmunology*, **58**, 81–8.

Zoritch, B. (1995). Nitric oxide and asthma. *Archives of Diseases in Childhood*, **72**, 259–62.

Zweier, J. L., Wang, P. & Kuppusamy, P. (1995). Direct measurement of nitric oxide generation in the ischemic heart using electron paramagnetic resonance spectroscopy. *Journal of Biological Chemistry*, **270**, 304–7.

Zygmunt, P. K., Persson, K., Alm, P., Larsson, B. & Andersson, K. E. (1993). The L-arginine/nitric oxide pathway in the rabbit urethral lamina propria. *Acta Physiologica Scandinavica*, **148**, 431–9.

Index

Page numbers in **bold** indicate illustrations; those in *italics* indicate tables